Biomedical Ethics for Engineers

Biomedical Ethics for Engineers: Ethics and Decision Making in Biomedical and Biosystem Engineering

Daniel A. Vallero, Ph.D.

AMSTERDAM • BOSTON • HEIDELBERG • LONDON • NEW YORK • OXFORD
PARIS • SAN DIEGO • SAN FRANCISCO • SINGAPORE • SYDNEY • TOKYO

Academic Press is an imprint of Elsevier

Academic Press is an imprint of Elsevier
30 Corporate Drive, Suite 400, Burlington, MA 01803, USA
525 B Street, Suite 1900, San Diego, California 92101-4495, USA
84 Theobald's Road, London WC1X 8RR, UK

This book is printed on acid-free paper. ♾

Library of Congress Cataloging-in-Publication Data
Application Submitted

British Library Cataloguing-in-Publication Data
A catalogue record for this book is available from the British Library.

ISBN 13: 978-0-7506-8227-5
ISBN 10: 0-7506-8227-2

For information on all Academic Press publications
visit our Web site at www.books.elsevier.com

Printed and bound by CPI Group (UK) Ltd, Croydon, CR0 4YY

Transferred to Digital Print 2011

Contents

Chapter 3

An Engineered Future: Human Enhancement

Chapter 4

The Bioethical Engineer

Chapter 5

Bioethical Research and Technological Development

Chapter 8

Justice and Fairness as Biomedical and Biosystem Engineering Concepts

Chapter 9

Sustainable Bioethics

Preface

Many thoughtful books have been written about ethics as it pertains to the practice of engineering. Most include the terms engineering ethics, environmental ethics, and medical ethics in their titles and subtitles. I have relied upon these texts in my courses and ethics forums at Duke University and in training programs at other institutions, as well as in seminars, symposia, and as an invited speaker. While these resources have enriched my own grasp of ethics, their eclectic nature has presented me with the challenge, first in my classes, and ultimately to this writing.

The texts that I have used to date have some drawbacks in trying to apply them to bioethics from an engineering perspective. The medical ethics texts tend to focus mainly on the role of the caregiver. This makes sense for engineers in clinical settings, but not so much for most engineering venues. The engineering ethics texts are mainly based on cases, most of them failures. They usually give short shrift to bioethics (some attention to genetically modified organisms, biomedical devices, and emerging technologies). The environmental ethics texts are steeped in philosophy and environmentalism.[1] This book addresses that challenge, namely, to consider life issues from the perspective of the engineer, written in a way that engineers think.

Engineers and engineering students are busy people. They are driven to learn as much as possible about good design. From a very young age, they have been selected by their mentors and teachers as having special, even unique, abilities. Society continues to increase its expectations for the engineer's skills in applying mathematics and the physical sciences to address the growing challenges of human development, with a simultaneous understanding of the social sciences to modulate designs accordingly. In other words, in recent years, we have added a new engineering onus.

I recently completed a project with P. Aarne Vesilind, delving into the matter of justice as it pertains to the engineering profession (*Socially Responsible Engineering: Justice in Risk Management*, 2006, John Wiley & Sons, Hoboken, NJ).[2] For some time (Aarne well before me), we have applauded the engineering profession's efforts to incorporate the social sciences into curricula and professional certification; this project accentuated the challenge. The engineer of the future (and the future is already here!) will increasingly have to find ways to eliminate and to reduce risks, and to will have to ensure that designs are inclusive of all members of the society.

Throughout my career, I have been blessed with mentors and role models who foresaw and actually helped to bring about this new engineering paradigm. They have included teachers, sanitary (later environmental) engineers, civil engineers, biomedical engineers, earth scientists, environmental scientists, physicians, psychologists, theologians, philosophers, sociologists, economists, mathematicians, chemists, biologists, decision theorists, statisticians, angry citizens, happy citizens, fellow students, and my students. Allow me to say a word about the last group.

Bioethics is many things, but chief among them is that it is multidimensional in scope and requires an integration of myriad perspectives. It also requires a heavy dose of humility. My students have always helped me with both, but especially the latter. I have had the privilege to teach (share and guide might be a better description) a diverse group of undergraduate and graduate students in engineering, urban planning, biology, the geosciences, and ethics. The two groups of students in my current courses, Green Engineering/Sustainable Design and Ethics in Professions, differ in many ways, but share a number of characteristics.

The Green Engineering class strikingly reflects the current challenge. I team-teach this course with a very able and gifted architect, Chris Brasier, who draws from his personal portfolio of environmentally friendly designs. The students are matriculated in one of the majors offered by the Pratt School of Engineering. The class comprises mostly mechanical and civil engineering students, along with a few biomedical and electrical engineering majors. Interestingly, we have had few environmental engineering majors, possibly because they need an elective course outside of the environmental arena. My role is complementary to Chris's design perspective; that is, to provide the science and engineering foundation to good design. I begin with thermodynamics and fluid mechanics, and carry these themes throughout, including the assurance that the class and individual student projects are scientifically sound.

When Chris and I envisioned this course, we were concerned yet hopeful that the students could manage to maintain their creativity and engineering intuition, while adhering to the laws of science. While we had a few "teachable moments" (e.g., in one homework assignment a number of students calculated an energy efficiency of 112%, necessitating a review of the second law of thermodynamics), the students have been quite facile in moving back and forth from objective science (e.g., homework calculations of carbon monoxide concentrations from indoor heating) to societal issues (e.g., global climate change and its ancillary human impacts). The students demonstrated an ability to be integrative and adaptive.

The Ethics in Professions students are even more diverse, majoring not only in engineering (most have been biomedical) but in the humanities and the social sciences. One of the largest groups is those seeking a certificate from Duke's highly successful Markets & Management program in the Sociology Department.

One of the exercises that I use in every course is to have each student find and describe the code of ethics for their expected career choice. What they share has been quite revealing. First, the engineering and medical codes are useful, at least when compared to most others. Second, a major difference between engineering and medical codes is the focus on the public *versus* the patient, respectively. This creates somewhat of a quandary for the biomedical engineering students. There is good discussion about when, if ever, the public perspective must supplant the needs of the patient. Interestingly, this almost always evolves into a discussion of the US Constitution, especially the First Amendment (*mea culpa!*). Third, there are big differences between business ethics and engineering ethics. Few businesses, for example, hold paramount the health, safety, and welfare of the public, as do engineers. This is an important potential source of professional conflict, since most engineers will be working in some type of hierarchical structure and all engineers will have clients working for these organizations. Fourth, a number of students come to the course with their own code of ethics and apply interesting approaches for reconciling this code with that of their chosen career.

I strongly challenge the students who hold the view that all ethics is situational (see Teachable Moment: Truth and Turtles in Chapter 5). Bioethics discussions always include terms that evoke visceral responses. Some are necessary, such as those used to differentiate medical treatments. Certain terms are terrifying, like cancer, central nervous system dysfunction, toxics, and ominous-sounding chemical names, like steroids, free radicals, dioxin, polychlorinated biphenyls (PCBs), vinyl chloride, and methyl mercury. In fact, these chemicals are ominous, but many chemicals that are less harmful can also elicit anxieties and associated increased perceived risk. In the course pre-test, I have asked my students for some years to now answer a number of questions as evidence. The first two questions on the exam are:

1. The compound dihydrogenmonoxide has several manufacturing and industrial uses. However, it has been associated with acute health effects and death in humans, as a result of displacement of oxygen from vital organs. The compound has been found to form chemical solutions and suspensions with other substances, crossing cellular membranes and leading to cancer and other chronic diseases in humans. In addition, the compound has been associated with fish kills when supersaturated with molecular oxygen, destruction of wetlands and other habitats, and billions of dollars of material damage each year. A prudent course of action dealing with dihydrogenmonoxide is to:
   a. Ban the substance outright
   b. Conduct a thorough risk assessment, then take regulatory actions
   c. Work with industries using the compound to find suitable substitutes
   d. Restrict the uses of the substance to those of strategic importance to the United States
   e. Take no action, except to warn the public about the risks
2. The class of compounds, PCBs, had several manufacturing and industrial uses during the twentieth century. However, PCBs were associated with acute health effects and death in humans. The compound has been found to form chemical solutions and suspensions with other substances, crossing cellular membranes, and leading to cancer and other chronic diseases in humans. In addition, the compound has been associated with contaminated sediments, as well as with wetlands and other habitats, and billions of dollars of material damage each year. A prudent course of action dealing with PCBs is to:
   a. Ban the substances outright
   b. Conduct a thorough risk assessment, then take regulatory actions
   c. Work with industries using compound to find suitable substitutes
   d. Restrict the uses of the substances to those of strategic importance to the United States
   e. Take no action, except to warn the public about the risks

The two questions were intentionally worded similarly and the answers worded identically. Everything in the question is factually correct. Most of the students taking the examination have earned A's in high school or college chemistry, physics, and biology, and several are well on their way to completing engineering and other technical degrees. Interestingly, the answers to the two questions have differed very little. Most students seem to be struck by the many negative effects to health and safety. The most frequent answer is "b," conduct a risk assessment. This may in part be due to the fact that we as educators have been relentless in reminding students to get their facts straight before deciding. That is the good news. The bad news is that many saw no difference between the two questions and several chose "a," outright bans on both chemicals, the first of which is water.[3]

Actually, the answers to the two questions should have been very different. I would recommend "e" for water and "a" for the PCBs (simply because they have already been banned since the 1970s following the passage of the Toxic Substances Control Act). Of course, water is not risk free. In fact, it is a contributing factor in many deaths (drowning; electrocution; auto accidents; falls, especially in its solid phase, i.e., ice; and workplace incidents, such as steam-related accidents). Water contributed to the worst chemical plant disaster in history at Bhopal, India. We blame methylisocyanate (MIC), but the explosion would not have occurred had MIC not reacted violently with water. However, none of us could survive if we banned or placed major restrictions on its general use!

Perceived risks often differ from actual risk, sometimes much greater and sometimes much less. Like so many engineering concepts, the timing and the scenarios are crucial parts of the factual premises upon which an ethical decision is based. What may be the right manner of saying or writing something in one situation may be very inappropriate in another. Approaches will differ according to whether we need to motivate people to take action, alleviate undue fears, or simply share our findings clearly, no matter whether they are good news or bad news. For example, some have accused certain businesses of using pubic relations and advertising tools to lower the perceived risks of their products. The companies may argue that they are simply presenting a counterbalance against unrealistic perceptions.

The engineer must be vigilant to give the highest priority to the principal client, the public.

DONE IS GOOD

Frequently for us mere mortals, "the perfect is the enemy of the good."[4], Engineers seek "perfection" by continuous improvement, but they must know the inflection point when the expected benefits of excellence are overtaken by the costs resulting from missing a deadline. I have to believe that the reason so many bright, motivated PhD students who fail to finish (i.e., known as ABDs – for "all but dissertations") do so because they continue to seek perfection in their chosen research long after their work has contributed substantial knowledge to advance the state of science and is good enough to defend. In engineering language, their factors of safety (i.e., ensuring success in defense of their research before the committee) were disproportionate to what was absolutely necessary to succeed. "Done is good" is the latest colloquialism from Voltaire's quote above.

So, it is with this project. There is so much more to say than what I have been able to share in the ensuing pages. For many of the issues, just when I felt confident that I had adequately addressed the topic, I would open the morning newspaper and find another compelling point of view. So, it is with such a dynamic and challenging endeavor as bioethics.

The path leading to this book has been a tortuous and circuitous one. My initial interest was to capture the many ethical commonalities and contrasts among the technical professions, especially engineering and medicine. For years, I have combined and integrated materials from excellent medical and engineering texts and cases. For example, I have had much success in my undergraduate Ethics in Professions course at Duke University using two required texts; most recently, *Classic Cases in Medical Ethics: Accounts of Cases that Have Shaped Medical Ethics, with Philosophical, Legal, and Historical Backgrounds* by Gregory Pence (McGraw-Hill, 2003) and *Ethics in Engineering* by Mike Martin and Roland Schinzinger (McGraw-Hill, 2004). I augment these with numerous other resources, including materials from:

- *Paradigms Lost*, D.A. Vallero, Butterworth-Heineman
- *Hold Paramount*, A.S. Gunn and P.A. Vesilind, Brooks-Cole

- *The Responsible Researcher: Paths and Pitfalls*, Sigma Xi
- *Pasteur's Quadrant: Basic Science and Technological Innovations*, Stokes, Brookings
- *Character is Destiny*, Gough, Primas
- *Engineering Ethics*, Fleddermann, Prentice-Hall
- *Engineering Ethics*, Mitcham and Duvall, Prentice-Hall
- *Engineering Ethics and the Environment*, Vesilind and Gunn, Cambridge
- *Hold Paramount*, A.S. Gunn and P.A. Vesilind, Brooks-Cole
- *Fundamentals of Ethics for Scientists and Engineers*, Seebauer and Barry, Oxford
- *The Responsible Researcher: Paths and Pitfalls*, Sigma Xi
- *Engineering the Risks of Hazardous Wastes*, Vallero, Butterworth-Heinemann
- *Pasteur's Quadrant: Basic Science and Technological Innovations*, Stokes, Brookings
- *Trust Us, We're Experts!* Rampton and Stauber, Tarcher/Putnam
- *Virtue of Prosperity*, D'Souza, Simon and Schuster
- *Character is Destiny*, Gough, Primas
- *The Perception of Risk*, Slovic, Earthscan
- Publications from the National Academies, university websites (notably Texas A&M and Penn State), and professional organizations, such as the American Society of Civil Engineers

The medical and engineering ethical writings share many concerns. As mentioned, their codes are similar in many ways. Conversely, the two professions differ substantially in important areas. For example, the principal interest of the physician is the patient and that of the engineer is the public. In fact, a major driving force behind professional registration of engineers is public safety. So, a text that could consolidate and integrate these common features while emphasizing the differences would be quite valuable to professional ethics courses.

This initial focus became increasingly refined, however, as I consulted with colleagues in the biomedical engineering profession, especially faculty at Duke and elsewhere. The biomedical engineers must straddle the ethical landscape of both professions, so it seemed logical that these experts could provide insights and suggest cases to inform future practitioners. In fact, they did. One of the principal insights was the diversity in ethical challenge, which ranged from ethical dilemmas common to any profession, such as conflicts of interests, intellectual honesty and integrity, and "dual use" considerations to larger issues like public health and social justice.

As I gained greater awareness of these microethical and macroethical issues, one common theme arose, "life." Engineers, physicians, veterinarians, nurses, attorneys, scientists, and numerous other technical professions have intimate experiences with protecting, enhancing, and preserving the quality of life.[5] Thus, the first working title of this book was *Biomedical Ethics for Engineers*. Biomedical ethics is a worthy topic and I give much attention to it in this book. However, "life issues" go beyond biomedical ethics, including topics that may best fall under topics like environmental ethics (e.g., ecological issues) and social justice (e.g., equitable land use decision making and planning). Thus, the book's working title changed to *Bioethics for Engineers* to reflect this more encompassing view.

After discussing many of the issues addressed in this book, I became convinced that there would be little chance that I would select the cases deemed to be most useful to all engineers, or even to all engineers within a single engineering discipline. So, humility drove me to a third title: *Bioethics: An Engineer's Perspective*. The final title reverted again to biomedical ethics, but with a subtitle that captures one of the book's major themes: systematic ethics. You undoubtedly would choose different cases and topics, depending on your needs and interests. However, I truly hope those that I have selected will trigger your thinking about the daunting ethical challenges facing the present and the future engineer. William Wulf of the National Academy of Engineering (NAE) put the "life" challenge of engineers in perspective:

> Frankly, one of the reasons I feel very proud to be an engineer is because of the strong ethical orientation.
>
> So, why do I want to talk about ethics? Why do I believe that ethics will be the greatest engineering challenge in the 21st century? Why do I think the NAE needs to start a new program? There are two reasons that are closely intertwined.
>
> First, the practice of engineering is changing; and, in particular, it's changing in ways that raise a different kind of ethical issue. Second, the issues that are arising from this particular nature of engineering practice are macroethical issues that the profession has not dealt with before.[6]

The reader, undoubtedly, will be uneasy with some of my assessments and opinions, but my intention is that the book will elicit discussions that are useful in practice and in the classroom. Before we can decide on the ethics of any decision or action, we must first comprehend the facts. Then, we must apply reasoning. All too often, we avoid the issues altogether, seeking the safety of our science. But, the issues are there, whether or not we choose to confront them. Hopefully, this text will ease this confrontation, since it uses the language and structure familiar to the engineer.

STRUCTURE AND PEDAGOGY

I have structured this book in a conversational style. I do not presume to present all the bioethical cases, not even those that are most relevant to every engineer. Most of the discussion is intended to invite the reader to consider bioethics in a practical way. Once the "pump is primed," I expect the reader to pursue those cases and discussions most pertinent to his or her specific needs and interests. Enjoy and embrace the quest. For example, I present debates, such as that between Descartes and Gassendi on animal sensibilities and humane treatment. This is certainly not the only type of debate, not necessarily the most memorable, to demonstrate the compartmental thinking of people about the dignity of creatures. However, I hope it will engender the reader's thinking about moral inconsistencies and whether we are comfortable with "defining away" issues, rather than dealing with them (e.g., if one takes Descartes' view, there is little concern about how an animal lives, so long as it provides some human utility, such as meat and cures for diseases).

Besides readability and ability to provoke thought about bioethical issues, I have also added some features in hopes of helping teachers and workshop discussion leaders to use this book in engineering courses, ethics courses, breakout discussions, symposia, and seminars. The thought experiments are the first such tool. These may be a bit unsettling to most students and participants, so there are few "right" answers given. Here, my hope is that the teacher and discussion leader will be a guide, more than an authority. This Socratic approach has worked well in my experience with engineers. They may not like it at first and there is a strong likelihood that they will resist the discovery approach at first (many would prefer an answer sheet and a "to do" list). But, have faith; as soon as a few brave souls give their point of view, others tend to join in.

A number of us engineers, especially engineering professors, are control freaks and have type "A" personalities (I must admit to these tendencies as well). We fear complete trust in the Socratic method and worry that the student's journey of self-discovery and the paucity of teachable moments will keep us from sufficiently covering every topic promised in the syllabus. My own teaching experience has led to an acknowledgment up front that the syllabus is a guide and that not every topic will automatically be addressed.[7]

That said, there really are some topics that need to be addressed in detail, especially those where our lack of thoughtfulness puts us at a career risk or where practical ethical considerations are not given to important bioethical issues. For example, you and I may disagree on an important bioethical issue, such as human enhancement. Probably, we both agree that some human enhancement is permissible and even obligatory. For example, most of us agree that technologies are needed to sustain certain "at risk" people, such as insulin pumps for diabetics. As engineers, we see little difference between an insulin pump needed to sustain life and pumping water from a stream to sustain life. They are both reasonable and necessary applications of available technologies. However, we may strongly disagree that if Barry Bonds enhanced his performance with anabolic steroids, he should be labeled a cheater and ought to be reproved by society. The latter is an example of something about which society in general and engineers specifically diverge. Such examples are common in bioethics. To help with such discussions, I have added some discussion boxes, called "Teachable Moments," throughout the text. These can be assigned as homework, as in-class discussions, or simply read and summarized along with a litany of possible perspectives. The last use could be in the form of a poll (open or confidential), much like the ethical and professional practice journals use for perplexing engineering cases. Similarly, the boxes can elicit some brainstorming, followed by multivoting (e.g., each student gets two votes for what they think are the "right" answers or what they deem most important). The ensuing discussion can lead to breakout groups or team discussions.

Another pedagogical approach is to let the students debate issues among themselves, with the professor serving as an information resource and, where necessary, a dispassionate referee. To help here, I have added several discussion boxes (they could have been called "Choosing Sides"). These are designed to work with individual students, such as a homework assignment, or for teams. The class can be divided, physically and ideologically, during class to square off. One tool would be to have them construct logical syllogisms to support their positions. However, depending on the subject matter and the group dynamics of the class, this approach can work with any of the case analysis tools, like line drawing, event trees, and net goodness analyses (see last section of Chapter 7).

I have found that my ability to be dispassionate depends on the actual subject being discussed. The challenge of whether to disclose fully one's personal perspective about an ethical issue is unsettled. For example, students need to know enough about you to allow them to gauge your competence and perspective. But, if you give away too much information about your personal beliefs or if you disclose up front your own position on a matter, you have biased the discussion. I have received useful advice in this regard: You may disagree with me or agree with me philosophically and ideologically, but this does not count with ethical analysis. What counts is the validity of your argument. Some of this is moot in this age of Google searches. Most students can "profile" you in a few minutes.

Copious troves of information about professors' triumphs and hopes are readily available on the Internet (some might say we are narcissistic; kinder folks might say that we are more "transparent"). So, in the interest of full disclosure and to save you a few minutes online, let me inform you that I am a Christian of the Roman Catholic tradition. I was born in East St Louis, IL, on 5 September 1953, the second of three sons of Berneice and Jim Vallero, an Italian-American coal miner. I married Janis over thirty years ago. Both of our children, Daniel and Amelia, are married. My respect for animals is closer to that of Gassendi than that of Descartes, and I am particularly fond of dogs. I eat meat, but I am increasingly troubled about pain and the quality of an individual animal's life (I am yet to have an experience as when Albert Schweitzer saw the hippos, leading to his "reverence for life" precept; mine is more evolutionary than revolutionary). I strongly believe that life is a gift from the Creator and we are entrusted to cherish this gift. This charge includes both the respect for the sanctity and dignity of all human life and protection of all creation. As such, I have worked for over three decades with dedicated engineers, scientists, and others to provide for a sustainable environment that supports humans and all other species. I am a technological optimist, but like all engineers, I know that the earth is a system with constraints. Technology will be a big part of the solution to many global problems, but the whole solution will require application of the social sciences as well. Given this profile, you will have to ask my students whether I am an effective "honest broker" during class discussions. They tell me that I am, but *hey, they want an A*.

I hasten to add one more disclosure. I believe risk is a critical component of engineering ethics. Holding paramount the health, safety, and welfare of the public gives the engineer no room for spin. However, the public often exaggerates risks. So, abating risks that are in fact quite low could mean unnecessarily complicated and costly measures. It may also mean choosing the less acceptable alternative, i.e., one that in the long run may be more costly and deleterious to the environment or to public health. Scientifically based analyses rely on problem identification, data analysis, and risk characterization, including cost–benefit ratios. Perception relies on thought processes, including intuition, personal experiences, and personal preferences. Engineers tend to be more comfortable using risk assessment processes, while the general public often employs less rigorous approaches. One can liken this to the "left-brained" engineers' attempts to communicate with a "right-brained" audience. This difference sometimes seems to be the "elephant in the living room" in many bioethical debates, where the scientific types tend to seek out those like themselves and generally reach consensus on a bioethical issue. They are later dismayed and flummoxed by the differing opinion of the general public and their elected representatives. While sometimes (all too often) the average citizen is taken in by "junk science," he or she has an uncanny ability to reach a correct conclusion. I have to agree with Elmo Roper, the famous pollster, who in 1947 said:

> [M]any of us make two mistakes in our judgment of the common man. We overestimate the amount of information he has; and underestimate his intelligence.

The public frequently has too little information to decide on important matters. Amazingly, in spite of this lack of sufficient information, Roper found that the common person's "native intelligence generally brings him to a sound conclusion." But, the world is a far more complicated place than it was in 1947. And, many would argue that we share less of a unified vision of what is right and wrong. Many of the ethical norms that were considered fundamental, both within the technical arena and in society as a whole, are now subject to debate and lack consensus. We, as experts, must provide people with ample information and credible science, while respecting their intelligence. This is especially crucial as we consider the many divisive bioethical issues that affect us as engineers and which the engineering community can help to resolve. We must discuss even the uncomfortable issues with reason and honesty. That is the overriding objective of this text.

DAV

NOTES AND COMMENTARY

[1] Environmental ethics is a worthy area of endeavor, but I draw a sharp distinction between environmentalism *versus* environmental science and engineering. My view is that the former is steeped in advocacy, is almost a religion (and sometimes is), is politically charged (some modern critics believe it to have been co-opted by socialist and anti-capitalist views), and sees science as optional (when it serves the environmentalist's needs). The latter are based on sound science, adhering to the scientific method, and differ from the other sciences in subject matter, not methodology. Advocacy has its place, a very important place, but many engineers grow weary with some of the junk science expounded by many environmentalists.

[2] I address justice and fairness as these concepts pertain to bioethics in Chapter 8.

[3] Note of advice to teachers: The formula of this compound can elicit a different response when spoken than that when it is written. For example, students who hear it may ask you to write the formula on the board (ruining the question). Also, some students are most attentive to the details that are orally transmitted (a minority), so if you use this in a verbal class exercise, one or a few students may feel compelled to advise the class that this is water. A way to counteract this is to set some ground rules before beginning the discussion, such as asking that any students who have expertise in these specific chemicals recuse themselves from the discussion by writing a note to you beforehand.

[4] *Le mieux est l'ennemi du bien.* This phrase has been attributed to *Voltaire*, pen name for François-Marie Arouet (1694–1778) in *La Bégueule* (1772).

[5] I realize that some have used "quality of life" as a euphemism for devaluing certain persons and certain life stages of human beings. That is not the meaning here. Engineers look for ways to make life better for everyone, or at least they should.

[6] W. Wulf, 2000, Sigma Xi Forum Proceedings: *New Ethical Challenges in Science and Technology*, Albuquerque, New Mexico, 9–10 November, 2000, Sigma Xi, Research Triangle Park, NC.

[7] However, I also agree that if a student believes that I have not covered a topic sufficiently or if the student fears that an important topic will not be covered, I will go into more detail on that topic.

Acknowledgments

I cannot acknowledge everyone whose ideas and work have served as the resources that are integrated within this book. Of course, they would include all of my teachers, colleagues, students, friends, and relatives. The great and challenging thing about engineering ethics is that it requires rich and robust diversity of not only those with technical expertise, but also those with philosophical insight. Indeed, many of the lessons that I have learned have come from those who would not consider themselves to be either an ethicist or a scientist, but merely people who want to do the right thing. With this proviso, let me thank especially Dean Kristina Johnson and Professor Tod Laursen of the Pratt School of Engineering, who for some years from now have placed enormous faith in me to establish the engineering ethics programs at Duke University. In addition, Tod is the principal investigator for a truly innovative program sponsored by the National Science Foundation to help to design new teaching and learning tools to address the social aspects of emerging technologies. We are focusing on the macroethics of groundbreaking nanotechnologies taking place in two of Duke's engineering centers: the Center for Biologically Inspired Materials and Material Systems and the Center for Biological Tissue Engineering. The Center Directors, Rob Clark and Monty Reichert, have been gracious in allowing us to use their researchers as a "laboratory" for testing these pedagogies. Thus, the macroethical and nanotechnology sections of this book have been supported in part by NSF grant EEC-0530053 to Duke University.

When I first conceived of the idea to write a book on bioethics, among the first experts that I contacted was George Truskey, Chair of Duke's Biomedical Engineering Department. George gave me some excellent leads and encouraged other members of the department to share their ideas. As a result, I started with some clear information about the major challenges in biomedical engineering.

Biomedical engineering is half of the book's challenge. The other half, biosystem engineering, is the direct and indirect outcome of my work for many years in the environmental engineering. What I have learned and what I continue to learn from my past and present colleagues at Duke (especially Jeff Peirce, Aarne Vesilind, Karl Linden, David Schaad, Andrew Schuler, Ken Reckhow, Lynn Maguire, Fred Boadu, Miguel Medina, Mark Wiesner, Henri Gavin, Ana Barros, Chris Brasier and Henry Petroski) are woven into every chapter. By the way, practically everything I know about the history of science started with my former teacher and present colleague, Sy Mauskopf. I also want to thank those on the ethical side of the ledger, especially Elizabeth Kiss, former Director of the Kenan Institute for Ethics. I congratulate her on the honor of now serving as President of Agnes Scott College. Duke's loss is surely Agnes Scott's gain.

I have greatly benefited from the insights of Paul Lioy and Panos Georgeopolous of the Environmental and Occupational Health Sciences Institute (Piscataway, NJ), Tom McKone, Randy Maddalena, and Agnes Lobscheid of Lawrence Berkeley National Laboratory, Steve Randtke of the University of Kansas, as well as Loran Marlow, Charles Hess, Melvin Kazak, Harry Kircher, Halsey Miller, Donald Strohmeyer, and Ron Yarborough of Southern Illinois University. My collaborators at the US Environmental Protection Agency are dedicated in sustaining a livable world. For over 30 years, I have had the privilege to work with literally hundreds of the brightest engineers and scientists in EPA offices and laboratories at Kansas City, Research Triangle Park, Washington, DC, Las Vegas, Athens, Georgia, and Cincinnati. Let me name one; Ed Vest was the first to put faith in me as a professional. If I name any more, I would need to name many, but you know who you are!

Although I have learned much from these experts, I have learned even more from my students. They have challenged and enlightened me. The mix of students from engineering, physical sciences, biological sciences, and even quite a few of those from the humanities and the social sciences has been a steady source of ideas found in these pages.

I greatly appreciate the patience of the editors at Elsevier and Integra. Biomedical ethics is a fluid and rapidly changing endeavor. Andrea Sherman, Lathika Rajan, and their colleagues allowed me sufficient leeway to include currency (e.g., amniotic stem cell breakthroughs at Wake Forest), while working feverishly at their end to meet publication dates.

Finally, I want to thank the person who has contributed the most to this book. My wife, Janis, is ever a font of knowledge and right thinking. She combines this with the rare quality of the rigor that comes from being Jesuit trained. Her insights and queries have enriched my thinking about the issues addressed in this book.

Bioethics Questions Posed in Text

1. When is an action morally permissible and what kinds of behavior are morally obligatory?
2. What must a professional do to be trusted?
3. What is the role of values in bioethical decisions?
4. What does it mean to be a good engineer?
5. Does biosystem engineering ethics go beyond utility in defining what is right?
6. How can medical and engineering ethics coalesce?
7. When is human enhancement ethically right and ethically wrong?
8. How can engineering problem-solving skills be applied to biethical issues like human enhancement?
9. Can intuition be a reliable guide to ethical decision making?
10. When does pushing the envelope become unethical and, conversely, when does not pushing the envelope become unethical?
11. Is it ethical to enhance future people?
12. Is all human life sacred?
13. How do engineers fail?
14. What characterizes a morally commendable engineer?
15. Is ethics rational?
16. What is the role of the engineer in sustaining the environment?
17. What is the bioethical responsibility of the engineering profession?
18. Is research using human pluripotent stem cells ethical?
19. What are the bounds on the engineer's right of conscience?
20. What is the moral standing of non-human species?

Prologue: Bioethics – Discovery through Design

Engineering is a creative process. It involves discovery and brilliance. Engineers are simultaneously asked to provide proven products for clients, while encouraged to think outside the box and to push the envelope. As such, engineering is always in the balance between flux and stability.

Today's frontiers have evolved from those forged by engineers in previous centuries. Indeed, we are still looking for better ways to design and to build more efficiently. But, now we are challenged by smallness and greatness. Our scale now ranges from the atom to the solar system. We are charged with finding solutions to humanity's greatest problems. Certainly, human health and biomedical systems are vital concerns for all engineers. Not only are we charged with a mandate to solve and to prevent biological problems, such as finding better medical devices, improving drug manufacturing and delivery, and modeling the life processes that dictate wellness, but we must do such things in a way that does not lead to unacceptable side effects, especially those that affect our vital life systems on this planet. Thus, every new biomedical and biosystem engineering innovation must incorporate ethics. Engineers have become increasingly adept at distinguishing between what we can do and what we should do. The latter falls within the realm of ethics.

Bioethics presents a particular challenge to engineers. How should we approach it? The typical step-wise approach of introducing the theories and applying it to cases is rather unsatisfying for a number of reasons. First, many of the cases are best applicable to physicians and medical caregivers. Second, the elegance of the engineering process is ignored or given little credence in the ethical decision-making analysis. Third, the intricacies of mathematical and the physical science aspects (i.e., the facts and systems) of the cases are merely summarized in an effort to get to the "meat" of the ethical dilemma, whereas to engineers, the scientific details often are the focus of what went right or wrong. Fourth, the creative process that is engineering is frequently omitted from the causal and consequence analysis. Fifth, the differences between engineering and science are rarely noted, with the "technical" aspects of a case lumped together.

The fourth and fifth shortcomings are especially difficult to address for an audience of engineers. Too often, we lose our engineering audience when we fail to appreciate the creative and intuitive problem-solving aspects of ethics. Thus, rather than make the engineer conform to the prototypical ethics approach, I have chosen to begin our journey in a more open-ended manner, using two thought experiments. Many problems and issues in bioengineering are not "case-friendly." We cannot look at a negative or positive example and directly apply it to the issue at hand. Sam Scott, a cognitive scientist at Carleton University, makes a strong case for the value of thought experiments:

> The thesis that I want to defend is that thought experimentation is a meta-activity – a duel between conflicting theories in which one appears to be a clear winner, thus challenging anyone who holds both theories. Thought experiments can never tell us something new about the world, because the world doesn't participate in the experiments. The objects of evaluation are theories of the world. On the other hand, thought experiments are a perfect device not only for revealing problems with various theories of the world, but in some cases, for making clear to us what our theories of the world actually are.[1]

The thought experiment is conducive to the way problems come at engineers. Our "objects of evaluation" indeed are theories of the world through which we must sort to find those which apply, and to determine how they ought to be applied under our design constraints. Bioethical challenges are not neatly organized within a well-defined framework. Often, they are not even recognized as ethical problems, but merely as design or research challenges.

Engineers and designers seldom are satisfied with simply pulling a handbook off the shelf and "plugging and chugging" for a singular answer. No, a principal reason that the public places such enormous trust in engineers is not so much attributed to the day-to-day, automatic answers that are provided, but the difficult quandaries that have subtle nuances that make them seemingly intractable. I contend that this is where scientists and engineers really earn their keep. The subtle differences, however, are what make engineering such a highly demanded profession, and which make engineering ethics so difficult. It is also why the engineering profession demands much postbaccalaureate experience before one can be considered a competent and trusted professional. Only by experience, interactive learning, and mentorship can the neophyte engineer develop into a responsible professional who can recognize the importance of seemingly inconsequential details and differences in design needs among very similar scenarios.

Bioethical challenges, it can be argued, are even more complex and unwieldy than most engineering problems, particularly those that arrive with little structure and organized in deceptively simple ways. For one thing, the engineers' knowledge about physical systems is plainly necessary, but not sufficient, to answer most ethical questions. This is because ethical questions always involve the social and human sciences.

Obviously, engineering is a human endeavor aimed at improving the lot of humans; so, it can rightfully be argued that "to engineer is human," as Henry Petroski, my Duke colleague, aptly puts it.[2] But, the human element is exaggerated in some cases, such as those where there is a debate about what is morally right and wrong. Bioethics is heaving with such debates. Sometimes, the ethics is hidden within the technological challenge or the technical question, and appearing like other more easily resolved and previously confronted problems. Such problems can be deconstructed using the thought experiment.

The thought experiment is a creative tool. In a way, engineers engage in thought experiments at the beginning of every project. The design is an act of faith. It provides the incremental steps needed to achieve something that previously did not exist, except in the mind of the designer. This creativity is invaluable, since the usual physical constraints have no power over ideas, allowing investigations to venture into previously unexplored territories. Einstein, Galileo, and all great thinkers have applied the thought experiment to advance the state of knowledge. Ethicists and philosophers have also found these to be useful. The sayings of Confucius begin with simple, observable phenomena that teach us much about life's profundities. Socrates started most quests for knowledge with simple questions. The parables of Jesus Christ are invitations to explore ideas by beginning with the seemingly mundane and progressing to the deepest spiritual concepts.

What makes the thought experiment intriguing to engineering decision making is that it is built upon the foundation of modern scientific thought, while maintaining the best of the Ancients. The "experiment" aspect follows the Renaissance science expounded by Paracelsus, Francis Bacon, Robert Boyle, Isaac Newton, and other giants who moved us toward *a posteriori* reasoning. The experiment became the central tool in advancing science. Facts had to be observed first, and then analyzed, before conclusions could be drawn. Thought experiments also share elements of *a priori* reasoning. The clarity and open-endedness of the human mind can be put to use to explore endless possibilities. What frustrates the experiential scientists (i.e., the lack of physical constraints) opens additional avenues of reasoning. Conversely, what would likely frustrate the *a priori* thinkers who follow the scientific approach of Thomas Hobbes and many of the Ancient Greeks (i.e., immutable adherence to physical constraints) is a real-world calibration against which empirical conclusions can be compared. Thus, the thought experiment is an ideal design tool for analyzing bioethical situations that provide an effective means for the present and the future engineers to begin to consider bioethics (For more information, visit: http://plato.stanford.edu/entries/thought-experiment/).

A DIFFERENT APPROACH TO BIOETHICS

Bioethics shares many of the same elements as other areas of engineering ethics. As such, these universal ethical foundations must be explored. However, other elements are unique to biomedical ethics in scope and kind. Like all ethics, bioethics operates between the realm of the clearly wrong and clearly right. This is evident by the words we use to compliment or to condemn the work of a professional. Consider the following adjectives:

- Although painful, the therapy was "worthwhile"
- He exposed his patient to "unnecessary" pain
- The stent's failure rate is "unacceptable"
- The defibrillator's reliability is "tolerable"
- The genetic modification is "desirable"
- Protecting human subjects and getting their informed consent are "essential"

These adjectives beg the question: "To whom?" The answer, at least in an overriding sense, is "society." The engineer's obligation to society is addressed throughout this text.

Bioethics addresses morality in medicine as well as in other living systems (biosystems), so it is unique from most other areas of engineering ethics. Infact, we will revisit these specific adjectives. It deals with life directly. All engineering codes demand that our foremost concern be for public health, safety and welfare; thus, what we conceive, design, and build indirectly incorporates a "life" factor. However, bioethics is directly concerned with moral decisions specifically related to the life sciences and their attendant technologies. As such, all the cases and discussions in this text will support engineering decisions, while most will introduce concepts unique to medicine and biological systems.

This text is concerned with conveying biomedical and biosystem ethics in a manner and form that is "engineer-friendly." Doing so leaves us with yet another challenge. Biomedical engineering is a rapidly growing field. Its gifted members must navigate the rigors of both engineering and medicine. Biosystems engineers address life systems from subcellular to biome scales. This challenge not only includes the technical subject matter and methods, but also the ethics. For example, biomedical engineers must understand the human body's functions and structures in the same way as a civil engineer and a chemical engineer understand the systems and constraints in designing a building or a chemical reactor, respectively. The human body is both of these and more. In fact, the function and the structure of each cell in various tissues must be fully understood before a proper (and ethical) design is possible. Beyond this, biomedical engineers must also understand the language of the colleagues with whom they will interact throughout their careers. These colleagues share the desire to improve health, but their vision, clientele, approach, and attitudes will vary from those of the biomedical engineer. Again, one noteworthy distinction is that engineers generally have clients and medical professionals have patients. While clients and patients may share numerous characteristics, the professional's onus is different. For example, physicians hold paramount the well-being of the patient and engineers hold paramount the well-being of the public.

So then, what is the best way to talk about biomedical ethics and to convey the lessons being learned? There are two popular approaches. We can explore cases to find examples of good and bad biomedical ethics. This is the most common pedagogical approach. We learn lessons from experiences (good and bad), and build from these lessons. Another approach, which is most familiar to engineers, is to treat biomedical ethics as a series of design problems. While this book uses both approaches, it is worthwhile to consider the advantages and downsides of each.

ARGUMENTS FOR AND AGAINST CASE ANALYSIS

The principal value of cases is that they really happened (or are presently happening in what educators call a "teachable moment"). Engineers have deconstructed cases successfully for failure analysis. In fact, some of these same tools, such as decision trees, Bayesian networks, and contingent probabilities, are also useful in ethical cases.

The major drawback of case analysis is that the process can oversimplify and detach the observer, who is placed in the position to judge right or wrong. Another disadvantage of case analysis is that the cases are the result of a unique confluence of decisions and events, which may never be exactly replicated.

Since case analysis places the reader in the position of judge, we may not be effectively learning the real lessons for practicing engineers. By simply asking whether the engineer acted appropriately, we may learn not to do something specifically wrong, but have we learned the larger professional lessons? For example, were there other alternatives, and how can such options be incorporated into general designs? Hopefully, the cases will demonstrate the proper approaches that should have been taken, but there is still much uncertainty as to how directly these approaches would apply in another similar, but not identical, situation.

Another disadvantage of the case approach is the premature exposure of these cases, especially to students and others who many not have a significant amount of experiential context. Telling someone that they should boldly "blow the whistle" on wrongdoing may mean little to a person who has yet to draw a paycheck as a professional. Indeed, they should do the right thing, but dispassionately considering it while taking a class or reading a text may be a wholly insufficient preparation when the ethical dilemma appears some years later. Put crudely, sometimes even the best cases are figuratively "pearls before swine" (and, we have all been "swine" at numerous stages in our careers). Put a bit more nicely, even when the facts are right in front of us, we fail to grasp their meaning. We simply do not have the context or the perspective to see the lesson that seems so obvious a few years hence.

DRIVER'S EDUCATION ANALOGY

The lack of readiness to translate cases into professional action can be likened to driver's education training, where the basics of driving a vehicle from a textbook (i.e., the "Rules of the Road") is augmented by hypothetical cases and scenarios to engage the student in "what ifs" (e.g., what factors led to a bad outcome, like a car wreck?). Society realizes that new drivers are at risk and are placing other members of society at risk. Teenagers are asking permission to handle an object with a lot of power (e.g., hundreds of horsepower), a large mass (greater than one ton), with a potential to accelerate rapidly, and travel at high speeds (you've seen the BMW commercials). To raise the consciousness (and we hope, their conscientiousness, as well), we show films of what happens to drivers who do not take their driving responsibilities seriously. Likewise, in ethics class, we also show films and discuss cases that scare the future engineers in hopes that this will remind them of what to do or not to do when an ethical situation inevitably arises. We do this in a safe environment (the classroom with a mentor who can share experiences), rather than relying on one's own experiences.

But, memory fades with time. Psychologists refer to this as extinction, which can be graphed much like a decay curve familiar to engineers (Figure P.1). Unless the event is very dramatic, it will eventually be forgotten. This may be why educators often use cases with extremely bad outcomes (e.g., the toxic gas release in Bhopal, India, the Hyatt walkway collapse, or the Tacoma Narrows bridge failure) as

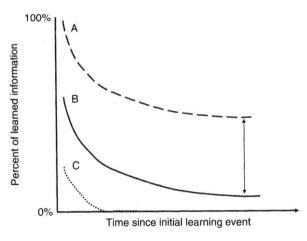

Figure P.1 Hypothetical extinction curves. Curve A represents the most memorable case and curves B and C represent less memorable ones. Curve C is completely forgotten with time. While the events in curves A and B are remembered, less information about the event is remembered in curve B because the event is less dramatic. The double arrow represents the difference in the amount of information retained in long-term memory.

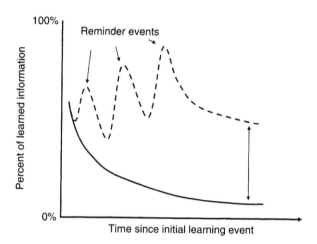

Figure P.2 Hypothetical extinction curves. The solid line represents a single learning event with no reminders (reinforcements). The dashed line shows reminder events in addition to the initial learning event. The double arrow is the difference in information retained in long-term memory as a result of adding reminders.

opposed to less extreme, yet more likely cases, such as the engineer who must decide whether to avoid a conflict of interest in selecting bids for a project.

If a bad thing happens to someone else, like the scenarios in the drivers' education films, they are not easily remembered, even if the results are gory. Much more memorable are events that occur to us personally. Anyone who has been in a car wreck will remember it for many years. The hope for young people is that the wreck that is sure to come will be severe enough to be memorable but not physically harmful to the driver.

There are at least two ways of trying to make sure that future professionals remember the importance of ethical decisions: (1) using powerful cases or (2) repeating the lessons. In the latter case, the extinction curve is bumped up periodically (Figure P.2). But, like driving, most ethical education occurs "out there." Only when a professional is confronted with the actual ethical problem will what has been learned be put to the real test. Learning engineering ethics, as is the case in learning to drive a car, results from a combination of formal study, interactive learning, and practice.

EXAMPLE CASE: PRIMING THE PUMP

The following case is a composite of numerous cases, using fictitious names and situations. We need not worry about "solving" it yet. It is provided merely to begin the conversation about responsible conduct and wise choices.

What about Bob?

A number of public health and social sciences researchers, from various universities, government agencies, and companies have been collaborating on interdisciplinary project that is trying to characterize the factors that lead to terrorism, especially in hopes of finding ways to prevent it. The United Nations is the major funding source for the research.

Bob, the only biomedical engineer on the research team, is charged with aiding the others in finding "low-tech" devices to extract and store fluid samples and to assist behavioral scientists in data management and informatics. Sid, one of Bob's colleagues, had an opportunity to visit Pakistan. While there, Sid interviewed young people in a terrorist training camp, using a questionnaire developed by the investigators. Sid also decided to retain a local health worker to draw blood and to take urine samples from the subjects to compare hormone levels with age-matched controls in the United States. The health care worker was a very valuable addition, since he had a rapport with and understood the cultural mores (especially since extracting blood and requesting urine specimens can be taboo in certain social groups). However, before Sid left, he neglected to ask the young people for their consent to participate in the research project. Sid had previously relied exclusively on anonymous databases from which he extracted data and he had not had experience in obtaining consent.

Sid asked Bob to take a look at the protocol for taking and handling the samples. Bob gave Sid some technical advice that greatly improved the local health worker's extraction process and that would double the shelf-lives of the samples to allow for lab analyses of better resolution and sensitivity upon the team's return to the United States. Bob, however, had enough experience to know that Sid should have gotten advance consent to obtain fluid samples, but did not think it was any of his business since this was an ancillary study separate from the larger study and Bob's role was clearly to provide engineering, not policy, advice.

After Sid returned home, he discussed his findings with the eight co-principal investigators at his university. Most of his colleagues wanted to publish *posthaste*, but a few thought that the results should be given to the UN in confidence, because of the timeliness and the nature of what was found. Disagreements began to arise about the appropriate journal where the data could be published, most wanting the journal that best fit their own respective academic discipline. When word got out that there was a new emphasis on endocrine differences, one key collaborator from another university became angry about being excluded.

In addition, major arguments and disagreements centered around who should be first author, who should be co-authors, and who should simply receive an acknowledgement. The grant project period has two more years remaining.

CASE ANALYSIS

1. What are the technical and management issues in this case?
2. What are ethical issues in this case?
3. What should Bob, the engineer, have done?
4. What should he do now?
5. What is different about ethical expectations of Bob compared to those of his colleagues?

The case presents the reader with an opportunity to assess one's own ethical "baseline." Of these questions, which are relatively straightforward? Which are complicated? Which seem to be outside at the bounds of engineering practice? All will be addressed in this book.

NOTES AND COMMENTARY

[1] S. Scott, 2000, "Dueling Theories: Thought Experiments in Cognitive Science," 22nd Annual Meeting of the Cognitive Science Society, Philadelphia, PA, 13–15 August 2000.

[2] H. Petroski, *To Engineer Is Human: The Role of Failure in Successful Design* (New York, NY: Vintage Books, 1992).

Chapter 1

Bioethics: A Creative Approach

The Brain—is wider than the Sky—
For—put them side by side—
The one the other will contain
With ease—and You—beside—
The Brain is deeper than the sea—
For—hold them—Blue to Blue—
The one the other will absorb—
As Sponges—Buckets—do—
The Brain is just the weight of God—
For—Heft them—Pound for Pound—
And they will differ—if they do—
As Syllable from Sound—

Emily Dickinson (1830–1886)[1]

Engineers do not have to be reminded of the importance of imagination. As Dickinson poetically and eloquently elucidates, the mind is almost limitless in its ability to reason. Given enough reliable information, humans are capable of amazing feats. We can see beyond the constraints of "what is" to "what can be." Imagination is arguably the greatest asset of the engineer. Engineers design because what they are able to see does not yet exist. That is the purpose of design. Likewise, imagination is a useful device to help us to understand the nature of bioethical challenges to practicing and future engineers, as well as an important means of designing ways to avert bioethical problems before they arise. So, what is it we are up against in the decades ahead?

Throughout this book, we consider bioethics questions relevant to the practice of engineering. Few, if any, of these will be answered to the complete satisfaction of every engineer. However, the very process of inquiry may help to give context and structure to some otherwise amorphous issues. So, let us begin with an essential query:

Bioethics Question: When is an action morally permissible and what kinds of behavior are morally obligatory?

Let us begin on a balmy Saturday morning in the North Carolina Piedmont, where a hundred or so new graduate students were gathered in the auditorium of Duke University's Chemistry Building. It was orientation week in late August, when Duke requires that every Ph.D.-seeking graduate student participate in a full day of ethics training.

The Responsible Conduct of Research training program developed by the Graduate School had had some inauspicious beginnings. Some years earlier, Duke, like other prominent research universities, risked losing funding from the National Institutes of Health because their research programs could not demonstrate that their researchers were receiving adequate training in ethics. So, as is often the case, Duke did not simply respond to the criticism and threat by instituting a *pro forma* program to keep the auditors happy, but decided to use this as a "teachable moment" and galvanizing event. Among the creative approaches to be adopted was to engage the students as soon as they arrive in Durham. Part of this commitment was to invite provocative speakers to evoke responses and to stir the imagination of the students. It was hoped that this first gathering would begin each student's self-directive, proactive, lifelong ethos of responsible research. On that Saturday, the students were asked to participate in two thought experiments.

In this vein, rather than to try the familiar ethical analytical approach of tediously (and often boringly) delineating, point by point, each possible bioethical aspect of engineering, we apply a tool used by philosophers to approach bioethics intuitively, using our imagination.

THOUGHT EXPERIMENTS

After the students had conducted a couple of projects and were given some advice on how to avoid ethical dilemmas, Duke University philosophy professor, Alex Rosenberg,[2] presented the following thought experiments:[3]

Thought Experiment 1

You are riding in a trolley car in San Francisco. You are the only passenger. Being the inquisitive type that you are, you watch the trolley driver's methods of controlling speed, changing tracks, and other aspects of handling the machinery. You ascend a hill, come to the top, and begin your descent. To your surprise and horror, during the downhill acceleration, the trolley driver jumps from the car, leaving you alone.

You move to the driver's seat and quickly find out why the driver jettisoned. The brakes have completely failed. The driver must have had a strong understanding of the laws of motions, since he jumped just before the car accelerated to the velocity at which anyone jumping from the car would suffer mortal injuries. In other words, you cannot leave the car. The horn and warning devices have also failed and do not operate. The only remaining control available to you is the lever that changes tracks.

The car continues to accelerate until you see on the track in front of you four workers (all the same size, sex, age, and demographics) who are standing on the tracks with their backs to you. There is no way that they can hear you due to the noisy equipment around them and the fact that your horn does not work. You also happen to notice that there is a switch track in front of you just before the workers that allows you to change tracks. However, a single worker (of the same size, sex, age, and demographics as the other four workers) is standing on the track in front of you to which you can switch.

What do you do?

Your ethical decision consists of only two choices. As an ethicist might ask: What is morally permissible? What is morally obligatory?

The group of students unanimously declared that switching the tracks was morally permissible since you would kill one person to save four. Most also agreed that you were morally obliged (not just permitted) to take the step.

Rosenberg then offered a second thought experiment:

Thought Experiment 2

You are a respected biomedical engineering major who has gone on to become a world-renowned surgeon at St. Bob, the Scientist Hospital, in San Francisco. You have kept up your skills in engineering and blended them with your surgical expertise to design a transplant device, based on nanotechnologies, that allows you to transplant any organ from one person to another in ten minutes with absolutely no rejection. (Hey, it's a thought experiment, so we can assume such things!) The transplant device is up and running at St. Bob's.

You are also a charitable physician. In fact, the reason you chose St. Bob's is that it has two distinct wings. Wing 1 is a world-class hospital that treats the rich and famous. Wing 2 is a public clinic that allows anyone to walk in and receive treatment. You practice in both wings, including dedicating *pro bono* services in Wing 2 every Tuesday from 9:00 p.m. to 1:00 a.m. Wednesday. Many of the patients you see in Wing 2 are completely destitute and have no family or friends. They are loners. Recently, a Wing 2 patient from a local homeless shelter was brought in to see you. You find in your interview with him that he has no family, no job, and no interest in bettering himself. You gave him a checkup, including a checkup of his lungs, liver, and pancreas, and found him to be in excellent physical health.

Earlier in the evening, in Wing 1, you saw a number of patients. Four of these patients were in dire need of transplants. One needed a pancreas, one a liver, and two needed a lung. In fact, if they do not receive them by midnight, all four will die. Your transplant machine could easily save them if the organs were available from a healthy donor, such as your Wing 2 patient.

What should you do? What is morally permissible in this case?

The Ph.D. students were flummoxed. Almost all of them said that it is morally impermissible to take the organs from the one person (and in the process kill him) to save the four. But, most could not say why or put their finger on the difference between the two scenarios.

Both situations involve an interface of technology with ethics. A crude technology in the first case (the level and switch track) and an advanced technology in the second case (the nanotechnology-based transplant machine) are part of the ethical decision. In both, we can change the situation, but the bioethical question is "Should we?"

Bioethics Question: What must a professional do to be trusted?

On the face of it, these cases are identical questions about what is moral; that is, are you permitted to kill one to save four? So, what is the big difference? If we are talking about a decision concerning anything but life, the question is almost silly. It would be merely arithmetic. But, when it comes to bioethics, such questions are often complicated and debatable. But, the difference basically comes down to trust.

Teachable Moment: Trust

Consider the people whom you can trust. Write down at least three reasons that they can be trusted. These can be intrinsic qualities, such as, to your knowledge they have never lied to you. They may also be extrinsic constraints, such as they are duty bound by law or a professional standard (e.g., a medical doctor). Often, they are combinations of the two (e.g., a trusted family doctor).

Explain how such trust applies to the practice of engineering. Discuss the types and degrees of controls that the engineering profession has over its members (extrinsic controls), and how much the social contract with engineering depends on characteristics of the individual engineer (intrinsic controls).

THE PRINCIPLE OF DOUBLE EFFECT

It is tempting to jump in and start analyzing the student's responses for right and wrong elements. That is the typical case-based approach. That is to say, philosophers and ethicists commonly apply "casuistry," an approach to determine right and wrong decisions or behavior by analyzing cases that illustrate general ethical rules. Among the problems with this approach is that it forces us to judge. Engineers, on the other hand, are problem solvers and designers, not judges. We are more comfortable about building a solution than about sitting in judgment (see Teachable Moment: The Engineer as Agent *versus* Judge). However, to prime the pump, let us begin to think about why the students responded as they did.

Teachable Moment: The Engineer as Agent *versus* Judge

 Philosopher Caroline Whitbeck is widely recognized as one of the first and strongest advocates of a design-based approach to professional ethics. She accentuated the synthetic as well as analytic elements in responses to moral problems.

In the 1980s and 1990s, she developed the parallel between ethical problems and design problems. She broke new ground in active learning methods in the teaching of engineering ethics and the responsible conduct of research, especially methods that place the learner in the position of the agent who must actually respond to the problem (rather than in the position of a judge who merely evaluates responses that have already been constructed). Among the ethical design criteria are collaboration, trust and trustworthiness, responsibility, and diligence (opposite of negligence).

Whitbeck draws four points of analogy between ethical problems and design problems:[4]

1. For interesting or substantive ethical problems, such as substantive problems of research design, there is rarely, if ever, a uniquely correct solution or response.
2. Some possible responses are unacceptable – there are wrong answers even if there is no unique right answer – and some are better than others.
3. However, solutions may have advantages of different sorts, such that where there are two candidate solutions, neither may be clearly better than the other.
4. A proposed solution must do all the following (in addition to being reasonably secure against accidents and miscarriages):
   - Achieve the desired performance or end – In the case of an ethical problem this might be to fulfill some moral responsibility, a professional responsibility, or a family responsibility.
   - Conform to given specifications or desired criteria – For an ethical problem, these specifications might include meeting the standards of care for one's profession, and not taking so much time that one fails in other particular commitments.
   - Be consistent with (usually unstated) background constraints, for example, that one not violate anyone's human rights and that one minimize the infringement of other rights.

Much like engineering design problems, ethical dilemmas are often ill-posed. Case-based approaches often erroneously assume that an engineer is confronted with a well-posed ethical problem; that is, one that is uniquely solvable (i.e., a unique solution exists) and one that is dependent upon a continuous application of data. That is to say, cases are best when a solution to the problem exists, the solution is unique, and the solution depends continuously on the data. So, if ethical cases were solved by something akin to the Laplace's equation,[5] casuistry is the way to go. By contrast, an ill-posed problem does not have a unique solution and can only be solved by discontinuous applications of data, meaning that even very small errors or perturbations can lead to large deviations in possible solutions.[6] Engineers are well aware of the complexity and challenge, as well as of the need to solve ill-posed problems. After all, most emergent problems are ill-posed. For example, an ill-posed problem may be solved by what are known as "inverse methods," such as restricting the class of admissible solutions using *a priori* knowledge. *A priori* methods include variational regularization using a quadratic stabilizer. Usually, this requires stochastic approaches, i.e., assumptions that the processes and systems will behave in a random fashion.

By extension, small changes for good or bad can produce unexpectedly large effects in an ethical decision. A seemingly small mistake, mishap, or ethical breech can lead to some dramatic, even devastating, results. Engineers are constantly warned to pay attention to the specific details of a design. The same admonition holds for ethical issues. In fact, the famous engineer, Norman Augustine has told us that "engineers who make bad decisions often don't realize they are confronting ethical issues."[7]

The design problem model of ethical problems represents problems as characteristically possessing more than one good (i.e., wise and responsible) solution. This allows the engineer to avoid the trap of situational ethics, where perpetrators are allowed to excuse their poor ethical choices since there are no right or wrong answers. This contradicts common sense and common morality. People know that there are indeed wrong answers. We also know that in most decisions, certain answers are more ethical than others, with some clearly wrong (the worst is known as the negative paradigm). Whitbeck's approach also obviates the over-simplistic and often flawed attempt to make an ethical decision into a multiple-choice question, selecting the best among two or more choices (this is often done in professional surveys, asking that as a practicing engineer "you be the judge"). This attempt at standardizing and modularizing professional ethics flies in the face of one of engineering's most important assets, creativity. The design problem model clarifies the character of what Whitbeck calls the agent's "synthetic" or constructive task of devising and improving responses, a much-preferred approach to the "analytic" approach of the judge.

An overarching advantage of the design approach is that it places the engineer squarely within the situation. From this vantage point, we can build solutions, rather than select them. Whitbeck has argued that moral problems are frequently mischaracterized as dilemmas; the roots of this misrepresentation and its relation to the failure of applied ethics are used to illuminate how to construct responses to moral problems. She argues that in many cases, there is a tendency to examine problems retrospectively, based on the medical case method approach, as opposed to placing the engineer in his or her most comfortable position of designer. The academic approach of the philosophers seldom matches what professionals and students need.

The medical case model tends to address acts, such as whether to participate in a specific medical action (e.g., use of genetic testing, withholding treatment, performing an abortion, or removing a feeding tube). Within enumerated constraints, society has granted the physician much latitude in treating a patient. The moral decision-making process in a medical situation depends on the general circumstances that may justify or fail to corroborate such acts. In other words, the attending physician in a hospital has inherited the patient's entire history and must make specific decisions about the act of patient care at that precise moment. Conversely, engineers are more adept at looking at problems from a life cycle perspective. Indeed, like the physician, part of our fact finding should include similar instances where a design has succeeded or failed, for instance, by using event and fault tree scenarios to understand the situation. However, the engineer goes well beyond the "act" to consider how the system can be better designed not only for damage control, but to prevent similar situations in the future. Whereas an attending physician may be limited to a decision of whether to remove a feeding tube for a specific patient, the design approach of the engineer is to look at the entire situation that led to the potentially tragic consequences of this decision. What could have been done to prevent the situation from occurring, such as concrete steps toward better technologies for health care and monitoring, and even better-designed roads and vehicles so that the accident that led to the coma and brain damage could have been wholly prevented? The life cycle design perspective calls for attention to all of the functions leading to an outcome, not simply the goodness or shortcomings of products.

One of the best examples of the life cycle viewpoint has been articulated by a nonengineer. Theologian Ronald Rolheiser[8] shares a parable of a community that dutifully, carefully, and honorably pulls a continuous stream of dead bodies from the town's river. Each day the moral people of the community give a proper burial to each deceased person represented by the body, as dictated by some type of social contract. In fact, these efforts led to a well-organized system to deal with the bodies, even providing jobs sorely needed by their citizenry. However, the community never makes the effort to travel upstream to find the source of the dead bodies! The case-based approach would look at the act of each person. Many of them are behaving quite morally by providing the burials and paying their respect to the dead. The agent-based design approach, however, requires a trip upstream. What is it about our systems in contemporary society that brings us to these bioethical dilemmas? Why does the "bottom-up" approach, with each person seemingly behaving morally, not lead to an overall ethical system? Stepping back and taking the comprehensive viewpoint clearly shows an overall societal ethical transgression. So long as we have a piecemeal, myopic view, no progress can be made in eliminating the core problem.

An example of this myopia is the recent argument for stem cell research because reproductive clinics are "just going to throw the embryos away anyway, so why not get something good out of them by using them in stem cell research?" The life cycle view would require that the whole process, including the ethics of the treatment of embryos and other aspects of reproductive technologies, be part of the whole argument. In this case the question is not limited to "What is my duty?" (known as the deontological view) or "What is the best result?" (known as the consequentialist view), but is really "What is going on here?" (i.e., a rational-relationship view, which considers the entirety of the issue, including duties, consequences, and the means toward these consequences).

Amy the Engineer

The design approach advantage is not limited to the big issues, but may be even more directly beneficial to the individual decisions (i.e., microethical decisions) of the practicing engineer.

As mentioned, according to Whitbeck, ethical questions are all too often incorrectly presented as "dilemmas." A dilemma is defined as a forced choice between two alternatives that are exactly and equally unfavorable. This leads to the representation of moral problems as though they were forced choices between two (or more) equally unwelcome alternatives. Again, this approach tries to make the solutions less messy by "pretending" that they are well-posed. We do this frequently in engineering, such as assuming "spherical chickens" when designing a poultry processing plant. Such a misrepresentation of moral problems as dilemmas implies that the only possible responses are the proposed courses of action (all of which are objectionable), stifling creative attempts to offer better alternatives. In a way, this is an over-prescribed design. Engineers perform best when they are not overly constrained by the client.

Whitbeck[9] shares the so-called Heinz dilemma recounted by Carol Gilligan in her book *In a Different Voice*.[10] Lawrence Kohlberg, a founder of moral development theory, posited to children a dilemma of whether a man named Heinz should steal a drug he cannot afford to save the life of his wife. Gilligan describes the performance of Jake, a child who does well by Kohlberg's criteria:

> Jake, at eleven, is clear from the outset that Heinz should steal the drug. Constructing the dilemma, as Kohlberg did, as a conflict between the values of property and life, Jake discerns the logical priority of life and uses that logic to justify his choice: For one thing, a human life is worth more than money,

and if the druggist only makes $1000, he is still going to live, but if Heinz doesn't steal the drug, his wife is going to die. (Why is life worth more than money?) Because people are all different and so you couldn't get Heinz's wife again. Jake understands the game to be one of finding the covering "principles," ordering them, and cranking out a solution, and this is the abstract exercise that he has successfully performed. What is notable is that Jake has not only learned the game but also recognizes it as an abstract puzzle; he aptly describes it as "sort of like a math problem with human beings." When ethical problems are constructed as abstract math problems with human beings it is no wonder that they have nothing much to do with moral life.

Another young respondent, Amy, refused to abide by the arbitrary conditions and constraints. As a result, she did not perform well by Kohlberg's criteria because she insisted on trying to work out a better response to Heinz's problem. When Amy was asked whether Heinz should steal the drug, she proposed new alternatives:

Well, I don't think so. I think there might be other ways besides stealing it, like if he could borrow the money or make a loan or something, but he really shouldn't steal the drug but his wife shouldn't die either.

Asked why he should not steal the drug Amy replies:

If he stole the drug, he might save his wife then, but if he did, he might have to go to jail, and then his wife might get sicker again, and he couldn't get more of the drug, and it might not be good. So, they should really just talk it out and find some other way to make the money.

Gilligan interprets Amy's solution as "a narrative of relationships that extends over time." Like the essay question's advantage over the multiple-choice test, the narrative can express "the dynamic character of a situation with its unfolding possibilities and resolutions." This avoids the need for abstraction.

Amy's answers follow an engineering paradigm for problem solving. She looks for numerous possibilities and alternative solutions. She refuses to be pigeonholed into predetermined solutions. She is put off by the arbitrariness of the design landscape, and seeks accommodation of her client's needs. Amy deserves more credit than Gilligan offers. Her response is not simply addressing relationships, as Gilligan asserts. In fact, as Whitbeck points out, Amy did not even identify the failure of Heinz and his spouse's relationship as the result of Heinz's jail sentence. She was more concerned that Heinz's wife may need the drug again. Amy was still seeking a solution to the root problem (need for the drug) and providing a sustainable solution. As an "engineer," Amy is more concerned that the process supports acceptable outcomes on an ongoing basis than merely solving a short-term problem.

The principal lesson seems to be that cases can be arbitrary and can limit engineering (and ethics) creativity, by imposing a forced choice between two (or a few) right and wrong answers. Amy, as most good engineers do, begins with an open-ended set of possibilities and brainstorms to arrive at an acceptable means of solving the problem. Next, she looks at the feasible options and explores one or more reasonable solutions, including attempts to persuade the pharmacist to help, obtaining a loan, and other practical actions.

Amy also reminds us that engineers need to keep thinking. We must analyze each possible outcome from a number of perspectives, including those that are not so easy for technical types to grasp, such as the sociological and psychological risks and benefits. From there, the engineer can follow a critical path to an acceptable solution.

Questions

1. Consider a case of bioethical importance. The case can be a negative or a positive paradigm.
2. What are the strengths and weaknesses of this case (as per Whitbeck)?
3. How might you be more of an ethical agent than a judge in answering these questions?
4. Identity a current bioethical issue that is being approached as an abstract math problem with human beings. What are the shortcomings of this approach?

The first thought experiment introduces us to an important consideration in bioethics, the so-called principle of double effect. Think of this as an equation with two constraints. The first constraint stems from Socrates' moral challenge to "first do no harm." The second constraint was fully first articulated by the thirteenth-century Italian theologian, Saint Thomas Aquinas:

Good is to be done and promoted and evil is to be avoided.[11]

In its most simple form, the doctrine says that an act that leads to negative side effects is permitted, but deliberate harm (even for good causes) is wrong. This seems similar in both scenarios, but the second thought experiment involves actively harming one person, whereas the first thought experiment involves harm as a side effect. In the trolley case, you are just redirecting the harm. However, in the second case you have to do something to the homeless man to save the four. In the first case, no person has more rights than anyone else not to be crushed, but in the second case, the homeless man has a right not to be killed, even though his organs would be put to good use.

These distinctions among professional decisions point to the fact that engineers will face risk trade-offs and double effects to some degree during their careers. For example, in designing a device, some persons (e.g., immunocompromised) may be harmed by its use. If all of a certain group are harmed at the expense of another group's benefit, this could be conceived as intentional harm. The way to mitigate this harm is to disclose fully the shortcomings of the device and to work on improvements that will decrease the likelihood of harm. The device must do inherent good (e.g., provide insulin to diabetics or deliver drugs to ill patients). In other words, the designer must not actually intend to accomplish the bad effect (harming immunocompromised people). The ill effect is simply unavoidable in the effort to do good. If another approach would avoid the bad and still accomplish the good, then such an option is the preferred and obligatory act.

Another provision of the double effect is that the good effect must be at least as directly an effect of the action as is the negative (bad) effect. In particular, the bad must not cause the good effect. Also, the benefit of the bad must not outweigh the benefit of the good. Finally, another ethically acceptable approach without the side effects must not be available.

A bioethical example of the double effect is that of the vaccine. A government and any manufacturer normally know the population risk of administering a vaccine to the public. Most are expected to benefit, with a small number with adverse side effects. And, from this small group, a subset of vaccine recipients will die. The lives are saved as a result of the vaccine, not as a result of the deaths of those who die of side effects. The side effects do not advance any goals of the drug manufacturer. Thus, the side effects are not intended as a means to any other outcome. Finally, the proportion of lives saved is very high compared to the lives lost, satisfying the requirement that benefits outweigh the negative outcomes. It would not be permissible to produce and administer the vaccine with these side effects (deaths) if another means of preventing the disease were available.

Teachable Moment: Who Was Van Rensselaer Potter?

The word "bioethics" was coined by Van Rensselaer Potter (1911–2001), an American biochemist. He wrote two important books. The first, *Bioethics: A Bridge to the Future* (Prentice-Hall), wherein he coined the term, was written in 1970 as a call to integrate many scientific and engineering disciplines to provide for an environment that was both livable and sustainable (that term as applied to the environment did not yet exist either). Bioethics was subsequently coopted by the medical community and was redefined specifically to address the morality of medical practice (including appropriateness of treatments and technological advances). Meanwhile, Potter grew more and more respectful of the "land ethic" espoused by Aldo Leopold, and attempted to recapture his original meaning and coined another phrase, "global bioethics," in this 1988 book, *Global Bioethics: Building on the Leopold Legacy* (Michigan State University Press). For our purposes, the term also harkened a new role for biosystem engineering.

Questions

1. Consider the prefix "bio" and how it applies to both terms: bioethics and global bioethics.
2. Contrast these terms from Potter and Leopold's perspectives with those of the contemporary definition of biomedical ethics.
3. Why has the term bioethics morphed in meaning so much since Potter coined it?

CREDAT EMPTOR

The thought experiments demonstrate, at least at some intuitive level, that trust is one of the distinguishing attributes of the professional. Engineering is no exception. Arguably, the engineer's principal client is the public. And, the clients need know little about the practice of engineering because, owing to the expertise, they have delegated the authority to the engineer. Just as society allows a patient to undergo brain surgery even if that person has no understanding of the fundamentals of brain surgery, so also does our society cede authority to engineers for design decisions. With that authority comes a commensurate amount of responsibility, and when things go wrong, culpability. The first canon of most engineering ethical codes requires that we "hold paramount" the health and welfare of the public. The public is an aggregate, not an "average." So, leaving out any segment violates this credo. Thus, even though most people favor an approach, it is still up to the engineer to ensure that any recommended approach is scientifically sound. In other words, no design, even a very popular one, is to be recommended unless it meets acceptable, scientifically sound standards. Conversely, the most scientifically sound approach may have unacceptable social side effects. The right and just decision is not a "popularity contest." Once again, the engineer is put in the position of balancing trade-offs.[12]

Bioethics Question: What is the role of values in bioethical decisions?

Recently, much debate has occurred within the legal community as to whether the use of lethal injection to execute convicted criminals is "cruel and unusual punishment." The debate includes positions across a wide spectrum, ranging from lethal injection being morally reprehensible to its being morally obligatory. The US Constitution prohibits cruel and unusual punishment, so it is not surprising that the debate reached the United States Supreme Court. In the spring of 2006, the Court refused to examine this issue, but a large controversy is brewing. A central bioethical issue important to engineering is whether the two-step system of injection may cause extreme pain if the first chemical, an anesthetic, is not properly injected. Thus, when the actual lethal dose of the second chemical is injected, the resulting pain puts the state in the position of a cruel executioner.

Perhaps another thought experiment will help us to begin to consider the debate rationally:

Thought Experiment

Dante is a chemical engineer employed by InfernoChem, Inc. (ICI). Dante's expertise is in small-scale reactors. Recently, he developed a manifold system that allows two different chemicals to be injected at specific preset times. His device is particularly useful in injecting catalysts into reactors. For example, a polymerization step in one of the reactors requires that the first catalyst be injected when the reactor temperature is $50\,^{\circ}C$ and the second catalyst injected 10 minutes later when the reactor temperature reaches $80\,^{\circ}C$. For the past 18 months, ICI has required that Dante continuously miniaturize the two-step injection system. In fact, the company gave Dante two monetary awards for his progress. Recently, the system was shrunk to 25 cubic centimeters of fluid volume.

Dante's perspective is that of a devout Roman Catholic. He opposes capital punishment for religious and personal reasons. He briefly mentioned these convictions to Tom, his immediate supervisor, on two occasions after executions were covered in the local news. Dante discussed this in a very matter-of-fact and dispassionate way.

Tom is aware that the reason for the short deadline to miniaturize Dante's injection system is that ICI has a contract with the State Bureau of Prisons to produce a single injection system that would ensure that prisoners receiving lethal injections get the entire dose of the anesthetic drug before the toxic dose is administered. However, based on Dante's opposition to the death penalty, Tom has chosen not to disclose the actual application of the new technology.

Was Tom morally permitted to withhold the company's plans from Dante? What is the appropriate thing for Dante to do?

Teachable Moment: Capital Punishment, Abortion, and the Definition of Human Life

The American Medical Association (AMA) policy on abortion is:

> The Principles of Medical Ethics of the AMA do not prohibit a physician from performing an abortion in accordance with good medical practice and under circumstances that do not violate the law. (III, IV)[13]

And

> (1) [A]bortion is a medical procedure and should be performed only by a duly licensed physician and surgeon in conformance with standards of good medical practice and the Medical Practice Act of his state; and (2) no physician or other professional personnel shall be required to perform an act violative of good medical judgment. Neither physician, hospital, nor hospital personnel shall be required to perform any act violative of personally held moral principles. In these circumstances, good medical practice requires only that the physician or other professional withdraw from the case, so long as the withdrawal is consistent with good medical practice. (Sub. Res. 43, A-73; Reaffirmed: I-86; Reaffirmed: Sunset Report, I-96; Reaffirmed by Sub. Res. 208, I-96; Reaffirmed by BOT Rep. 26, A-97; Reaffirmed: CMS Rep. 1, I-00)[14]

Compare and contrast this policy that the AMA policy on capital punishment:

> An individual's opinion on capital punishment is the personal moral decision of the individual. A physician, as a member of a profession dedicated to preserving life when there is hope of doing so, should not be a participant in a legally authorized execution. Physician participation in execution is defined generally as actions which would fall into one or more of the following categories: (1) an action which would directly cause the death of the condemned; (2) an action which would assist, supervise, or contribute to the ability of another individual to directly cause the death of the condemned; (3) an action which could automatically cause an execution to be carried out on a condemned prisoner.

(See the rest at http://www.ama-assn.org/apps/pf_new/pf_online?f_n=browse&doc=policyfiles/-HnE/E-2.06.HTM&&s_t=&st_p=&nth=1&prev_pol=policyfiles/HnE/E-1.02.HTM&nxt_ pol=policyfiles/HnE/E-2.01.HTM&.) Compare this policy to an engineering code of ethics. Identify commonalities and any inconsistencies between the two policies, especially the differences between the engineer's paramount focus on the public good and the physician's focus on the patient.

In particular, consider that for lethal injection executions, the AMA prohibits a physician aiding the execution by:

- selecting injection sites;
- starting intravenous lines as a port for a lethal injection device;
- prescribing, preparing, administering, or supervising injection drugs or their doses or types; inspecting, testing, or maintaining lethal injection devices; and
- consulting with or supervising lethal injection personnel.

Some have argued that, in an abortion, the doctor does not "kill" the unborn child intentionally, but makes the "choice" to evict the child from the womb. This position argues that the unborn child's right to life does not "entail that the child *in utero* is morally entitled to the use of the mother's body for life support."[15] Tragically, this position is tantamount to subjugating the unborn child to a "parasite."

Questions

1. Compare the technical and moral differences and similarities between the prohibitions against participating in an execution and the applications of the technologies of abortion.
2. Is the child *in utero* living? Is it human? Be specific in these definitions. If so, explain the bioethical distinction between calling abortion an act of killing and a choice to "evict" the child.

One of the means of determining whether an ethical decision is sound is to see if we are being morally consistent in our values. For example, some have pointed out inconsistencies in those who support the death penalty, yet ideologically oppose abortion, embryonic stem cell research, euthanasia, and assisted suicide, as well as other acts against humans at the beginning and end of life. The counterargument is that in these other situations, innocents are being killed, while in the case of capital punishment, if due process is followed, the person being executed is being treated fairly. Again, it comes down to what is most cherished, what has the most value. If human life is sacred, as most believe (however they define it), can it ever become unsacred, as in the case of a convicted murderer?

. The bottom line is drawn with values. What we perceive to be precious is protected and cherished. What we see as having less value becomes commodified. Less value translates to less protection and greater expendability.

THE GOOD ENGINEER

Engineering is an active process of solving problems and building new things using computers and other technologies.

Duke University's K-PhD Program[16]

Bioethics Question: What does it mean to be a good engineer?

From the Ancient Greeks, excellence is a dichotomous phenomenon. Good was considered to be given and almost undefinable (although Socrates can be credited with starting the process of defining it). Excellence requires skill and character. Our academic and professional preparation is designed to give us the former and some of the latter. For example, the engineering curriculum continues to place high demands on the student's grasp of mathematical and scientific concepts. This has never let up. However, the curriculum of today increasingly requires an understanding of the social sciences and humanities and demands core competencies in interpersonal skills.

Engineering is active. It is a process that takes the raw materials and energy that exist at a given time and creates new things and improves things that already exist. Engineers solve problems and build. Engineers do things.[17] We are seldom satisfied merely in possessing information or mastering skills.

The real test of engineering is when we put our knowledge and aptitude into practice. I noticed this in the Duke engineering students who recently returned from Indonesia and Uganda after participating in the Engineers without Borders projects abroad. They were joyous about their opportunities to apply the theoretical information in real projects. It is this eagerness to apply what we know that is a gift that has truly made the world a better place. Many of the improvements in public health, safety, and quality of life are largely attributed to engineers.

In North America and Europe, public works projects designed by engineers have given us clean water,[18] which has prevented many of the diseases responsible for the majority of premature deaths and disabilities so common 100 years ago. Pollution control equipment has allowed for cleaner air. Vehicle and transportation designs continue to improve the safety of travel. Chemical engineering advances have improved product manufacturing, leading to higher quality and safer consumer products and pharmaceuticals. And, biomedical devices and systems have lengthened and extended an improved quality of life to millions.

The engineering call will be even stronger in the future. The knowledge and creativity of the engineer will have to grow in proportion to the increased societal expectations. Kristina Johnson, Dean of Duke University's Pratt School of Engineering, characterized the engineer's future obligation to society:

> [A]s an engineering dean, I'd argue that it is our responsibility as good citizens of the planet to solve many of our global problems, such as developing renewable energy sources, purifying water, sustaining the environment, providing low-cost health care and vaccines for infectious diseases, to mention a few. Coupled with global climate issues, transportation and urbanization, we need all the technical horsepower we can educate.[19]

As in all aspects of engineering, however, there are challenges and obstacles. We are familiar with design challenges, such as the unique conditions of situations that make us think outside of the box. An equation or model seems to work well in most cases, but those few instances that fail can be the difference between good engineering and poor design. Or, we commonly experience the challenge of the lessons to be learned when we move from a prototype to an actual application. Slight changes in scale or complexity (i.e., effects of some unknown variable) limit the application of a design. One specific challenge, the subject of this book, is the ethical challenge, specifically the challenge of how engineering decisions and actions affect human life, for good or ill.

FEEDBACK AND ENHANCEMENT OF DESIGN

> Engineering, like poetry, is an attempt to approach perfection. And engineers, like poets, are seldom completely satisfied with their creations.
>
> Henry Petroski (1985)[20]

Engineering is not only active, but, as Petroski reminds us, it is a system filled with feedbacks. We are frequently told where we fall short, but we are sufficiently optimistic to recognize our progress. This

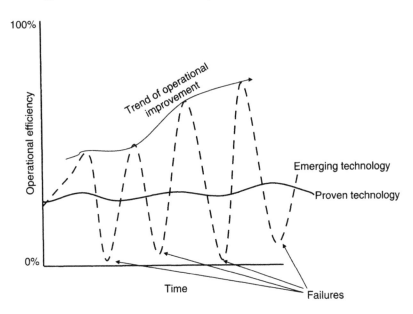

Figure 1.1 Hypothetical failure rate of new *versus* proven technologies.

continuous improvement was formalized by another engineer, W. Edwards Deming, who established the total quality management (TQM) process. TQM is defined as "a management approach of an organization, centered on quality, based on the participation of all its members and aiming at long-term success through customer satisfaction, and benefits to all members of the organization and to society."[21] This is a strong mandate. However, it should not imply that engineering advances are linear. In fact, growth is incremental and is often accompanied by setback, hypothetically shown in Figure 1.1.

Especially in biomedical and biosystematic applications, engineers push the envelope. They must. The growth in medical science is directly attributable to new technology, pharmacology, and systems. The unforeseen setbacks depicted in Figure 1.1 go with the territory, but the public does not take kindly to those that could have been prevented if reasonable steps had been taken. It hurts the professional and the profession because a preventable failure delays the overall advance in engineering solutions to problems, as shown in Figure 1.2. It is even worse if the setbacks are the results of mishaps and mistakes. And, since engineering requires intensive and extensive training and experience, most situations ripe for mistakes and mishaps are only detectable to fellow engineers. This was well put by Petroski[22]:

Engineers . . . are not superhuman. They make mistakes in their assumptions, in their calculations, in their conclusions. That they make mistakes is forgivable; that they catch them is imperative. Thus it is the essence of modern engineering not only to be able to check one's own work but also to have one's work checked and to be able to check the work of others.

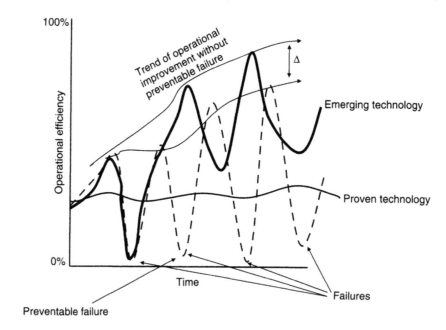

Figure 1.2 Hypothetical effect of preventable failure on overall advance of a new technology. Solid line is the technological advance had the preventable failure not occurred. The delta (Δ) is the difference in technology development attributed to the preventable mistake. Without the failure, the technological progress would be greater by delta.

Teachable Moment: The Good Engineer

The famous physicist Freeman Dyson (born 1923) said:

> A good scientist is a person with original ideas. A good engineer is a person who makes a design that works with as few original ideas as possible. There are no prima donnas in engineering.

He also said:

> If we had a reliable way to label our toys good and bad, it would be easy to regulate technology wisely. But we can rarely see far enough ahead to know which road leads to damnation. Whoever concerns himself with big technology, either to push it forward or to stop it, is gambling in human lives.

Questions

1. Give two examples in bioengineering that demonstrate Dyson's first point and two emerging technologies in which we presently need to heed the warning of his second quote.
2. What does this tell us about risk taking, especially the optimum between necessary and irresponsible risks?
3. Give an example of when avoiding a risk leads to bigger problems than taking the risk.

In addition to the aesthetic aspects of design, it must be underpinned by sound science. Herbert Hoover, one of only two US presidents trained as an engineer (the other is Jimmy Carter), eloquently captured this balance in his memoirs (see The Profession of Engineering).

The Profession of Engineering[23]

By Herbert Hoover

It is a great profession. There is the satisfaction of watching a figment of the imagination emerge through the aid of science to a plan on paper. Then it moves to realization in stone or metal or energy. Then it brings jobs and homes to men. Then it elevates the standards of living and adds to the comforts of life. That is the engineer's high privilege.

The great liability of the engineer compared to men of other professions is that his works are out in the open where all can see them. His acts, step by step, are in hard substance. He cannot bury his mistakes in the grave like the doctors. He cannot argue them into thin air or blame the judge like the lawyers. He cannot, like the architects, cover his failures with trees and vines. He cannot, like the politicians, screen his shortcomings by blaming his opponents and hope that the people will forget. The engineer simply cannot deny that he did it. If his works do not work, he is damned. That is the phantasmagoria that haunts his nights and dogs his days. He comes from the job at the end of the day resolved to calculate it again. He wakes in the night in a cold sweat and puts something on paper that looks silly in the morning. All day he shivers at the thought of the bugs which will inevitably appear to jolt his smooth consummation.

On the other hand, unlike the doctor his is not a life among the weak. Unlike the soldier, destruction is not his purpose. Unlike the lawyer, quarrels are not his daily bread. To the engineer falls the job of clothing the bare bones of science with life, comfort and hope.

No doubt as years go by people forget which engineer did it, even if they ever knew. Or some politician puts his name on it. Or they credit it to some promoter who used other people's money with which to finance it. But the engineer himself looks back at the unending stream of goodness that flows from his successes with satisfactions that few professions may know. And the verdict of his fellow professionals is all the accolade he wants.

Cogito, ergo sum. (Latin: "I think, therefore I am.")[24]

René Descartes (1637)

In the broadest sense, engineering is an outgrowth of rationalism; that is, engineers apply reason to solve problems. Descartes' observation can be extended to say: "I think, therefore I design." But, on what do we base such designs, and if these are flawed, what makes an engineer choose another path? According to physicist-philosopher Thomas S. Kuhn, there are essentially two types of paradigm shifts: those that result from a discovery caused by encounters with anomaly and those that result from the invention of new theories brought about by failures of existing theories to solve problems the theory defines. In the case of a paradigm shift brought about by discovery, the first step in shifting the said paradigm is the discovery of the anomaly itself.

Engineers are constantly interpreting theory to provide realistic examples. In this regard, Louis Pasteur can be considered to be among the first "modern era engineers." He was concerned with advancing the state of the science of diseases in parallel with practicing private and public health care. Thus, engineering cannot rely on a single-minded paradigm – in this case, Pasteur's work found that the standard medical practice paradigms would fall short (e.g., in treating anthrax). He also found that a new foundational paradigm was needed (e.g., an enhancement of germ theory).

Thus, the good engineer explores the anomalies. Once the paradigm has been adjusted so that the anomalous becomes the expected, it is said that the paradigm change is complete. In the case of a paradigm shift that results from the invention of new theories caused by the failure of existing theory, the first step is the failure itself (when the system in place fails, the creation of a new system is necessary). Several things can bring about this failure: observation of discrepancies between the theory and the fact, changes in the surrounding social or cultural climate, and academic and practical criticism of the existing theory. While these problems have generally been known for a long time, Kuhn noted that on numerous occasions the scientific community has been highly resistant to change in paradigms. Kuhn also noted that in the early stages of a paradigm, it is easy to invent theoretical alternatives that can be placed on a given set of data. However, once the paradigm is more well established, these theoretical alternatives are strongly resisted.[25]

Similarly, the process through which advances are made, and paradigms developed, may also be examined through the lens of the social sciences, especially economics. Although there are numerous examples of the differences in thinking between economists and engineers, this is an area of agreement. For example, economist John Maynard Keynes wanted to understand the dynamics of economics fundamentally. At the same time, he wanted to solve the problem of economic depression. Thus, Keynes was taking the engineering perspective, seeing the fusion of these goals as a joint desire to extend basic knowledge and reach applied goals.

ENGINEERING BIOETHICS AND MORALITY

I not only acknowledge but insist upon the fact that morality only limits the range of morally acceptable answers, it does not always provide a unique solution to a moral problem. I hold that it is very rare that any ethical theory, including mine, can resolve any controversial ethical disagreement.

Bernard Gert[26] (twentieth-century ethicist)

The more complicated the problem, the less certain and unique is the solution. Why does engineering attract the best and brightest young minds? Surely, the attraction goes beyond the mathematical and scientific accolades. It goes beyond the sorting process of high analytical and quantitative scores on college admission exams. It must have something to do with a calling. That calling is an integration of a myriad of factors and variables. This integration is an uneasy one. It is simultaneously rewarding and risky, as are all great callings. It incorporates the most basic and most complex algorithms, given that it involves science and people, respectively.

Humans are quite complex. Since engineering is a human enterprise, it should come as no surprise that applying the sciences to solve human problems is a complicated business.

Mathematics has evolved to deal with such complications. In 1902, the mathematician Jacques Hadamard defined a well-posed problem (*un problème bien pose*) as one that is uniquely solvable (*déterminé*). A year earlier he defined ill-posed problems (*questions mal posées*) as those without a unique solution; that is, such problems depend in a discontinuous way on the measurements so that tiny errors can create very large deviations in the solution. The modern rendition of this phenomenon is the "butterfly effect" (see Discussion Box: Ethics and the Butterfly Effect). Hadamard believed anything that is physically important must be well posed. We now know better that numerous engineering, medical, and physical science problems are ill-posed. It can be argued that engineering ethics cases may often be even less uniquely solvable than some of the most complicated physical and engineering problems, such as the tragic case of Jesica Santillan, where a usually well-managed variable (blood type) was mistakenly mismatched, leading to a cascade of events that ended in tragedy (rejection of the transplanted heart and subsequent death).

Discussion Box: Ethics and the Butterfly Effect

The Butterfly Effect is the name for "sensitive dependence upon initial conditions,"[27] as a postulate of chaos theory. To engineers, the effect can mean that a small change for good or bad can reap exponential rewards or costs.

Edward Lorenz, at a 1963 New York Academy of Sciences meeting, related the comments of a "meteorologist who had remarked that if the theory were correct, one flap of a seagull's wings would be enough to alter the course of the weather forever." Lorenz later revised the seagull example to that of a butterfly in his 1972 paper "Predictability: Does the Flap of a Butterfly's Wings in Brazil Set off a Tornado in Texas?" presented at a meeting of the American Association for the Advancement of Science, Washington, DC. In both instances, Lorenz argued that future outcomes are determined by seemingly small events cascading through time. Engineers and mathematicians struggle mightily to find ways to explain (and to predict) such outcomes of so-called ill-posed problems. As engineers, we generally like orderly systems so we prefer a well-posed problem; that is, one that is uniquely solvable (i.e., a unique solution exists) and one that is dependent upon a continuous application of data. By contrast, an ill-posed problem does not have a unique solution and can only be solved by discontinuous applications of data, meaning that even very small errors or perturbations can lead to large deviations in possible solutions.[28] Finding the appropriate times and places to solve ill-posed problems is a promising area of mathematical and scientific research.

By extension, small changes for good or bad can produce unexpectedly large effects. Upfront considerations of possible losses of privacy in designing information systems or possible malfunctions under certain physiological constraints when designing medical devices can prevent large problems down the road. Allowing for and dutifully considering the ideas from any member of the design team, no matter how junior, can lead to a successful study and help avoid costly mistakes. We sometimes hear after the fact how someone had noticed a disturbing trend in a laboratory study or an apparent cluster of effects in a group of early adopters, but whose questions and complaints were ignored until a larger data set eventually showed that the device or drug caused unnecessary ill effects. Many catastrophic failures, upon analysis, had at least one internal memorandum from an engineer presciently stating misgivings.

Engineers ignore this information at their peril. Ignoring small details can lead to big problems.

"SMALL" ERROR AND DEVASTATING OUTCOMES

Duke University is blessed with some of the world's best physicians and medical personnel. As a research institute, it often receives some of the most challenging medical cases, as was the previously mentioned case for Jesica Santillan, a teenager in need of a heart transplant. Although the surgeon in charge had an impeccable record and the hospital is world renowned for such a surgery, something went terribly wrong. The heart that was transplanted was of a different blood type than that of the patient. The heart was rejected, and even after another heart was located and transplanted, Jesica died due to the complications brought on by the initial rejection. The logical question is how could something so vital and crucial and so easy to know as blood type be overlooked? It appears to be a systematic error. The system of checks and balances failed. And, the professional, i.e., the surgeon, is ultimately responsible and primarily accountable for this or any other failure on his watch.

What can we learn from the Santillan case? One lesson is that a system is only as good as the rigor and vigilance given to it. There is really no such thing as "auto pilot" when it comes to systems. Aristotle helps us here. He contended that the whole is greater than the sum of its parts. This is painfully true in many public health disasters. Each person or group may be doing an adequate or even superlative job, but there is no guarantee that simply adding up each of the parts will lead to success. The old adage that things "fall through the cracks" is a vivid metaphor. The first mate may be doing a great job in open waters, but may not be sufficiently trained in dire straits when the captain is away from the bridge. A first response team may be adequately trained for forest fires (where water is a very good substance for firefighting), but may not properly suited for a spill of an oxidizing agent (where applying water can make matters considerably more dangerous). Without someone with a "global view" to oversee the whole response, the perfectly adequate and even exemplary personnel may contribute to the failure.

Systems are always needed and these systems must be tested and inspected continuously. Every step in the critical path that leads to failure is important. In fact, the more seemingly "mundane" the task, the less likely people are to think a lot about it. So, these small details may be the largest areas of vulnerability. Like the butterfly effect in chaos theory, the chain of events or critical path of one's decision will ultimately determine whether it is a good one or a bad one. One must wonder how many presurgery meetings before the Santillan case had significant discussions on how to make sure that the blood type is properly labeled. One can venture that such a discussion occurs much more frequently now in pre-op meetings (as well as hospital board meetings) throughout the world. Although, it is quite likely even this focus has attenuated in the years since the tragedy. The public and our peers will judge whether we apply due diligence or whether our designs and projects are impaired by something we should have known and considered.

TECHNOLOGY, ENGINEERING, AND ECONOMICS

Although engineers are a diverse lot, most are more than a little utilitarian. They strive to provide the greatest amount of goods and services to the greatest number (however these terms are defined and constrained by the design specifications). The National Academy of Engineering recently declared that "engineers and their inventions and innovations have helped shape the changes that have made our lives more productive and fruitful."[29] But what does it mean to become more fruitful? It must be something beyond pure utilitarianism. To begin, we can consider the economic implications.

Bioethics Question: Does biosystem engineering ethics go beyond utility in defining what is "right?"

Technology is the obvious indicator of biosystem engineering. It is the "workhorse" that delivers on biosystematic designs. In economics, technology can be considered a tool of empowerment, in that it empowers producers to generate more output from given levels of the two inputs, labor and capital. In this sense, as a catalyst in chemistry is a substance that increases the rate of reaction, technology is a catalyst for production of output – as it increases the amount of output we get from given inputs.[30] Technology allows for the use of more advanced capital, which results in better and faster ways to create output. Producers are rendered more efficient, as they are able to produce more output, given the same amount of input. This results in greater profit, which results in economic growth. For example, if we consider the basic supply–demand model, we see that improvements in technology cause the supply curve to shift out to the right, meaning that producers will create more output for any given price (Figure 1.3). This raises equilibrium output and lowers equilibrium price. Equilibrium, in this case, is used to refer to the price level at which the aggregate supply curve (an upward sloping line that illustrates how much producers would be willing to supply at any given price) crosses the aggregate demand curve (the downward sloping line that shows how much of a good consumers would demand at any given price). This price level is called the "equilibrium" price because here the amount supplied equals the amount demanded. It is a "balance," as any further increase in supply would disrupt the equilibrium. The same connotation of "balance" is seen in the usage of "equilibrium" in chemistry and thermodynamics, which is the study of work, heat, and energy on a system.

An object is said to be in thermodynamic equilibrium when it is in thermal, chemical, and mechanical equilibrium. It is observed that some properties of an object can change when the object is heated or cooled. Should two objects be brought into physical contact, an initial change in the property of both

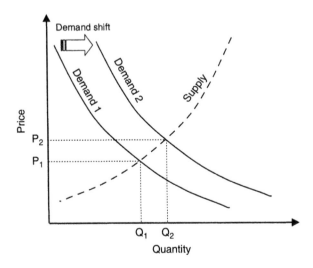

Figure 1.3 Supply–demand economic model. The increase in demand results in a relative increase in price and quantity needed to reach a new equilibrium point on the supply curve.

objects results. For instance, heat, a form of energy, is transferred between these two objects when they are brought into contact with one another. Eventually, this transfer of energy stops. At this point the objects are said to be in thermodynamic equilibrium.[31] Thus in both the economic and the thermodynamic definition of equilibrium, the system is said to be in balance; nothing changes unless acted upon by an exogenous force. In the case of market equilibrium, these endogenous forces act to shift the supply curve or the demand curve, subsequently changing where they cross to form the equilibrium price. Some examples of exogenous forces are increased costs for producers, changes in the regulatory environment, or changes in wealth of consumers.

This supply–demand relationship is analogous to the engineers paramount ethical demand for safety, as shown in Figure 1.4. Like price in the economic model, the safety of an engineered device or system translates into costs. However, this is not a linear relationship. An unsafe product is not cheap if all the factors are included in the cost calculations. Primary costs of production are only part of the equation. Premature obsolescence and failures can lead to significant costs of lawsuits, lack of public trust, and recalls and warranty costs.[32] A well-designed device may still be a failure if its useful life is too short or too uncertain.

Technological advances often improve the use of scarce resources, which are then allocated through the economic system. Therefore technology broadens the horizon through which economics operates. When firms invest, they increase capital; and increasing input means more output, i.e., more economic growth. Depreciation on capital stock yields less output. The term sustainability in economics is similar to the concept of environmental sustainability. In economics it is used to describe capital withstanding time and continuing to function to facilitate the production of output. Technology allows higher levels of sustainability for capital, including a predictable and reliable engineered system.

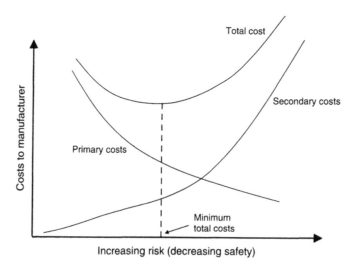

Figure 1.4 Safety and risks associated with primary and secondary costs. Increased safety can be gained by considering secondary costs in product and system design.
Adapted from: M. Martin and R. Schinzinger's, 1996, *Ethic in Engineering*, McGraw-Hill, New York, NY.

Teachable Moment: The Dismal Scientist *versus* the Technological Optimist

British political economist Robert Malthus[33] (1766–1834) predicted that starvation would result as projected population growth exceeded the rate of increase in the food supply. His forecast was based on the population growth at an exponential rate whereas the food supply growth rate would be linear. These predictions greatly underestimated the role of technology in increasing the global food supply, causing many to doubt such economic models. In a later incarnation of Malthusian thinking, Paul Ehrlich's *Population Bomb* gives an exceedingly grim prognosis for the future:

> Each year food production in underdeveloped countries falls a bit further behind burgeoning population growth, and people go to bed a little hungrier. While there are temporary or local reversals of this trend, it now seems inevitable that it will continue to its logical conclusion: mass starvation.[34]

Not only does Ehrlich state that the world is headed toward calamity, he is convinced that there is nothing anyone really can do that will provide anything more than temporary abatement. Ehrlich's attitude toward technology as part of the solution to the impending problem is technological pessimism, so to speak. Ehrlich's lack of confidence in technology to deal with the problems plaguing the future is perhaps seen most explicitly in his statement: "But, you say, surely Science (with a capital "S") will find a way for us to occupy the other planets of our solar system and eventually of other stars before we get all that crowded."[35] Ehrlich was sure that "the battle to feed humanity is over." He insisted that India would be unable to provide sustenance for the 200-million-person growth in its population by 1980. He was wrong – thanks to biotechnologists like Norman Borlaug. Borlaug and his team engaged in a program that developed a special breed of dwarf wheat that was resistant to a wide spectrum of plant pests and diseases and that produced two or three times more grain mass than the traditional varieties. His team then taught local farmers in both India and Pakistan how to cultivate the new strain of wheat. This astonishing increase in the production of wheat within a few years has come to be called the green revolution, its inception credited to Borlaug.[36] Since 1968, when Ehrlich published his frightful predictions, "India's population has more than doubled, its wheat production has more than tripled, and its economy has grown nine-fold."[37] Pakistan has progressed from harvesting 3.4 million tons of heat each year to around 18 million tons, and India has made similar impressive movement from 11 million tons to 60 million tons.[38]

Malthus' pessimistic predictive framework earned for economics the nickname of "dismal science." Conversely, engineers look upon the same problems with more of a technical optimism. Engineers "mess up" the Malthusian curve by finding ways to accomplish this (e.g., Borlaug's spoiling Ehrlich's predictions). And, such positions lead to moral arguments as when Ehrlich criticizes the Pope for saying "You must strive to multiply bread so that it suffices for the tables of mankind and not, rather, favor an artificial control of birth, which would be irrational, in order to diminish the number of guests at the banquet of life."[39] Ehrlich feels that this representative bread cannot indeed be multiplied: "Can we expect great increases in food production to occur through the placing of more land under cultivation? The answer to this specific question is most definitely no."[40] The truth is, however, that Borlaug did in fact foresee ways to multiply the bread and, that engineers feel that they can continue to design technologies that help to support the growing population and the expanding needs of mankind. In *The Engineer of 2020*, the NAE describes various ways that engineers in the future will help solve the very same problems Ehrlich (and the Malthusian model in general) is concerned with. Where Ehrlich felt technology's role in solving the problem would only be seen through how "improved technology has greatly increased the potential of war as a population control device,"[41] engineers look toward technology not as a "means for self-extermination"[42] but rather as an option for supporting and improving life in the future.

Having examined Ehrlich's prognosis from 1970s and 1980s, consider now a prognosis made by engineers for new future. According to *The Engineer of 2020*, the world's population will approach 8 billion people; much of this increase will be seen in groups that are today considered underdeveloped countries, mainly in Asia and Africa. Apparently, "by 2015, and for the first time in history, the majority of people, mostly poor will reside in urban centers, mostly in countries that lack the economic, social, and physical infrastructures to support a burgeoning population."[43] Engineers, however, see an opportunity for "the application of thoughtfully constructed solutions through the work of engineers"[44] in the challenge posed by the highly crowded and densely populated world of 2020. Likewise, engineers look upon the necessity for improved health care delivery in the world of the future with confidence. They feel that they will be able to make advanced medical technologies accessible to this ever-growing global population base. In the developed world of twenty years from now they see positive implications on human health, due to improved air quality and the control and cleanup of hazardous waste sites, and focused efforts to treat diseases like malaria and AIDS.[45]

Engineers believe that they can solve the problems posed by the future, as opposed to views like the one posed by Ehrlich who sees a future where "small pockets of *Homo sapiens* hold on for a while in the Southern Hemisphere, but slowly die out as social systems break down, radiation poisoning takes effect, climactic changes kill crops, livestock dies off, and various man-made plagues spread. The most intelligent creatures ultimately surviving this period are cockroaches."[46] Indeed, this dramatic example serves to illustrate the differences between the engineer's technical optimism and the doomsayer's technical pessimism, so to speak. Of course, this discussion has considerd the extremes. Many economists are quite optimistic and many engineers are rather pessimistic. But the ethos of each discipline does travel in different directions, since the nature of engineering is to solve problems. If one has little or no hope at finding a solution, why ever bother? But what does the future call for – the proverbial idealist or the modern-day skeptic? Mankind can either resign itself to failure and deem the big problems unsolvable or it can press forward, attempting to solve the problems it faces and overcome tomorrow's challenges. While the question is subjective, it becomes clear that in order to progress, most engineers choose the latter option. Thus, it can be strongly asserted that utility is but one measure of biosystematic success albeit an important one. Reason and duty are also important, as we will investigate throughout this text.

Questions

1. Is the moniker "dismal scientist" truly applicable to the economist? Explain.
2. What are some downsides to technological optimism?
3. What is the difference between high density and crowding (or "overcrowding")?

4. Explain the bioethical aspects of population controls proposed by two recent Canadian political agendas:
   a. A private member's bill, C-407, introduced to the Canadian Parliament that proposed to allow any person, under certain conditions, to aid a person close to death or suffering from a debilitating illness to "die with dignity" if that person has expressed the free and informed consent to die.
   b. Health Canada's plans to introduce preimplantation genetic diagnosis (PGD) regulations in May 2006.

   Defend your agreement or disagreement of the following statement:

   Both proposals establish legally binding mandates as to when, how, and why innumerable innocent human beings shall be consigned or abandoned to death, reminiscent of the early "slippery slope" actions that led to the eugenic[47] movement embraced by many scientists in the United Kingdom, the United States, and Germany in the 1930s, which ultimately led to the heinous actions of the Nazis to produce a "master race." Thus, physicians who participate in euthanasia or in prenatal diagnosis often think in terms of "weeding out" the unfit, thus denying them their inalienable right to life. Stemming overpopulation is simply the environmental rationalization for eugenics. (Hint: Read varying views about "Social Darwinism".)

We must not forget that everything engineers do is to protect and to enhance life, no matter the specific discipline. Thus, even for those of us in engineering specialties other than biomedical, we must be mindful of our work's impact on life. And, while our keen interest is usually in human life, we must be aware of how our work affects other species. In fact, some of the greatest resistance against scientific and engineering research is the result of the perceived indifference to pain and suffering in nonhuman animals. Even microbial life must be respected in our laboratories. One of the scariest scenarios is that of unchecked self-replication of unicellular species. Many engineering researchers are presently pushing the envelopes of nanotechnology, for example. This includes using microbes to manufacture complex pharmaceuticals and chemical products that either cannot be manufactured or can be manufactured much less efficiently in abiotic chemical reactors. However, a number of credible researchers fear the attendant risks of such nanomachinery, including the creation of unintended toxic by-products and dangerous new strains of microbes.

The life aspects of engineering are, for the most part, the major success stories of engineering. For example, more than any other profession, engineers have prevented (in some cases, eliminated) devastating diseases with their public works projects, such as wastewater treatment, sanitary landfills, hazardous waste facilities, air quality controls, and drinking water supplies. Engineers not specifically practicing in biomedical engineering sometimes need to be reminded that their work serves life.

The "medical" part of biomedical engineering implies a strong link between the medical and the engineering professions. Like engineers, medical practitioners apply the basic sciences to achieve results. Thus, our designing of devices and structures is part of the larger health care provision. It is quite interesting to watch the growth and evolution of professions. They seem to oscillate between stages of specialization and contraction. Presently, both seem to be occurring in biomedical engineering. Medical doctors have become highly specialized, but their responsibilities have increasingly called for broader accountability. While the individual practitioner may lead one area of health care for the patient, they must build systems to ensure that all of the other specialties effectively work together to provide the best care for each patient. In this sense, medicine is part of a larger biosystem (human health).

Some of the most dramatic ethical and legal failures have resulted not from the practitioners' incompetence in their area of specialization, but in their lack of oversight and quality assurance of others who are part of the comprehensive care. Engineers are part of this system of care. In fact, some of the major advances in devices have come about through the close relationships with medical practitioners, such as the collaborations between teaching hospitals and biomedical engineering programs (like those at Duke, Johns Hopkins, and Stanford, to name a few). The engineer brings a number of assets to the team, including practicality, creativity, adaptability, and a long-term view.

The next term, ethics, has been summed up by Socrates as the way we ought to live. For engineers, this can be modified a bit. Engineering ethics is the way we ought to practice. The fundamental canons of the National Society of Professional Engineers (NSPE) code of ethics[48] captures what engineers "ought" to do (see Appendix 1). It states that engineers, in the fulfillment of their professional duties, shall:

1. Hold paramount the safety, health and welfare of the public.
2. Perform services only in areas of their competence.
3. Issue public statements only in an objective and truthful manner.
4. Act for each employer or client as faithful agents or trustees.
5. Avoid deceptive acts.
6. Conduct themselves honorably, responsibly, ethically, and lawfully so as to enhance the honor, reputation, and usefulness of the profession.

Let us consider these canons as they relate to biomedical and biosystem ethics for engineers. The canons are the professional equivalents to "morality," which refers to societal norms about acceptable (virtuous/good) and unacceptable (evil/bad) conduct. These norms are shared by members of society to provide stability as determined by consensus.[49] Philosophers consider professional codes of ethics and their respective canons to be normative ethics, which is the philosophical study of ethics concerned with classifying actions as right and wrong without bias. Normative ethics is contrasted with descriptive ethics, which is the study of what a group actually believes to be right and wrong, and how it enforces conduct. Normative ethics regards ethics as a set of norms related to actions. Descriptive ethics deals with what "is" and normative ethics addresses "what should be."

Philosopher Bernard Gert categorizes behaviors into what he calls a "common morality," which is a system that thoughtful people use implicitly to make moral judgments.[50] According to Gert, humans strive to avoid five basic harms: death; pain; disability; loss of freedom; and loss of pleasure. Arguably, the job of the engineer is to design devices, structures, and systems that mitigate against such harms in society. Similarly, Gert identifies ten moral rules of common morality:

1. Do not kill.
2. Do not cause pain.
3. Do not disable.
4. Do not deprive of freedom.
5. Do not deprive of pleasure.
6. Do not deceive.
7. Keep your promises.

8. Do not cheat.
9. Obey the law.
10. Do your duty.

Most of these rules are proscriptive. Only rules 7, 9, and 10 are prescriptive, telling us what to do, rather than what not to do. The first five directly prohibit the infliction of harm on others. The next five indirectly lead to prevention of harm. Interestingly, these rules track quite closely with the tenets and canons of the engineering profession (see Table 1.1).

The Gert model is good news for engineering bioethics. Numerous ethical theories can form the basis for engineering ethics and moral judgment. Immanuel Kant is known for defining ethics as a sense of duty. Thomas Hobbes presented ethics within the framework of a social contract, with elements reminiscent of Gert's common morality. John Stuart Mill considered ethics with regard to the goodness of action or decision as the basis for utilitarianism. Philosophers and ethicists spend much effort and

Table 1.1
Canons of the National Society of Professional Engineers compared to Gert's rules of morality

Engineers shall:	Most closely linked to rules of morality identified by Gert
1. Hold paramount the safety, health, and welfare of the public	• Do not kill • Do not cause pain • Do not disable • Do not deprive of pleasure • Do not deprive of freedom
2. Perform services only in areas of their competence	• Do not deceive • Keep your promises • Do not cheat • Obey the law • Do your duty
3. Issue public statements only in an objective and truthful manner	• Do not deceive
4. Act for each employer or client as faithful agents or trustees	• Do not deprive of pleasure • Keep your promises • Do not cheat • Do your duty
5. Avoid deceptive acts	• Do not deceive • Keep your promises • Do not cheat
6. Conduct themselves honorably, responsibly, ethically, and lawfully so as to enhance the honor, reputation, and usefulness of the profession	• Do your duty • Obey the law • Keep your promises

energy deciphering these and other theories as paradigms for ethical decision making. Engineers can learn much from these points of view, but in large measure, engineering ethics is an amalgam of various elements of many theories. As evidence, the American Society of Mechanical Engineers (ASME)[51] has succinctly bracketed ethical behavior into three models:

Malpractice, or Minimalist, Model – In some ways this is really not an ethical model in that the engineer is only acting in ways that are required to keep his or her license or professional membership. As such, it is more of a legalistic model. The engineer operating within this framework is concerned exclusively with adhering to standards and meeting requirements of the profession and any other applicable rules, laws, or codes. This is often a retroactive or backward-looking model, finding fault after failures, problems, or accidents happen. Any ethical breach is assigned based upon design, building, operation, or other engineering steps that have failed to meet recognized professional standards. This is a common approach in failure engineering and in ethical review board considerations. It is also the basis of numerous engineering case studies.

Reasonable Care, or Due Care, Model – This model goes a step further than the minimalist model, calling upon the engineer to take reasonable precautions and to provide care in the practice of the profession. Interestingly, every major philosophical theory of ethics includes such a provision, such as the harm principle in utilitarianism, the veil of ignorance in social contract ethics, and the categorical imperative in duty ethics. It also applies a legal mechanism, known as the reasonable person standard. Right or wrong is determined by whether the engineer's action would be seen as ethical or unethical according to a "standard of reasonableness as seen by a normal, prudent nonprofessional."[52]

Good Works Model – A truly ethical model goes beyond abiding by the law or preventing harm. An ethical engineer excels beyond the standards and codes and does the right thing to improve product safety, public health, or social welfare. An analytical tool related to this model is the net goodness model, which estimates the goodness or wrongness of an action by weighing its morality, likelihood, and importance.

This model is rooted in the moral development theories such as those expounded by Kohlberg,[53] Piaget,[54] and Rest,[55] who noted that moral action is a complex process entailing four components: moral awareness (or sensitivity), moral judgment, moral motivation, and moral character. The actor must first be aware that the situation is moral in nature; that is, at least that the actions considered would have consequences for others. Second, the actor must have the ability to judge which of the potential actions would yield the best outcome, giving consideration to those likely to be affected. Third, the actor must be motivated to prioritize moral values above other sorts of values, such as wealth or power. Fourth, the actor must have the strength of character to follow through on a decision to act morally.

Piaget, Kohlberg, and others (e.g., Duska)[56] have noted that the two most important factors in determining a person's likelihood of behaving morally are age and education; that is, of being morally aware, making moral judgments, prioritizing moral values, and following through on moral decisions. These are strong indicators of experience[57] and seem to be particularly critical regarding moral judgment: A person's ability to make moral judgments tends to increase with maturity as they pursue further education, generally reaching its final and highest stage of development in early adulthood. This theory of moral development is illustrated in Table 1.2.

Kohlberg insisted that these steps are progressive. He noted that in the two earliest stages of moral development, which he combined under the heading "preconventional level," a person is primarily motivated by the desire to seek pleasure and avoid pain. The "conventional level" consists of stages

Table 1.2
Kohlberg's stages of moral development

Preconventional level	1. Punishment-obedience orientation
	2. Personal reward orientation
Conventional level	3. "Good boy"–"nice girl" orientation
	4. Law and order orientation
Postconventional level	4. Social contract orientation
	6. Universal ethical principle orientation

Source: L. Kohlberg, 1981, *The Philosophy of Moral Development (Vol. 1)*, Harper & Row, San Francisco, CA.

three and four: in stage three, the consequences that actions have for peers and their feelings about these actions; in stage four, considering how the wider community will view the actions and be affected by them. Few people reach the "postconventional" stage, wherein they have an even broader perspective: Their moral decision making is guided by universal moral principles;[58] that is, by principles that reasonable people would agree should bind the actions of all people who find themselves in similar situations.

The moral need to consider the impact one's actions will have on others forms the basis for a normative model. Pursuing an activity with the goal of obeying the law has as its driving force the avoidance of punishment, and pursuing an activity with the goal of improving profitability is a goal clearly in line with stockholders' desires; presumably customers', suppliers', and employees' desires must also be met at some level. And finally, pursuing an activity with the goal of "doing the right thing," behaving in a way that is morally right and just, can be the highest level of engineering behavior. This normative model of ethical engineering can be illustrated as in Figure 1.5.

There is a striking similarity between Kohlberg's model of moral development (Table 1.2) and the model engineering professional growth. Avoiding punishment in the moral development model is similar to the need to avoid problems early in one's career. The preconventional level and early career experiences have similar driving forces.

At the second level in the moral development model is a concern with peers and community, while in the professionalism model the engineer must balance the needs of clients and fellow professionals with those of society at large. Engineering services and products must be of high quality and be profitable, but the focus is shifting away from self-centeredness and personal well-being toward external goals.

Finally at the highest level of moral development a concern with universal moral principles begins to govern actions. The driving force or motivation is trying to do the right thing on a moral (not legal or financial) basis. These behaviors set the example for the whole profession, now and in the future.

Professional growth is enhanced when engineers and technical managers base their decisions on sound business and engineering principles. Ethical content is never an afterthought, but is integrated within the business and design decision-making process: That is, the engineering exemplars recognize the broad impact their decisions may have, and they act in a way such that their actions is in the best interest of not only themselves and the organization they represent, but also the broader society and even future generations of engineers.

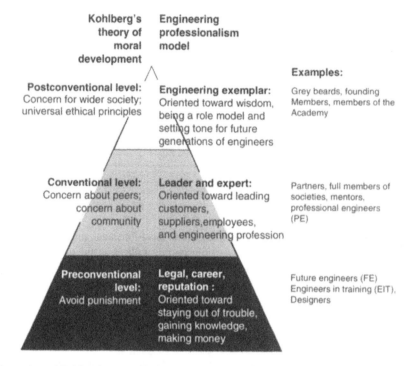

Kohlberg's
theory of
moral
development

Engineering
professionalism
model

Examples:

Postconventional level:
Concern for wider society;
universal ethical principles

Engineering exemplar:
Oriented toward wisdom,
being a role model and
setting tone for future
generations of engineers

Grey beards, founding
Members, members of the
Academy

Conventional level:
Concern about peers;
concern about
community

Leader and expert:
Oriented toward leading
customers,
suppliers,employees,
and engineering profession

Partners, full members of
societies, mentors,
professional engineers
(PE)

**Preconventional
level:**
Avoid punishment

**Legal, career,
reputation :**
Oriented toward
staying out of trouble,
gaining knowledge,
making money

Future engineers (FE)
Engineers in training (EIT),
Designers

Figure 1.5 Comparison of Kohlberg's moral development stages to professional development in engineering.

Much of the ethics training to date has emphasized preconventional thinking; that is, adherence to codes, laws, and regulations within the milieu of profitability for the organization. This benefits the engineer and the organization, but is only a step toward full professionalism, the kind needed to confront bioethical challenges. We who teach engineering ethics must stay focused on the engineer's principal client, "the public." One interpretation of the "hold paramount" provision of this ethical canon is that it has primacy over all the others. So, anything the professional engineer does cannot violate this canon. No matter how competent, objective, honest, and faithful, the engineer must not jeopardize public safety, health, or welfare. This is a challenge for such a results-oriented profession.

Bioethics Question: How can medical and engineering ethics coalesce?

In our zeal to provide the best technologies, devices, services, and plans, we cannot treat the general public as a means to such noble ends. Here is where the primary focus of the physician and that of the engineer begin to diverge. The medical practitioner's principal client is the patient, whereas the primary client of the engineer is the public. Nothing is more important to the engineer than the health and safety of the public.

The engineer, especially the biomedical engineer, must navigate both professional codes. As evidence, the Biomedical Engineering Society recently recognized this with its approval of a new code of ethics in 2004 (see Appendix 2). The code recognizes that the biomedical engineer practices at the confluence of "expertise and responsibilities in engineering, science, technology, and medicine."[59] Mirroring the

NSPE code, the biomedical engineering community reminds its members that "public health and welfare are paramount considerations."[60]

Public safety and health considerations affect the design process directly. Almost every design now requires at least some attention to sustainability and environmental impacts. Biomedical designs are not excluded. For example, there is a recent requirement for changes in drug delivery to decrease the use of greenhouse gas propellants like chlorofluorocarbons (CFCs) and instead using pressure differential systems (such as physical pumps) to deliver medicines. This may seem like a small thing or even a nuisance to those who have to use them, but it reflects an appreciation of the importance of incremental effects. It also combines two views, that of the patient (drug therapy) and the public (environmental quality).

One inhaler does little to affect the ozone layer or threaten the global climate, but millions of inhalers can produce enough halogenated and other compounds that the threat must be considered in designing medical devices. Environmental quality and sustainability are public virtues. To the best of our abilities, we must ensure that what we design is sustainable over its useful lifetime. This requires that the engineer think about the life cycle not only during use but when the use is complete. Such programs as "design for recycling" (DFR) and "design for disassembly" (DFD) allow the engineer to consider the consequences of various design options in space and time. They also help designers to pilot new systems and to consider scale effects when ramping up to full production of devices.

Like virtually everything else in engineering, best serving the public is a matter of optimization. The variables that we choose to give large weights will often drive the design. Treating cancer, providing devices to aid cardiovascular functioning, and delivery of efficacious drug therapies are noble and necessary ends. The engineer must continue to advance the state of the science in these high-priority areas. But, the public is a complicated entity and the human body is uniquely exquisite. Thus, any possible adverse effects must be recognized. These should be incorporated and properly weighted when we optimize benefits. We must weigh these benefits against possible hazards and societal costs.

ENGINEERING COMPETENCE

Engineering is a technical profession. It depends on scientific breakthroughs. Science and technologies are drastically and irrevocably changing. The engineer must stay abreast of new developments. This is particularly challenging for biomedical science and biosystem technology, where the scale of interest continues to decrease. It was not that long ago when organs were the most refined scale of interest, giving way to the organelles and cells. Now, the "nanoscale" is receiving the most attention, with structures and systems having design units of but a few angstroms.

ENGINEERING: BOTH INTEGRATED AND SPECIALIZED

Professional specialization has both advantages and disadvantages. The principal advantage is that the practicing engineer can focus on a specific discipline more sharply when compared to a generalist. The principal disadvantage is that integrating the different parts can be challenging. For example, in a very complex design only a few people can see the overall goals. Thus, those working in specific areas may not readily see duplication or gaps that they assume are being addressed by others.

A classic example of the shortcomings of overspecialization can be found in the video *Professional Ethics and Engineering* produced by Duke's Center for Applied Ethics. In it, a seasoned engineer is being interviewed by a professional review panel and asked whether he knew that the foundation of a building being constructed was mismatched to the soil type. He said that he did, but it was none of his business, since it was the job of the soil engineers. The panel reminded him that people died as a result of this failure. As the Santillan case reminds us, medical scenarios can also suffer since the whole is greater than the sum of its parts. This is the essence of biosystem engineering.

The work of technical professions is both the effect and the cause of modern life. When undergoing medical treatment and procedures, people expect physicians, nurses, emergency personnel, and other health care providers to be current and capable. Society's infrastructure, buildings, roads, electronic communications, and other modern necessities and conveniences are expected to perform as designed by competent engineers and planners. But how does society ensure that these expectations are met? Much of the answer to this question is that society cedes a substantial amount of trust to a relatively small group of experts, the professionals in increasingly complex and complicated disciplines that have grown out of the technological advances that began in the middle of the twentieth century and grew exponentially in its waning decades.

Professions, including engineering, are not neatly subdivided as they once were. A visit to the hospital shows that not only do many of the physicians specialize in particular areas of medicine (e.g., neuromedicine, oncology, and geriatrics), but all of these physicians must rely on chemists, radiologists, and tomographic experts to obtain data about their patients (e.g., from serum analysis, magnetic resonance and CT scans, and sonography). In fact, many of the solutions (cures?) to health problems require an intricate cacophony among doctors, biomedical engineers, and technicians, as well as public and community health professionals, epidemiologists, and environmental engineers to prevent and control many of the diseases, making treatment unnecessary. For example, a drug delivery system requires the understanding of the biochemical needs of the patient, the fluid mechanics of the pharmacology, and the actual design of the apparatus. This is a continuum among science, engineering, and technology.

Within this highly complex, contemporary environment, practitioners must ensure that they are doing what is best for the profession and what is best for the patient and client. This best practice varies by profession and even within a single professional discipline, so the actual codified rules (codes of ethics, either explicit or implicit) must be tailored to the needs of each group. However, many of the ethical standards are quite similar for most technical professions. For example, people want to know that the professional is trustworthy. The trustworthiness is a function of how good the professional is in the chosen field and how ethical the person is in practice. Thus, the professional possesses two basic attributes: subject matter knowledge and character. Maximizing these two attributes enhances professionalism.

WHO IS A PROFESSIONAL?

There is some debate about just who is a professional. I often ask the students enrolled in my Professional Ethics course to give examples of professionals. The list always includes physicians, engineers, airline pilots, and lawyers, and usually includes accountants. A few students consider clergy and military officers (not usually enlisted personnel) to be professionals. Some include businesspersons, teachers, and scientists. Only a small minority includes professional athletes, although many admit this is because the group includes the term "professional." Several other disciplines are included, but support diminishes after the first few. I approach the query quite unscientifically, simply asking their opinions, but this is interesting since I give them no criteria from which to label something as a professional; yet they are readily equipped to answer. I simply ask whom they would identify as professionals, and a list is generated.

I am often amazed by the intuitive powers of students (actually, of people in general). They usually can differentiate some very complicated subject matter (e.g., pollutant types, risk, values, and obligatory moral behavior), but they often cannot tell you why. In other words, they cannot explain their methodology, but they clearly use one. So, I delve a little more deeply by asking the students to tell me why one group is professional and another is not. They usually note readily that it is not that one is necessarily more "valuable" than the other or that ease or difficulty is a determining factor. Certain highly technical, critical, difficult, and respected "jobs," such as that of an aircraft mechanic, are not generally considered "professional."

Most of us will admit that a certain threshold of expertise is needed to ascribe the label "professional" to someone. However, as our aircraft mechanic example demonstrates, clearly expertise is a necessary but insufficient quality of professionalism. All professionals must be experts in the field, but not all experts are professionals. One distinguishing characteristic of a professional is the level of accountability and degree of responsibility. In our aircraft example, the mechanic is a highly trained expert in a particular area, but with a tightly defined span of control and realm of responsibility. However, the airline pilot is responsible for all aspects of the plane; that is, everything the mechanic does, everything the copilot does, and everything about the plane, the weather, the flying conditions, and whatever it takes to transport the plane, passenger, and cargo safely to the destination are the responsibility of the pilot. The captain is also accountable for everything that transpires on his or her "watch." Any organization, like any system, has a set of norms and mores that are distributed throughout its membership. Indeed, the mechanic who does not follow protocol will be reprimanded, usually severely, but the captain shares the blame, if for no other reason than because the "system" being led by the captain does not adequately ensure high-quality performance by the mechanic. Nor can the pilot defer and deflect blame to the airline company. No company policy or business decision should detract from the professional responsibilities of the pilot. In a word, the pilot remains responsible. The pilot is accountable for the whole flight experience.

WHAT IS TECHNICAL?

With a better idea of what it means to be a professional, we can now endeavor to characterize certain professionals as "technical." The technical professional must have a mastery of technical subject matter. But, what does this mean? Is the ability to play a video game or listen to an I-Pod a technical skill? Most of us would not think so. Is the ability to run sonigraphic software and hardware a technical skill? Most would agree. However, is this ability enough to be a considered a professional? Many would say: "No, it simply means the person is a skillful technician."

SYSTEMATICS: INCORPORATING ETHICS INTO THE DESIGN PROCESS

The key to engineering successes is ensuring that all of the right factors are considered in the design phase and that these factors are properly implemented and monitored throughout the project.

Integrated engineering approaches require that the engineer's responsibilities extend well beyond the construction, operation, and maintenance stages. Such an approach has been articulated by the ASME. One way to visualize a systematic view, such as design for the environment (DFE) recommended by the ASME is to use an integrated matrix[61] (Table 1.3). This allows for the engineer to see the technical and ethical considerations associated with each component of the design, as well as the relationships among these components. For example, health risks, social expectations, and environmental impacts and other societal risks and benefits associated with a device, structure, product, or activity can be visualized at various stages of the manufacturing, marketing, and application stages. This yields a number of two-dimensional matrices (Figure 1.6) for each relevant design component. And, each respective cell indicates both the importance of that component and the confidence (expressed as scientific certainty) that the engineer can have about the underlying information used to assess the importance (see legend in Figure 1.6).

The matrix approach is qualitative, but it allows comparisons of alternatives that would otherwise be incomparable, which is often the case in bioethics. To some extent, even numerical values can be

Table 1.3
Functions that must be integrated into an engineering design

1. Baseline studies of existing conditions
2. Analyses of project alternatives
3. Feasibility studies
4. Environmental impact studies and other macro-ethical, societal considerations
5. Assistance in project planning, approval, and financing
6. Design and development of systems, processes, and products
7. Design and development of construction plans
8. Project management
9. Construction supervision and testing
10. Process design
11. Start-up operations and training
12. Assistance in operations
13. Management consulting
14. Environmental monitoring
15. Decommissioning of facilities
16. Restoration of sites for other uses
17. Resource management
18. Measuring progress for sustainable development

Source: Adapted from American Society of Mechanical Engineers, http://www.professionalpractice.asme.org/communications/sustainability/2.htm; accessed 23 May 2006.

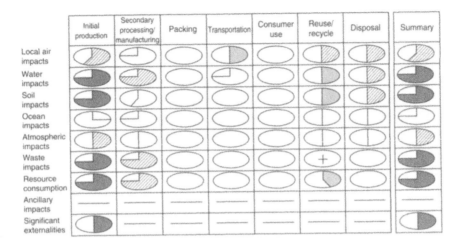

The indium environmental matrix for printed wiring board assembly.

Legend:

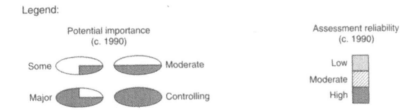

Figure 1.6   An example of an integrated engineering matrix; in this instance applied to sustainable designs.
Adapted from: American Society of Mechanical Engineers; http://www.professionalpractice.asme.org/communications/sustainability/
2.htm; accessed 25 May 2006.

assigned to each cell to compare them quantitatively, but the results are at the discretion of the analyst, who determines how different areas are weighted. The matrix approach can also focus on design for a more specific measure, such as energy efficiency or product safety, and can be extended to view corporate activities systematically.

Contemporary society demands many things from its professionals. For those in technical fields there are two prominent expectations: trust and competence. These expectations are built into the codes of practice and ethics of each technical discipline. They are two elements that the engineer keeps throughout his or her career.

NOTES AND COMMENTARY

[1]   E. Dickinson, 1862, *Wider than the Sky*.
[2]   I have modified Rosenberg's experiments a bit to make them more specifically relevant to bioethics and engineering, but the results reported are those observed in the Responsible Conduct of Research workshop, August 2004.

3 The original version of this thought experiment was conceived by philosopher Philippa Ruth Foot (born 1926). She is one of the leaders of the contemporary movement to virtue ethics. Virtue ethics centers around what a person should become; that is, the goal of ethics is "eudaimonia" (the Aristotelian concept of "success" sometimes translated as "happiness" but more correctly as "blessedness."). The twentieth-century movement is known as the "aretaic" turn, from the Greek term *arête*, meaning "excellence." This is in contrast to utilitarian and other "consequentialist" ethical models, where ethics is predominantly determined by outcome.

4 "Teaching Ethics to Scientists and Engineers: Moral Agents and Moral Problems," *Science and Engineering Ethics* **1**, no. 3 (1995): 299–308.

5 Laplace's equation on the rectangular region, $0 < x < a, 0 < y < b$, is subject to the Dirichlet boundary conditions:

$$u(x, 0) = x^2 u(x, b) = x^2 - b^2$$

$$u(0, y) = -y^2 u(a, y) = a^2 - y^2$$

The unique solution to this BVP is $u(x, y) = x^2 - y^2$, making it well posed.

6 J. Hadamard, *Lectures on the Cauchy Problem in Linear Partial Differential Equations* (New Haven, CN: Yale University Press, 1923).

7 N.R. Augustine, Ethics and the second law of the thermodynamics. *The Bridge*, 32(3) Fall, 2002.

8 R. Rolheiser, *The Holy Longing: The Search for a Christian Spirituality* (New York, NY: Doubleday, 1999).

9 This case is found at http://onlineethics.org/bib/appcw-pt3.html (accessed 30 June 2006).

10 C. Gulligan, *In a Different Voice: Psychological Theory and Women's Development* (Cambridge, MA: Harvard University Press, 1993).

11 T. Aquinas, *Summa Theologica*, I-II Q94 Art 2.

12 Often, texts, manuals, and handbooks are valuable, but only when experience and good listening skills are added to the mix can wise (and ethical) decisions be made. First-century thinking linked maturity to "self-control" or "temperance" (Greek *kratos* for "strength"). St Peter, for example, considered knowledge as a prerequisite for temperance. Thus, from a professional point of view, he seemed to be arguing that one can really only understand and appropriately apply scientific theory and principles after one practices them (I realize he was talking about spirituality, but anyone who even casually studied Peter's life would see that he fully integrated the physical and the spiritual). This is actually the structure of most professions. For example, engineers who intend to practice must first submit to a rigorous curriculum (approved and accredited by the Accreditation Board for Engineering and Technology), then must sit for the Future Engineers (FE) examination. After some years in the profession (assuming tutelage by and intellectual osmosis with more seasoned professionals), the engineer has demonstrated the *kratos* (strength) to sit for the Professional Engineers (PE) exam. Only after passing the PE exam does the National Society for Professional Engineering certify that the engineer is a "professional engineer" and eligible to use the initials PE after one's name. The engineer is, supposedly, now schooled beyond textbook knowledge and knows more about why in many problems the correct answer is "It depends." In fact, the mentored engineer even has some idea of what the answer depends on (i.e., beyond "knowing that one does not know" as Socrates would say).

13 American Medical Association (AMA), 1997, Policy E-2.01 Abortion, http://www.ama-assn.org/
    apps/pf_new/pf_online?f_n=browse&doc=policyfiles/HnE/E-2.01.HTM&&s_t=&st_p=&nth=1&
    prev_pol=policyfiles/HnE/E-1.02.HTM&nxt_pol=policyfiles/HnE/E-2.01.HTM& (accessed 13
    July 2006).
14 AMA, H-5.995, Abortion.
15 P. Lee and R.P. George, "The Wrong of Abortion," in *Contemporary Debates in Applied Ethics*,
    ed. A. Cohen and C.H. Wellman (Malden, MA: Blackwell Publishing, 2004).
16 Duke University, 2006, http://www.k-phd.duke.edu/purpose.htm (accessed 13 April 2006). Accord-
    ing to the site, the K-PhD program "provides opportunities for children to learn to think critically
    and analytically while developing a passion for understanding the world and an appreciation for
    improving the quality of all living things." Its mission "is to increase significantly the number of
    children, particularly female and under-represented groups, who choose to pursue science related
    careers."
17 This profundity is actually a quote by P.A. Vesilind, RL Rooke Professor of Engineering, Bucknell
    University, at the conference, "Engineers Working for Peace, Lewisburg, Pennsylvania, 15 Novem-
    ber 2003." Vesilind made the comment as a reminder of the practicality of the profession and the
    need to respect the ethos of engineers when addressing societal issues, like peace and justice.
18 The concept of "clean" is subject to debate within the engineering community. It parallels the
    questions about safety. When is a device or drug sufficiently "safe" to move to the production
    stage? Environmental engineers ask a similar question, "How clean is clean?" We wonder when we
    have done a sufficient job of cleaning up a spill or a hazardous waste site. It is often not possible
    to have nondetectable concentrations of a pollutant, especially since analytical chemistry and other
    scientific disciplines continue to improve. Commonly, a threshold for cancer risk to a population is
    one in a million excess cancers. In cleanup situations, the tolerable risk may be much higher (e.g.,
    one in ten thousand). However, one may find that the contaminant is so difficult to remove that we
    almost give up on dealing with the contamination and put in measures to prevent exposures, i.e.,
    fencing the area in and prohibiting access. This is often done as a first step in site remediation, but
    is unsatisfying and controversial (and usually politically and legally unacceptable). Thus, even if
    costs are high and technology unreliable, the engineer must find suitable and creative ways to clean
    up the mess and meet risk-based standards.
19 K. Johnson, "We Need to Keep Leading Students into Science, Math," Editorial in *The Durham
    (NC) Herald-Sun*, p. A-11, 12 February 2006.
20 H. Petroski, *To Engineer Is Human: The Role of Failure in Successful Design* (New York, NY:
    St. Martin's Press, 1985).
21 T.J. Albrecht, "ISO 9002 Implementation: Lessons Learned," *Quality Digest* 14 (1994): 55–61.
22 Petroski, *To Engineer Is Human*.
23 *The Memoirs of Herbert Hoover 1874–1920: Years of Adventure*, vol. 1, Library of Congress E
    802.H7 (Washington, DC: 1951), 132–3.
24 Translated from "*Je pense, donc je suis*," in R. Descartes, 1637, *Discourse on Method*.
25 T.S. Kuhn, *The Structure of Scientific Revolutions* (Chicago, IL: University of Chicago Press, 1962).
    Much of the economics and engineering comparative discussion benefits from the work of Duke
    undergraduate student, Rayhaneh Sharif-Askary's research.
26 B. Gert, Letter to the Editor, *The Ag Bioethics Forum* (Iowa State University) 5, no. 2 (Novem-
    ber 1993).

[27] R.C. Hilborn, *Chaos and Nonlinear Dynamics* (UK: Oxford University Press, 1994).

[28] Hadamard, *Lectures on the Cauchy Problem.*

[29] National Academy of Engineering, *The Engineer of 2020: Visions of Engineering in the New Century* (Washington, DC: National Academy Press, 2004), 48.

[30] This analogy does not hold completely to the economics of technology, since in chemistry the catalyst is a chemical substance that increases the rate of a reaction without being consumed. We know that technologies do indeed become consumed (antiquated and in need of replacement).

[31] National Aeronautics and Space Administration Thermodynamic Equilibrium, 15 March 2006, http://www.grc.nasa.gov/WWW/K-12/airplane/thermo0.html.

[32] M. Martin and R. Schinzinger, *Ethic in Engineering* (New York, NY: McGraw-Hill, 1996).

[33] His actual given name was Thomas Robert Malthus.

[34] Paul R. Ehrlich, *The Population Bomb* (New York, NY: Ballantine Books, 1968).

[35] Ibid., 20.

[36] Salil Singh, 17 April 2006, "Norman Borlaug: A Billion Lives Saved." A World Connected, http://www.aworldconnected.org/article.php/311.html.

[37] Ibid.

[38] Ibid.

[39] Ehrlich, *The Population Bomb*, 95.

[40] Ibid., 96.

[41] Ibid., 69.

[42] Ibid.

[43] National Academy of Engineering, 27–8.

[44] Ibid.

[45] Ibid., 28–9.

[46] Ehrlich, *The Population Bomb*, 78.

[47] In 1883, Francis Galton, Charles Darwin's cousin, coined the term "eugenics." He reportedly objected to charity because it encouraged the poor to have more children. Such elitism is an example of social engineering run amok.

[48] National Society for Professional Engineering, 2003, NSPE Code of Ethics for Engineers, http://www.nspe.org/ethics/eh1-code.asp (accessed 8 January 2006).

[49] T.L. Beauchamp and J.F. Childress, "Moral Norms," in *Principles of Biomedical Ethics*, 5th ed. (New York, NY: Oxford University Press, 2001).

[50] B. Gert, *Common Morality: Deciding What to Do* (New York, NY: Oxford University Press, 2004).

[51] American Society of Mechanical Engineers, 2006, Professional Practice Curriculum, "Engineering Ethics," http://www.professionalpractice.asme.org/engineering/ethics/0b.htm (accessed 10 April 2006).

[52] Note that this is not the "reasonable engineer standard." Thus, the reasonable person standard adds an onus to the profession, i.e., not only should an action be acceptable to one's peers in the profession, but to those outside of engineering. An action could very well be legal, and even professionally permissible, but may still fall below the ethical threshold if reasonable people consider it to be wrong.

[53] L. Kohlberg, *The Philosophy of Moral Development*, vol. 1 (San Francisco, CA: Harper & Row, 1981).

[54] J. Piaget, *The Moral Judgment of the Child* (New York, NY: The Free Press, 1965).

55 J.R. Rest, *Moral Development: Advances in Research and Theory* (New York, NY: Praeger, 1986); and J.D. Rest, D. Narvaez, M.J. Bebeau, and S.J. Thoma, *Postconventional Moral Thinking: A Neo-Kohlbergian Approach* (Mahwah, NJ: Lawrence Erlbaum Associates, 1999).

56 R. Duska and M. Whelan, *Moral Development: A Guide to Piaget and Kohlberg* (New York, NY: Paulist Press, 1975).

57 Hence, the engineering profession's emphasis on experience and mentorship.

58 J.A. Rawls, *A Theory of Justice* (Cambridge, MA: Harvard University Press, 1785); and I. Kant, *Foundations of the Metaphysics of Morals*, trans. L.W. Beck, 1951, (Indianapolis, IN: Bobbs-Merrill, 1959).

59 Biomedical Engineering Society, 2004, Biomedical Engineering Society, 2004, "Biomedical Engineering Society Code of Ethics," http://www.bmes.org/pdf/2004ApprovedCodeofEthicsShortForm.pdf (accessed 8 January 2006).

60 This wording is quite interesting. It omits "public safety." However, safety is added under professional obligations that biomedical engineers "use their knowledge, skills, and abilities to enhance the safety, health, and welfare of the public." The other interesting word choice is "considerations." Some of us would prefer "obligations" instead. These compromises may indicate the realities of straddling the design and medical professions. For example, there may be times when the individual patient needs supersede those of the general public and *vice versa*.

61 American Society of Mechanical Engineers, 2005, Sustainability: Engineering Tools, http://www.professionalpractice.asme.org/business_functions/suseng/1.htm (accessed 10 January 2006).

Chapter 2

Bioethics and the Engineer

> While engineering is a rapidly evolving field that adapts
> to new knowledge, new technology, and the needs of society, it
> also draws on distinct roots that go back to the origins of
> civilization. Maintaining a linkage of the past with the future is
> fundamental to the rational and fact-based approaches that
> engineers use in identifying and confronting the most difficult issues.
>
> National Academy of Engineering (2004)[1]

After reading the preface, prologue, and first chapter, the reader may be wondering when a specific definition of "biomedical ethics" would appear. After all, it is in the book title Sometimes, biomedical ethics is shortened to "bioethics." In fact, bioethics has numerous definitions. Here is mine:

> Bioethics is the set of moral principles and values (the *ethics* part) needed to respect, to protect, and to enhance life (the *bio* part).

Engineers, medical practitioners, and all technical professionals must be clear about meanings. A stray mark on a blueprint or a misreading of a prescription can lead to harmful outcomes, even death and destruction. So it is with each term in this book's title. Let us consider each.

Upon review, there may be a few parts of the definition that are missing, such as words like medicine, health, and biotechnologies. They are certainly embedded, but bioethics is much more. In light of the risk that my definition may appear overly simple and obvious, let us try to go back to the origins.

The term was coined by Van Rensselaer Potter II (1911–2001). Although Potter was a biochemist, he seemed to think like an engineer; that is, in a rational and fact-based manner. In fact, his original 1971 definition of bioethics was one rooted in integration and systematics. Potter considered bioethics to bridge science and the humanities to serve the best interests of human health and to protect the environment. In his own words, Potter describes this bridge:

> From the outset it has been clear that bioethics must be built on an interdisciplinary or multidisciplinary base. I have proposed two major areas with interests that appear to be separate but which need each other: medical bioethics and ecological bioethics. Medical bioethics and ecological bioethics are non-overlapping in the sense that medical bioethics is chiefly concerned with short-term views: the options open to individuals and their physicians in their attempts to prolong life.... Ecological bioethics clearly has a long-term view that is concerned with what we must do to preserve the ecosystem in a form that is compatible with the continued existence of the human species.[2]

Biomedicine, engineering, and the development and application of emerging biotechnologies all share a common feature; they call for balance. Society demands that the state of the science be advanced as rapidly as possible and that no dangerous side effects ensue. Most engineers appreciate the value of pushing the biotechnological envelopes. Engineers are adept at optimizing among numerous variables for the best design outcomes. However, most of these emergent areas are associated with some degree of peril. A recent query of top scientists[3] addressed this very issue. Its focus was on those biotechnologies needed to help the developing countries. Thus, the study included both the societal and the technological areas of greatest potential value (Table 3.2). Each of these international experts was asked the following questions about the specific technologies:

- *Impact.* How much difference will the technology make in improving health?
- *Appropriateness.* Will it be affordable, robust, and adjustable to health care settings in developing countries, and will it be socially, culturally, and politically acceptable?
- *Burden.* Will it address the most pressing health needs?
- *Feasibility.* Can it realistically be developed and deployed in a time frame of 5–10 years?
- *Knowledge gap.* Does the technology advance health by creating new knowledge?
- *Indirect benefits.* Does it address issues such as environmental improvement and income generation that have indirect, positive effects on health?

The top three areas require major advances in biomedical engineering. The fourth area is within the domain of environmental and civil engineering. The fifth area is a challenge for genetic and tissue engineers. The sixth area falls within biomedical engineering research and clinical engineering. The seventh area combines the work of computer engineers and biomedical engineers, while the eighth area is a blend of agricultural and biomedical engineering with food sciences. The ninth area will require advances in biomedical, clinical, and tissue engineering, and the tenth area will call on computational pharmacological modeling (e.g., compartmental models), material sciences, biomedical engineering, and chemical engineering. This is evidence that bioethics is a growing concern for all engineering disciplines. Notably, each of these technological areas is associated with bioethical issues, but in very unique ways.

Regarding biomedical engineering, both parts of the term "biomedical" are important. Again, "bio" connotes life. The dictionary definition of this combination form denotes "life or living organisms, or systems derived from them."[4] This is an engineering-friendly definition, since it incorporates systems. In fact, the discipline of "biosystem engineering" relates to the "operation on industrial scale of biochemical processes . . . and is usually now termed biochemical engineering."[5] Interestingly, this appears to be a distinction between what molecular biologists and biochemists do and what engineers do with the same

Table 2.1
Ranking by global health experts of top ten biotechnologies needed to improve health in developing countries

Final ranking	Biotechnology
1	Modified molecular technologies for affordable, simple diagnosis of infectious diseases
2	Recombinant technologies to develop vaccines against infectious diseases
3	Technologies for more efficient drug and vaccine delivery systems
4	Technologies for environmental improvement (sanitation, clean water, and bioremediation)
5	Sequencing pathogen genomes to understand their biology and to identify new antimicrobials
6	Female-controlled protection against sexually transmitted diseases, both with and without contraceptive effect
7	Bioinformatics to identify drug targets and to examine pathogen–host interactions
8	Genetically modified crops with increased nutrients to counter specific deficiencies
9	Recombinant technology to make therapeutic products (e.g., insulin, interferons) more affordable
10	Combinatorial chemistry for drug discovery

Source: Data from survey conducted in: A.S. Daar, H. Thorsteinsdóttir, D.K. Martin, A.C. Smith, S. Nast, and P.A. Singer, 2002, Top Ten Biotechnologies for Improving Health in Developing Countries, *Nature Genetics*, 32, pp. 229–32.

information. Bioengineering is the "application of the physical sciences and engineering to the study of the functioning of the human body and to the treatment and correction of medical conditions."[6] This closely tracks with the definition of "biomedical engineering."

Thus, engineers as agents of technological progress are at a pivotal position. Technology will continue to play an exponentially increasingly important role in the future. The concomitant societal challenges require that every engineer fully understands the implications and possible drawbacks of these technological breakthroughs. Key among them will be biotechnical advances at smaller scales, well below the cell and approaching the molecular level. Technological processes at these scales require that engineers improve their grasp of the potential ethical implications. The essence of life processes is at stake.

MAJOR BIOETHICAL AREAS

Engineering practice and research is deeply committed to and involved in the advancing technologies that will benefit humankind. However, this commitment and involvement calls for deliberate and serious considerations of actual and potential ethical issues. The President's Council on Bioethics[7] has summarized the dichotomy between the promise and the ethical challenges:

> For roughly half a century, and at an ever-accelerating pace, biomedical science has been gaining wondrous new knowledge of the workings of living beings, from small to great. Increasingly, it is providing precise and sophisticated knowledge of the workings also of the human body and mind. Such knowledge of how things work often leads to new technological powers to control or alter these workings, powers generally sought in order to treat human disease and relieve suffering. But, once available, powers sought for one purpose

are frequently usable for others. The same technological capacity to influence and control bodily processes for medical ends may lead (wittingly or unwittingly) to non-therapeutic uses, including "enhancements" of normal life processes or even alterations in "human nature." Moreover, as a result of anticipated knowledge of genetics and developmental biology, these transforming powers may soon be able to transmit such alterations to future generations.

So, let us consider some of these technological areas important to engineers and their attendant bioethical concerns.

CLONING AND STEM CELL RESEARCH

[C]loning represents a turning point in human history—the crossing of an important line separating sexual from asexual procreation and the first step toward genetic control over the next generation. It thus carries with it a number of troubling consequences for children, family, and society.

Leon R. Kass, Chair, The President's Council on Bioethics[8]

Researchers in the United States asserted in late 2001 that they were able to produce the first cloned human embryos, merely reaching a six-cell stage before the cells stopped division and died. In the meantime, a number of fertility specialists had declared a strong intent to clone human beings. In response to the technical and societal uncertainties and anxieties, the United States Congress had already begun in 1998 to consider these issues, with the House of Representative in July 2001 passing a strict ban on all human cloning, including the production of cloned human embryos. Since then, a number of cloning-related bills have been considered in the US Senate and several state legislations. Numerous nations have banned human cloning, with the United Nations considering an international convention on the issue. It suffices to say that the political and societal aspects of cloning are as challenging as the technical demands.

The biology of cloning begins with the ovum (egg). A human ovum consists of a single gamete cell, having only 23 active chromosomes. This means that the sex has not yet been determined. Once the ovum contains a complete nucleus from any species that is activated and developing – whether that has occurred by sexual fertilization or by asexual somatic cell nuclear transfer – an embryo of that species (*Homo sapiens*, sheep, cat, dog, etc.) is produced. Cloning is a type of reproduction wherein offspring result not from the chance of the union of ovum and sperm (sexual reproduction); rather it results from the deliberate replication of the genetic makeup of another single individual (asexual reproduction).

Human cloning is accomplished by introducing the nuclear material of a human somatic cell (donor) into an oocyte (egg) that has had its own nucleus removed or inactivated. This yields a product with a human genetic constitution virtually identical to the donor of the somatic cell. This technique is known as somatic cell nuclear transfer (SCNT). Since SCNT uses human genetic material, the developing embryo is of the species *H. sapiens*.

Bioethics Question: Is research using human pluripotent stem cells ethical?

A "pluripotent" cell can be differentiated into more than one alternative type of mature cell; so it is able to produce all the cell types of the developing organism's body. Thus, a pluripotent stem cell has the

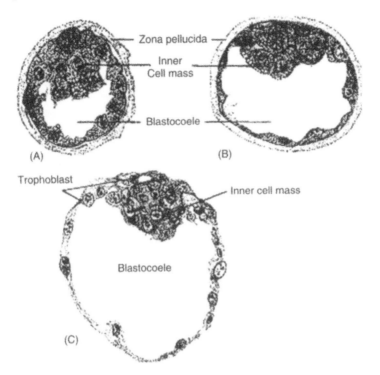

Figure 2.1 Three stages of blastocyst formation in the pig, drawn from sections to show the formation of the inner cell mass. Adapted from: The President's Council for Bioethics, 2004, *Monitoring Stem Cell Research*, Appendix A, Washington, DC.

same functional capacity (i.e., pluripotency) as an embryonic stem cell, though it does not necessarily share the same origin. "Stem cell research" involves isolating human embryonic stem cells from embryos at the blastocyst stage (Figure 2.1) or from the germinal tissue of fetuses (Figure 2.2). As of this writing, such harvesting kills the donor. The embryonic stem cells have been harvested from *in vitro* fertilization (IVF). Human adult stem (i.e., multipotent) cells have been isolated from a variety of tissues. These stem cell populations can be differentiated *in vitro* into various cell types, and are currently being extensively and intensively investigated for potential applications in regenerative medicine. Many scientists believe that embryonic and adult stem cells can lead to treatments for many human maladies. However, much of this is conjecture at this point.

The Chairman of the President's Council on Bioethics succinctly characterized the raging stem cell debate in a letter to the President:

> While they may well in the future prove to be of considerable scientific and therapeutic value, new human embryonic stem cell lines cannot at present be obtained without destroying human embryos. As a consequence, the worthy goals of increasing scientific knowledge and developing therapies for grave human illnesses come into conflict with the strongly held belief of many Americans that human life, from its earliest stages, deserves our protection and respect.[9]

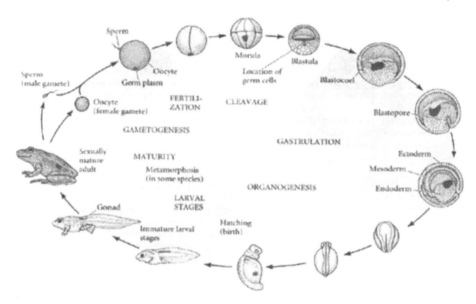

Figure 2.2   Stages of stem cell development in a frog. Note the continuity of germplasm.
Adapted from: The President's Council for Bioethics, 2004, *Monitoring Stem Cell Research*, Appendix A, Washington, DC.

The diametric opposition lies between those who see embryos as living human beings entitled to protection and those who consider them merely as "potential" humans that may be used as researchers see fit in an effort to advance the state of knowledge. This use includes the destruction (killing) of embryos to harvest stem cells.

One of the major ethical issues of stem cell research revolves around the first stage, especially the means of harvesting the embryonic stem cells by IVF. In particular, the argument contends that since fertilization has occurred, we are actually conducting research on a living human being. And, the IVF process itself is cavalier about the value of the individual, fertilized ova, destroying and discarding many in the process.

In a recent undergraduate seminar at Duke, students shared why IVF seems to "get a pass" morally, at least in comparison to the scrutiny given to stem cell and cloning research, even though all these are beginning of life (BOL) issues. One student pointed out that IVF is not new. Others noted that the technologies are more mainstream and understandable by the public. One student suggested the fact that many people know people who were conceived via IVF, so the mystique is gone. In the course of the discussion, it became obvious that most of the students do not even think about the morality of IVF. It simply exists as an alternative means of reproduction. And, the IVF industry is considered rather positively. The discarded embryos are just considered by many to be a byproduct of the process that allows people who could not ordinarily do so to have children. Thus, few seem to question the morality of IVF, so the discarding of embryos is not often seen as an immoral act of the fertility clinics. Further, those who oppose the use of embryos that would otherwise be destroyed are seen as "wasteful" and "myopic," or even as Luddites who stand in the way of progress and betterment of humankind.

As the President's Council on Bioethics puts it: "All extractions of stem cells from human embryos, cloned or not, involve the destruction of these embryos."[10] A rather common paradox is that research may either exacerbate or ameliorate ethical issues. For example, if emerging research supports the contention that embryonic stem cells are essential to provide the treatment and cures of intractable diseases, many scientists and ethicists will push the "greater good," utilitarian viewpoint, while others will see this as malicious means (sacrificing embryos and living human beings) toward the end (treatment).

Further, ethicists will be faced with the "slippery slope" dilemma. This argument sees a progression from diminished respect for embryos to a lack of respect for all unborn human beings, which leads to less respect for weaker members of the society. This last step in the progression is an argument for an incremental acceptance of eugenics, the belief that information about heredity can be used to improve the human race. The dreadful truth is that the genocide and other inhumane attempts at eugenics, such as the Nazis' attempted purging of Jewish people in the twentieth century, were clumsy and inefficient compared to emergent genetic manipulations available now and in the foreseeable future. The slippery slope advocates fear that the next Hitler will have subtle tools and a "rationale" supported by many to accomplish similar ends (i.e., eradicating selected human subpopulations).

In fact, one view extends this devaluing of certain human beings to even *after* they are born (a type of "postnatal abortion").

Peter Singer holds the title of Ira W. Decamp Chair of Bioethics at Princeton University's Center for the Study of Human Values. In his book, *Rethinking Life and Death: The Collapse of Our Traditional Ethics*, Singer asserts:

> Human babies are not born self-aware or capable of grasping their lives over time. They are not persons. Hence their lives would seem to be no more worthy of protection than the life of a fetus.[11]

Ironically, many pro-life advocates would agree with Singer's statement because they believe that a fetus is indeed worthy of protection as is any human person. However, they strongly oppose Singer's contention that self-awareness is an essential characteristic of personhood, so those who are not self-aware are, by Singer's definition, not persons. They may well agree with the definition for the species, but not for individuals. In other words, *H. sapiens* are indeed unique in their self-awareness, but each individual of the species varies in awareness of self and the awareness changes over the lifetime of the individual.

To Singer, those creatures at any given time that lack full self-awareness (the unborn, the newly born, those in vegetative states, the infirm, and those nearing the end of life) are merely human nonpersons. In fact, Singer, who is an emblematic advocate of animal rights, contends that unaware babies are less valuable than self-aware nonhuman animals. Singer's position points to a problem of unadulterated utilitarianism; that is, value is based entirely on the elitist's view of philosophical and scientific naturalism (materialism). In such a strident view, nonmaterial phenomena, such as a soul, are dismissed since they cannot be measured empirically (this also eliminates most human qualities, such as love, true happiness, and meaning). Such rigid scientism misrepresents the scientific data and even its own materialistic philosophy, since even the template for materialism cannot be proven by observation. That is, the premise that only that which can be measured is real is itself an unmeasurable and untestable premise.

Thus, such moral relativism is an easy and sloppy way to deal with personhood. Applying a "greatest good" defense (i.e., decisions wherein the most pleasure or desire is gained) allows for the rest of society to sacrifice certain members. Such sacrifice is easiest and scientifically erroneous if we call them "nonhuman" and it is philosophically erroneous if we call them "nonpersons." Singer and his ilk advocate both. Common sense and, hopefully, common morality would disagree. Such convenience is highly unscientific and grossly unethical. Theologian Michel Schooyans[12] has characterized this quite well:

> Men cast doubt on the human character of certain beings whenever they sought arguments to exploit or exterminate their fellow human beings. In antiquity slaves were considered as things and barbarians as second class men. In the sixteenth century, some conquerors considered the Indians as "beasts in human appearance." The Nazis looked upon some men as "non-men", as *Unmenschen*. To these arbitrary classifications dictated by the masters corresponded real discrimination and this, in turn, "legitimized" exploitation or extermination.

Scientific advances can wreak havoc with social values. What appears to be advanced thinking at times turns out to be retrograde attempts at dehumanization.

Advances in technology can be used to commoditize human beings. Genetic fetal testing, for example, can be used as a screen against the "unfit." Conversely, scientific advances can improve the bioethical landscape. Many of us hope that the perceived demand for embryonic stem cells will soon be obviated by scientific advances that allow the same cures and treatments using unquestionably morally acceptable tools, such as adult stem cells. Recent advances cause us to be hopefully optimistic (see Teachable Moment: Nanog).

Teachable Moment: Nanog

In Celtic legend, Nanog (short hand for *Tir Nan Og*) was a land of the eternally young. Fittingly, the Nanog molecule is at the threshold of what could be the unraveling of the need for embryonic stem cells to advance medical science. Scientists have argued that embryonic stem cells are unique in their ability to develop into any type of cell in the body, and further that they show great, yet still unfilled, potential to replace and to mend damaged tissue. Thus, the argument goes, embryonic stem cells are essential for such medical treatments.

Recently, researchers have become aware that the Nanog may someday allow ordinary cells to make use of that wonderful attribute.

Quite recently, working from the premise that amniotic fluid is known to contain multiple cell types derived from the developing fetus, researchers at Wake Forest University have advanced that state of science to show that these cells are able to produce an array of differentiated cells, including those of adipose, muscle, bone, and neuronal lineages.[13] The amniotic fluid-derived stem (AFS) cells are different from both adult and embryonic stem cells but share qualities of both. According to the lead researcher, Anthony Atala, AFS cells, like human embryonic stem cells, can double every one and a half days and can extensively differentiate. However, like adult cells, they do not seem to form tumors when implanted. As a scientific and potentially ethical bonus, AFS cells are readily gathered from amniotic fluid or placenta, making way for a potentially vast supply of cells available for therapeutic purposes.

Questions

1. What are some of the scientific obstacles that need to be overcome in Nanog research?
2. If the science supports this advance, what may be some of the reasons that scientists and others may still be skeptical of these findings?
3. Compare this type of skepticism to Kuhn's descriptions of the scientific community during a paradigm shift.

HUMAN ENHANCEMENT

Engineering is all about enhancing our environment and ourselves. However, the tools that have emerged in recent decades are changing the nature of what we can do. Consequently, ethical issues are emerging or are becoming increasingly complicated by biomedical and biosystem breakthroughs. One area that should be discussed here, though, is "genetic engineering."

The first question that engineers must ask is whether so-called genetic engineering is really engineering. At its most basic level, genetic engineering is artificially modifying the genetic code of an organism. But what is artificial? Obviously, an active approach like inserting DNA into a cell is artificial, but what about passive techniques like choosing potential mates with certain traits as the spouse and the future parent of one's children?

Since engineering is the application of sciences to address societal needs, then genetic engineering seems to be a type of engineering. However, few engineers actually perform human genetic engineering. In fact, most of it is accomplished by biologists. And, these are mainly physicians (e.g., those in IVF clinics) and biomedical researchers (e.g., those conducting DNA work under the auspices of biochemistry). So, the lay public and clergy work in passive genetic engineering (e.g., choosing spouses) and biologists work in active genetic engineering (e.g., cellular biochemists). Where does that leave the engineering profession?

Arguably, engineers' work is affected by that of other professions. Thus, we need at least some preparation for the bioethical issues that are certain to arrive in our individual specialties.

The emergence of genetic manipulation has been rapid. Scientists made huge gains throughout the 1900s in discovering DNA and its structure. In 1977, Genetech became the first company to use recombinant DNA technology. This served as a catalyst in the industry by being the first of many opportunities and discoveries throughout the 1980s and 1990s. In 1988, the Human Genome Project started with the goal of determining the entire sequence of DNA in humans. Since then, genetic engineering has been the focus of a myriad of ethical questions and debates.

Chapter 3 is devoted to the bioethical issues surrounding human enhancement.

PATENTING LIFE

Bioprospecting, the search for natural substances of medicinal value, is a very divisive topic in bioethical debates. In November 1999, the US Patent and Trademark Office rescinded a patent on the plant species *Banisteriopsis caapi* held by a Californian since 1986. The plant is sacred to tribal communities living in the Amazon basin and is the source of the hallucinogen *ayahuasca*, used in their religious rituals. In addition to being a harbinger of the complications of religious and cultural respect, it presages the looming, bitter debates about the extent to which biological materials can be "owned."

In fact, in one form or the other, humankind has been in the bioprospecting business for millennia. Like many bioethical issues, emerging technologies and research have changed the landscape (literally and figuratively) dramatically. And, powerful interests, such as pharmaceutical companies, see natural materials (including certain genes) as lucrative ventures that need to be harnessed for profit. For example, the biotechnology firm, Diversa Inc., entered into an agreement with the National Park Service to find efficacious and beneficial microbes in the geysers and springs in Yellowstone National Park. However, this met with much resistance and ultimately the agreement was suspended by a federal court ruling.

As controversial as patents on plant genetic material are, they pale in comparison to the bioethical debates surrounding that of animals. This is in part because patenting animals' genetic materials is linked to cloning. The larger bioethical issue is captured well by the Church of Scotland's Society, Religion and Technology Project:

> Many people would also say that knowledge of a genetic sequence itself is part of the global commons and should be for all to benefit from. To patent parts of the human genome as such, even in the form of "copy genes," would be ethically unacceptable to many in Europe. In response it is argued that patenting is the legal assessment of patent claims, and should not be confused with ethics. But patenting is already an ethical activity, firstly in that it expresses a certain set of ethical values of our society; it is a response to a question of justice, to prevent unfair exploitation of inventions. Secondly a clause excluding inventions "contrary to public order and decency" is part of most European patent legislation – an extreme case of something like a letter bomb would be excluded as immoral. But now we have brought cancerous mice and human genetic material in the potential frame of intellectual property that ethics has moved to a much more central position, where it sits uncomfortably with the patenting profession. They do not like the role of ethical adjudicator to be thrust upon them by society.[14]

Teachable Moment: Patenting Germplasm

Consider the following statement from Keith Douglas Warner of Environmental Studies Institute at Santa Clara University:

> The privatization of germplasm formerly considered the common heritage of humankind is incompatible with notions of the common good and economic justice. The scrutiny that life industries have been receiving is well deserved, although most of this attention has been focused on the potential threats to human and ecosystem health. The economic implications of the biotechnology patent regime are less obvious because they do not impact individuals, but rather social groups. The pubic appears less interested in this dimension of the biotechnology revolution. Nevertheless, addressing this patent regime through the lens of the common good is a better strategy for critics of agricultural biotechnology, who will likely be more successful in slowing down the expansion of corporate control over germplasm by addressing economic issues.[15]

Questions

1. The biotechnical revolution has improved crop yields and has greatly increased the world's food supply in recent centuries. What have been the "human and ecosystem health" tradeoffs associated with these benefits?
2. Is it morally preferable to engage in "slowing down the expansion of corporate control over germplasm" and other genetic materials? Why or why not?
3. Compare any opportunities lost with the risks prevented if germplasm ownership by private concerns is halted.

NEUROETHICS

Ethicists and scientists have continuously struggled with defining just what constitutes personhood, but virtually every definition includes the human mind. So, the mind–brain connection includes elements of both ethics and neuroscience. Nanotechnology and other emergent applications of neurotechnologies affect who a person is. The medical definition of neuroethics is a bit pedestrian; that is, ethical aspects of neuromedicine. However, neuroethics is more than a subset of biomedical ethics.

Manipulations of neural tissue affect who we are, including our free will. This has intrigued ethicists for millennia. Neuroscientists Dai Rees and Steven Rose[16] consider neuroethics more broadly to include aspects of responsibility, personhood, and human agency. These issues are already upon us:

> How will the rapid growth of human brain/machine interfacing – a combination of neuroscience and informatics (cyborgery) – change how we live and think? These are not esoteric or science fiction; we aren't talking about some science-fiction prospects of human cloning, but prospects and problems that will become increasingly sharply present for us and our children in the next ten to twenty years.

Every neuroethical issue is entangled with the concept of "self." Indeed, self is more than an aggregation of neurons, or even their synaptic functions. The challenge is how to characterize the mind and the self, while merging the scientific method with ethics. But this is nothing new. Rene Descartes tried but died before completing this project, followed by the rationalists (closest to Descartes' view), the empiricists (especially David Hume), until the thoroughly modern view of John Stuart Mill's utilitarianism. All of these views fall short in dealing with the self, so they also fail in providing an ethical framework for neuroethics. For example, Hume's logical positivism contended that all knowledge is based on logical inference from simple "protocol sentences" grounded in observable facts. This position is self-defeating since even this postulation cannot be observed! Perhaps, this argues that mind and brain are separate but interrelated concepts.

Thus, neuroethics must reconcile two perspectives: (1) the doctrine of psychophysical parallelism, which hold that a mental state is always accompanied by a neural activity, and (2) the concept of self, which holds that the self or person is more than the mental functions, and also includes temperament or motivation, which are affected by the physiological systems other than neural (e.g., endocrine) and, some would argue, nonphysiological factors (e.g., the soul). The first perspective is most widely held by the neuroscience community, but the second is the one that is embraced by most people, as well as many faith traditions. This difference is fodder for ethical conflict between the professional (e.g., the treating physician or hospital staff) and the patient. It may also help to explain the intense conflict in coma and other cases, such as those where a person is in a permanent vegetative state (PVS).

ORGAN TRANSPLANTATION

The debate over organ transplantation touches on many of the deepest issues in bioethics: the obligation of healing the sick and its limits; the blessing and the burden of medical progress; the dignity and integrity of bodily life; the dangers of turning the body, dead or alive, into just another commodity; the importance of individual consent and the limits of human autonomy; and the difficult ethical and prudential judgments required when making public policy in areas that are both morally complex and deeply important. It is no exaggeration to say that our attitudes about organ transplantation say much about the kind of society that we are, both for better and for worse.

President's Council on Bioethics, 2003[17]

Most scientists consider the human body to be like that of any other organism's body or even no different than any other system of matter. Thus, many do not see any moral relevance to organ transplants. The controversy arises when scientists and others set aside moral attitudes and strongly held convictions about the sacredness. This is in opposition to the beliefs of many people and of a number of faith traditions that see the human body as something much more than a bunch of cells. This is not much different from the psychophysical parallelism between the body and the mind, especially as it conveys a sense of self. Are we devaluing humanity if we have a free exchange of organs? The moral question asked by the President's Council on Bioethics is whether it is possible that in ignoring societal taboos have we diminished humanity, has it "lessened us, dehumanized us, and corrupted us?" This leads to the Council's more practical questions[18]:

1. What is the most ethically responsible and prudent public policy for procuring cadaver organs?
2. Should the current law be changed, modified, or preserved?

The engineer should be aware that such questions are being asked and should not assume that all, or even the majority, of people share the scientific community's perspective on organ transplantation. In fact, this may be the first time that a number of readers have heard about there even being an ethical issue associated with transplants. Suffice to say that the engineer should be aware of the diversity of opinion and of the likelihood that a number of his or her clients at a minimum are uneasy with the current system of transplants and may even be in outright opposition to the procedure.

RESPONSIBLE CONDUCT OF HUMAN RESEARCH

All human beings deserve respect. Those who are subjects of research must provide informed consent. Whereas philosophers for many centuries have extolled the virtues of respect, it really was not until 1979, with the publication of *the Belmont Report: Ethical Principles and Guidelines for the Protection* of Human Subjects of Research,[19] that the inviolable principles were articulated on how to treat people as research subjects. Basically this consists of three requirements: (1) respect for persons; (2) beneficence; and (3) justice. These topics will be considered in much detail throughout this text, especially in Chapter 5.

ANIMAL TESTING

Respect for animals has increasingly been integrated into the Western ethos, including rethinking the majority of the utilitarian perspectives within the medical and biological research communities. Animals have played a key role in biomedical research and technological development (see Chapter 5).

Animal research points to some of the problems of the popularly held utilitarian ethical framework, especially how the model influences compartmentalization and objectification.

Compartmentalization in thinking is not only a division of labor, but it is often a survival mechanism for many professionals engaged in having to live with an ethical decision. This was best demonstrated in my recent discussion with author and veterinarian, Richard Orzeck,[20] who candidly shared the following case:

During my second tour of duty through Cornell's pathology department, I was working late one weekday afternoon in the hospital's postmortem room performing an autopsy on a young Labrador who had died unexpectedly from no obvious cause, when one of my friends from the junior class walked into the room leading a dapple-gray, quarter horse mare. The pathology professor and I had just finished a long and detailed study of the poor dog's lungs and heart when he decided that we both needed a short break. Having nothing else to do while he was out smoking his cigarette, I went over to chitchat with my underclassman colleague.

As she stood there holding the silent and unusually well-behaved horse by its lead, we talked about all of the exciting things that vet students talk about when they're able to find the time: how our classes were going, how my rotations were going, and how demanding all of the professors were. After a couple of minutes of this small talk, I casually asked her what she was doing with the horse here in this room normally set aside for studying the dead.

Her eager and straightforward answer surprised me. She said that the animal was a donation to the college by an owner who, for whatever reason, no longer wanted it. She said that she had been working with a research professor after classes on a project focused on degenerative joint diseases in racing and performance horses, and that the animal was part of their study.

Reaching over to stroke the horse's neck, I looked into eyes of my future colleague and smiled as I congratulated her on being asked to be part of this research project. As a student research assistant, even though the job mostly involves doing all the "dirty work" such as mucking out the stalls and feeding and caring for the research subjects, you do get to interact on a higher level with the doctors and professors in charge. And it can all be pretty exciting stuff, especially if a scientific or medical breakthrough is discovered. Still curious as to why she was here in the postmortem room with this obviously healthy horse, I asked her again why they were here.

A few of seconds of awkward silence ensued as I waited for her to answer. Reaching up with her free hand to pet the horse on its muzzle, she finally told be that she was waiting for the professor to arrive. After he arrived, they would *euthanize* the animal in order to harvest its healthy joint cartilage. These tissues were needed as a *positive control* in their research. And then she said no more. She seemed quite excited by the whole thing, and I remember, just briefly, being a little surprised at her enthusiasm.

I congratulated her again, and not giving it any more thought at the time, returned to continue my autopsy of a Labrador retriever. Several minutes later I saw that the research professor had indeed arrived, and my friend and he carried out their grim task (I won't go into the details) to advance the "noble cause of medical science." My surprise at her enthusiasm was because I knew her to be—just like all my vet school colleagues—a dedicated and compassionate person who loved all animals, especially horses. She was subjecting herself to a rigorous, unrelenting, and expensive eight years of college and medical school to fulfill her life's dream of saving the lives and improving the welfare of her animal patients. But here she was, ready to end the life of a perfectly healthy horse and eager about the opportunity to do so.

Orzeck's case indicates the extent to which scientists can rationalize behaviors and decisions within a "research" context that we would not otherwise do. This is truly an example of the ends justifying the means. Orzeck continues:

I can't say for sure, but by the words that my fellow student used allowed her to justify and to absolve herself of what she was doing. "She and the researcher were going to euthanize the animal in order to harvest its healthy joint cartilage so that they could use the tissue as a positive control." Animal? Euthanize? Harvest? Positive control? To make what she was doing more bearable, she transformed this living, breathing horse into an object of exploitation through the use of these very specific words.

Animal: The creature was no longer a horse; it was now an animal object. It was easier to accomplish what she had to do to an animal or research subject than it was to a living horse.

Euthanize: Even though I and all other veterinarians (and, quite sadly nowadays, even the human medical profession) use the term all the time, the word *euthanize* also is quite sneaky. The term has been bastardized and twisted and applied to many situations less ennobling. It's nicer to say "euthanize the research subject" than it is to say "kill the innocent horse."

Harvest: The term brings to mind comforting images of strong and hardy farmers gathering up sheaves of wheat and corn, or perhaps little red-cheeked children helping their dear grandpa pick apples from well-trimmed trees in his ancient New England apple orchard. But the word also has recently been adopted by various groups of people to rationalize some despicable behavior. Credit and banking agencies "harvest" and "mine" the data that they extract and collect from their clients; various government agencies "harvest" mountains of covertly obtained information on its citizens for their so-called common good; and, most recently, scientists have sought to have the right to "harvest" stem cells from aborted babies to advance the causes of medical research. In summary, it was more pleasant to say "harvest the tissue" than it was to say, "We're exploiting this helpless horse to steal its healthy joint cartilage for our research."

Positive control: The use of positive controls is an honorable and long-used scientific technique that allows researchers to make accurate comparisons of information they are trying to obtain for whatever study they are trying to make. The best example I can think of involves human medical research and a study in which my father was asked to participate. As a frequent hospital patient suffering from heart disease, he had been asked to participate in a study of a new drug, which, if it worked, would improve the strength of his heart. On the permission sheet, he was offered free care during the length of the study, but he would not be told whether he would receive the new drug or a placebo. (A placebo is something that looks like a drug but is actually an inactive substance.) The placebo group would then be referred to as the positive controls. Their purpose was to act as a baseline, or starting point, against which any effects of the new drug could be compared.

The problems with using positive controls are many. The individuals, whether human or animal, don't receive any potential benefits of the scientific study that's being performed. In the worst case, the positive controls could be deprived of a life-giving procedure. But the biggest problem with regard to medical science and the use of positive controls is that by using the term *positive control*, researchers are able to make objects out of these patients. And making an object out of living and breathing people now makes them easier to use and take advantage of, all in the name of science. It is easier to justify and to appear heroic to sacrifice a positive control than it is to put to death an innocent horse.

I hasten to add that before we pass judgment on these otherwise compassionate and loving persons, we must be reminded that nearly all of us are guilty of objectification. For example, when clients bring their pets to my practice for euthanasia, I always ask them what they would like me to do with their remains; probably because the word *remains* is easier to deal with and just sounds nicer than *dead pet's body*.

Orzeck's case is yet another example of how terminology is not simply a neutral conveyance of information, but it is often steeped in ideology and perspective. Medical and engineering professionals, as Feynman reminded us, must be diligent and vigilant in using the correct terms to communicate. "Junk science" is fraught with the loose usage of language. In fact, strategically designed redefinitions and omission or selective use of actual data are common fallacies in junk science.

Is the Research Worth It?

Debates about animal research elucidate a number of ethical issues. First, as illustrated in the debate between Descartes and Gassendi (see discussion in Chapter 4), the difference between humans and animals is an important distinction. To most biologists, the difference is merely a continuum, as indicated by the development of the nervous system and other physiological metrics. These physiological complexities translate into sensory differences that differentiate the species' sentience (especially self-awareness), one of the variables that distinguish "humanness." In fact, sentient-centered ethics falls between "anthropocentric" (human-centered) and "biocentric" (i.e., all living creatures have inherent worth, e.g., Schweitzer's "reverence for life"[21]) ethical frameworks. That is, it calls us to appreciate the perspective of other creatures in a personal sense. With advances in the understanding of neurological processes, for example, we must assume that the more highly developed animals (and possibly even many of the less advanced) experience pain. That said, the bioethical view would cause us to want to do whatever we can to prevent or at least reduce suffering in these other species.

The ethical problem ensues when the utility of animal-derived knowledge is presented as a dichotomy. This can cause the utilitarian view to come to the fore and we are forced to choose between preventing and curing some human malady *versus* harming animals. However, this viewpoint is inherently weak, and such an argument is invalid. This illogical argument is referred to as the fallacy of *non sequitur* (Latin: "does not follow") since the outcome (cure) does not really depend on the condition (animal suffering). This particular *non sequitur* argument is known as "denying the antecedent."

1. If A then B. (Animal research leads to cures, so people benefit from animal research.)
2. Not A. (Not allowing animal research would end animal suffering.)
3. Therefore, not B. (Not allowing animal suffering would prevent cures.)

Not all animal research leads to animal suffering (at least it varies substantially). Another type of *non sequitur* is affirming the consequent:

1. If A is true, then B is true. (If animals suffer in research, cures are developed.)
2. B is stated to be true. (Animal research has resulted in cures.)
3. Therefore, A must be true. (Therefore, animals must suffer.)

In this instance, the argument derives from a false dichotomy that denies the possibility that animal research can be conducted in ways that are humane. It also denies that even if the animal research was necessary in the past (or at least beneficial to the advancement of medical knowledge), it ignores alternate approaches that can now displace such research (e.g., *in vitro* and *in silico*).

Moral concern grows with respect to what we value. For example, we are less tolerant of harm to our own pet than to a "generic" lab animal. Also, concern usually increases as an animal is considered more "human-like." This value can be seen even in people who believe that animals differ from humans both in degree (i.e., a continuum of self-awareness and cognitive processes) and in kind (i.e., inherently different, especially because many believe humans have a soul and animals do not). In both instances, people value certain animals more than others. In the first case, the species (i.e., the cat or the dog) may have more value to the pet owner than to species other than that of the pet; however, this value can be transferred to other animals. The second type of valuation increases with the complexity of the species. Greater concern may result because people ascribe human characteristics to nonhuman species (known as "anthropomorphism").

Primates are particularly important indicators of moral value, since they share many more "human" traits than most other animals. From a biomedical perspective, the fact that nonhuman primates have similar anatomical and physiological features makes them ripe for research. From a bioethical perspective, their similarities argue against much of the research, especially that which is invasive, painful, and unpleasant. The case of psychological researcher Edward Taub and his George Washington University student Alex Pacheco (who later became cofounder of People for the Ethical Treatment of Animals, PETA) dramatically illustrates the bioethical challenge.[22]

From 1981 to 1991, the National Institutes of Health provided $1.2 million to the Institute of Behavioral Research in Silver Spring, Maryland for Taub's attempt to regenerate severed nerves in 17 Rhesus Macaque (*Macaca mulatta*), popularly known as the Rhesus Monkey. Sensory nerves were cut in the monkeys' limbs (deafferentation), then stimuli such as electric shock, physical restraint, and food or sight deprivation were applied to compel animals to regain the use of crippled limbs. Taub's rationale for this research was to aid stroke victims and the mentally handicapped.

In 1982, Pacheco, who took a summer job at the institute, visited the laboratory and took photographs of the starved, uncared for animals. He reported the lab to the state police, who then raided the lab, removed the monkeys from the lab, and the case went to court. The monkeys were euthanized at the end of the case. Taub was convicted of six counts of animal cruelty.

The case illustrates some of the ethical aspects of animal research: definition of human life *versus* animal life, the application of precautionary principles, dignity in any research methodology, the question of whether animals have rights, and whether useful results truly justify research. On this last count, Taub's research appears to have advanced the state of neurological science. But was it worth it?

Animal research is of three types: pure research, applied research, and testing. Pure research strives to advance the understanding of biological structures, processes, and systems for the sole benefit to medical and scientific knowledge. Applied research is a research specifically done to address a biomedical need. Testing studies are special types of applied research that test the effects of a procedure, device, or drug to determine efficiencies and efficaciousness. Thus animal data provide a utilitarian purpose. Because animal researchers hold human life to be paramount, they see animals as a means to an end and view this research as justifiable based solely on possible benefits to the human race.

Viewed exclusively from this perspective, the Silver Spring monkey is a morally and ethically sound case. However, there must be other nonutilitarian aspects to the case that are immoral. Let us consider how the court may have arrived at a guilty verdict, and arguably, a pronouncement of unethical practices (Figure 2.3). Numerous consequences of this case are positive. Whereas the animals may have been abused and living in deplorable conditions, there were medical advances.

It appears that Pacheco made his decisions based mainly on three factors: legality (animal rights, guidelines), duty to provide health/safety of animals, and politics/public opinion. He showed less regard for the potential of these studies to aid human victims of stroke and mental retardation as Taub proposed. His position seems to have been that not only was the fruitful research not morally obligatory, but was also not morally permissible. This seems to be a more deontological view than that of Taub, who was most concerned with the benefits to science/medical research, finding cures for humans, and finance/economics of funding. In fact, many such whistle-blowing cases exhibit a duty-based view. Some are indeed utilitarian, for example, if a laboratory is doing bad science that will lead to negative consequences (e.g., dangerous device or drug), but others are in spite of good science if immoral activity

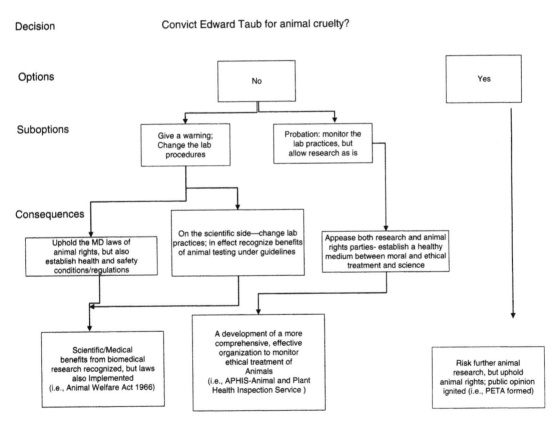

Figure 2.3 Event tree showing possible decision flow in Silver Spring Rhesus Monkey case.

is ongoing (e.g., cruelty and harassment). In the end, the courts sided with Pacheco and convicted Taub. We can show this logical progression as a syllogism:

Factual premise	Animal research is a powerful tool for scientific and medical research, and Taub's monkey-based studies were intended to study nerve regeneration and recovery of crippled limbs.
Fact-value premise	The animal subjects lived in deplorable conditions and were abused.
Evaluative premise	Even though animal research is a powerful tool for scientific and medical research, abuse of animal subjects in studies is morally wrong.
Evaluative conclusion	Therefore, Taub's studies, while intended for research, violate animals' rights and are morally wrong.

Therefore, animal research goes beyond "whether the nonhuman nature of other animals is morally relevant, and if it is relevant, what humanity and justice permit us to do with animals."[23] As we learn more about animal neurophysiology and psychology it becomes harder to argue that the values of justice and humanity are not applicable to animals, and do not merit ethical consideration. One does not have to be a hardcore member of the Animal Liberation Front to agree that some cases go beyond the pale. Few would disagree that some studies are clearly immoral, such as the studies done at the University of Pennsylvania wherein baboons were rendered brain dead from having their heads cemented to pistons that whipped the brain at accelerations up to 2,000 times the force of gravity, while researchers laughed while the primates were restrained, alert, and writhing before impact.[24] However, most animal welfare cases are not so clear.

SYSTEMATIC REALITY CHECK

Biomedical and biosystem engineers depend on animal studies, and they are more likely than many to support such research as a critical means toward a noble end. So, then, how can engineers be objective and realistic about matters where we are likely to have such a bias and conflicts of interests? One way is to do what engineers do best; use the systematic approach.[25] Start by gathering the pertinent facts and identifying the uncertainties and data gaps. Ethical problem solving is fraught with ambiguities and assumptions, so it is crucial that decisions are based on the best available factual information and that the decision maker be honest about what is not known.

Next, use the gathered information to identify realistic options and alternate solutions. Conduct a sensitivity analysis, i.e., compare how each alternative performs in terms of how sensitive it is to slight changes in scenarios needed to reach the decision. For example, one alternative may produce excellent outcomes (high on the utility scale), but in the process it leads to injustices or disproportional harm (low on the fairness scale). The alternative of withholding scientific findings that may promote bioterrorism (i.e., "dual use"), for instance, could be seen as one where one utility is optimized (i.e., security), but at the expense of another value (i.e., scientific freedom). Based on these analyses, a number of plans for addressing the problem can be considered.

All the reasonable plans can be compared and assessed against moral metrics, such as those embodied in the moral theories (e.g., utilitarianism, Rawlsism, deontology, and rational models). Key to these assessments is characterizing all the potentially affected parties and the stake each has in the decision. What are the risks and the benefits to each party from each option?

Finally, make a well-informed decision. However, even after the decision is made, keep seeking feedback and revisiting the options and alternatives to ensure that the decision continues to be the right one. Ethical decision making can be messy and chaotic. Unanimous decisions are the exception. Even consensus can be difficult. Sometimes, the right decision is made in the face of a resistant majority. So, the process is never static and the actions may need to be adapted as new information becomes available. For example, the constraint may disappear. The choice of keeping some lead (Pb) in gasoline even though it was known to have neurotoxic effects eventually went away in the 1980s when alternative fuel additives became available and when engines were redesigned. Or, when a sufficient repository of data becomes available, the need for even important animal testing can be replaced by computational methods (i.e., *in silico* studies and informatics). Thus, we need to continue to look for alternatives to animal research.

Animal welfare is sometimes categorized as the "3 Rs:"

1. *Reduction.* Methods that result in the use of fewer animals to obtain scientifically valid information.
2. *Refinement.* Methods to reduce stress or discomfort to the animals involved and to improve animals' overall well-being and environment.
3. *Replacement.* Methods other than animal studies that can provide robust biomedical information and modeling.

The bioengineering researcher must constantly rethink the research to adhere to these three methods.

GENETICALLY MODIFIED ORGANISMS

Genetic modification of organisms is a very old endeavor. Humans have changed the characteristics of numerous plants and animals through selective breeding techniques, beginning with attempts to encourage offspring from organisms with favorable traits, such as size, color, texture, and taste. However, in recent decades the process has become supercharged with the onset of direct genetic manipulations of DNA and RNA. Like a number of bioethical issues, genetic manipulation is associated with the slippery slope fear.[26] The slippery slope occurs when allowing an act makes it "impossible to hold the line and prevent extension to a less justifiable situation."[27] The goal of genetic modification is to delete specific phenotypic characteristics from or introduce new characteristics to an organism's progeny. Prior to the late twentieth century, this was done externally, but now it is increasingly accomplished internally within the cell's genetic material. The resulting progeny is known as a "transgenic organism." Transgenesis occurs when DNA from another organism is introduced using artificial gene transfer techniques.

Such techniques allow researcher to understand the interactions of certain genes more completely. The major ethical issues involved in genetically modified organisms (GMO) often center on animal welfare, and risks to human health and environment. For example, what if a new creature is so different in kind that it has such a competitive advantage (i.e., no effective predators) or an ability to self-replicate that it would pose risks to public health and welfare, in violation of the engineer's first ethical canon. And, how many animals' lives are worth an important discovery? Are we decreasing the genetic diversity of our wildlife or destroying the habitats of other animals?

Such concerns come from within and outside of the scientific community and have at least a basis in utilitarianism (i.e., disagreement about the utility *versus* risks). Others oppose the research based on religious and moral concerns, arguing that the researchers and the biotechnological companies are immorally attempting to "play God" by creating entirely new beings and unnaturally altering the genetic makeup of progeny.

Furthermore, GMOs are generally supported due to the dual effects principal. First, it is believed that such research could lead to more profitable or productive animals. Second, scientists hope that this experimentation could aid humans by developing treatments to deadly diseases or methods to assist in the creation of tissues and organs.

TRANSGENIC SPECIES

An organism that has been genetically modified (GM) is known as a "transgenic species." There are at least two ethical considerations of transgenic species. First, what impact might research and marketing on these species have on society, such as threats to health? Second, what are the ethical considerations needed on behalf of the creatures themselves? We address the first issue next in discussing GM food. The second is predominantly about animals. For example, is it ethical to modify a monkey so that its fur glows in the dark?

One issue with transgenic species is about who benefits. For example, Harvard University modified the genetic structure of a mouse, known as the "oncomouse," to produce a human cancer in the animal. From an anthropocentric viewpoint, this can be justified so long as it helps to advance cancer research, but it does not stand against a biocentric view or even a deontological view if one considers the protection of lower sentient species obligatory. It is also an example of bioprospecting, as Harvard has sought protection of their ownership of the oncomouse genetic information.

FOOD

The European Union has taken a number of recent actions, many precautionary, to address GMOs. These actions can be treated as a bioethical case. In 1999, a four-year ban on all GM crops was the result of several studies indicating that GM food could be harmful to humans. One study of the effects of GM potatoes on the digestive tract of rats found that there were large differences between the intestines of rats fed GM potatoes and those fed unmodified potatoes.[28] Other studies have indicated that introducing a new gene to a plant could create new allergens or cause reactions in susceptible people (e.g., a gene from peanuts transferred to soybeans could cause those allergic to peanuts to have reactions when they eat soybeans).

Based on these studies, the European Union announced a ban on all GM crops. At the end of 2002, the ban was lifted, but producers were forced to label all GM goods with a special DNA sequence that identified the origin of the crops. This made it easier for regulators to spot contaminated crops and feed. Later, in June 2003, the European Parliament agreed upon a UN biosafety protocol regulating the international trade of GM food. This protocol allows countries to ban the importation of GM organisms. However, there must be credible evidence about the dangers of the product. This protocol, therefore, states that goods developed from new technologies should be regulated based on the precautionary principle.

A number of factors surround GMOs. The safety and welfare of consumers appear to be the biggest concern. The European nations feared that their inhabitants would experience sicknesses similar to the mad cow epidemic (bovine spongiform encephalopathy). However, GM crops could lead to future discoveries in other fields. It is possible that certain technologies could be first tested on crops before animals, or that methods for altering plant genetics could be transferable to humans and/or animals. Opportunity costs, such as precautions slowing possible improvements in health, need to be considered. Enhanced nutritional value, disease prevention, and other health-related positive outcomes should be considered. Another possible negative outcome that should be investigated is the effects of the crops on the environment. Will new plant species interrupt predator–prey balances or be opportunistic, nuisance species?

A flow chart (Figure 2.4) shows the decision-making process in this case. At the beginning, Europe would like to ban the GM food. So, they approach their first question: Is there evidence to support such an action? The countries then must determine if there is sufficient evidence to support such an action. If they do have the evidence, Europe must next ensure that the studies meet all scientific requirements (i.e., they are replicable, etc.). If the studies are not sufficient, the ban cannot be established until further research has been conducted.

If all the scientific evidence is credible, then the countries encounter the feasibility issues. Can the countries be sustained without GM foods? The majority of the food produced has some form of genetic alteration, so such a ban would be dramatic.

This question also entails a political aspect. Moreover, with such a ban, it is undeniable that many of Europe's trading partners will be infuriated. So, will all the nations be able to handle the political backlash of such a regulation?

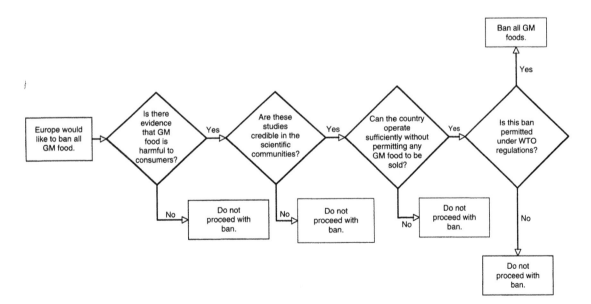

Figure 2.4 Decision flow chart for GM food in Europe.

By answering yes to the above questions, Europe is able to move on the final question – would this regulation be permitted under World Trade Organization guidelines? If the answer is yes, then they ban the food. If the answer is no, then they have no choice but to allow the goods to be sold in their nations.

The initial decision posed was whether Europe should ban GM foods. Europe had two options, to ban or not to ban the food. If they chose to ban, they had two suboptions. First, they could attempt to get studies and research to support the decision. Second, they could avoid the research step and proceed to ban the GM food. This demonstrates that the GMOs have macroethical and geopolitical ramifications, so no matter how comfortable the GMO researcher may feel, there will be opposition to the research.

ENVIRONMENTAL HEALTH: THE ETHICS OF SCALE AND THE SCALE OF ETHICS

The ultimate measure of a man is not where he stands in moments of comfort and convenience, but where he stands at times of challenge and controversy.

Martin Luther King, Jr. (1963)[29]

Environmental problems are often characterized by scale, a concept familiar to most engineers. In fact, we often describe phenomena by their dimensions and by when they occur; that is, by their respective spatial and temporal scales. Engineers are comfortable with dimensional analysis. But, can we "measure" ethics in a similar way? Of course, physical analogies do not completely hold for metaphysical phenomena, but they can be instructive. King's advice is that we can measure ethics, especially in our behavior during worst cases. How well can we stick to our principles and duties when things get tough? Philosophers and teachers of ethical philosophy at the university level frequently subscribe to one classical theory or another for the most part, but most concede the value of other models. However, they all agree that ethics is a rational and reflective process of deciding how we ought to treat each other.

TEMPORAL ASPECTS OF BIOETHICAL DECISIONS: ENVIRONMENTAL CASE STUDIES

An important consideration in making bioethical decisions is the amount and type of effects that will result from an action. This is the beginning of rational ethics, that is, forming a factual premise. A particularly difficult aspect of a decision or an activity is predicting the cascade of events and their future impacts.

From a teleological perspective, an event can represent a means or an end. Indeed, an end can actually be a means toward another end. So let us consider a few bioethical problems of various scales. Manufacturing and commercial decisions about material use, such as metallic pigments in paint, lead-based fuel additives, and industrial processes that generate carcinogenic byproducts, can have lasting effects for generations. Such decisions are quite complex. A case in point is the comparison of short- and long-term effects of using coal *versus* nuclear fission to generate electricity. Combusting coal releases particle matter and damaging compounds like sulfur dioxide and toxic substances like mercury. Nuclear power presents a short-term concern about potential accidental releases of radioactive materials and long-lived radioactive wastes (sometimes with half-lives of hundreds of thousands of years). Nuclear events have also been extremely influential in our perception of pollution and threats to public health. Most notably, the cases of Three Mile Island, Dauphin County, Pennsylvania in 1979 and the Chernobyl nuclear power plant disaster in the Ukraine in 1986 have had an unquestionable impact on not only nuclear power, but on other aspects of environmental policy, such as community "right to know" and the importance of risk assessment, management, and communication.

Similarly, decisions regarding armed conflict must consider not only the tactical warfare, but the geopolitical and public health changes wrought by the conflict. Furthermore, the psychological and medical effects on combatants and noncombatants must be taken into account. Prominent cases of these effects include the use of the defoliant Agent Orange in Vietnam and decisions to prescribe drugs and to use chemicals in the Persian Gulf War in the 1990s. The World War II atomic bombings on the Japanese cities of Hiroshima and Nagasaki in August 1945 not only served the purpose of accelerating the end of the war in the Pacific arena, but also ushered in the continuing threat of nuclear war. In addition, the bombings were the world's first entrees to the linkage of chronic illness and mortality (e.g., leukemia and radiation disease) that could be directly linked to radiation exposure. Following are a few cases where the effects did not become apparent until after a protracted lag period.

Agent Orange

The use of Agent Orange during the Vietnam War (used between 1961 and 1970) dramatically demonstrates the concept of "latency period," where possible effects may not be manifested until years or decades after exposure. Agent Orange is a defoliant and weed-killing chemical that was used by the US military during the Vietnam War. It was sprayed to remove leaves from the trees behind which the enemy troops would hide. Agent Orange was dispersed by airplanes, helicopters, trucks, and backpack sprayers. In the 1970s, years after the tours of duty in Vietnam, some veterans became concerned that exposure to Agent Orange might be the cause of delayed health effects. One of the chemicals in Agent Orange contained small amounts of the highly toxic compound 2,3,7,8-tetrachlorodibenzo-*para*-dioxin (TCDD).

The US Department of Veteran Affairs (VA) has listed a number of diseases, which could have resulted from exposure to herbicides like Agent Orange. The law requires that some of these diseases be at least 10% disabling under VA's rating regulations within a deadline that began to run the day a person left Vietnam. If there is a deadline, it is listed in parentheses after the name of the disease as follows:

- Chloracne or other acneform disease consistent with chloracne (must occur within one year of exposure to Agent Orange)
- Chronic lymphocytic leukemia
- Diabetes mellitus, type II
- Hodgkin's disease
- Multiple myeloma
- Non-Hodgkin's lymphoma
- Acute and subacute peripheral neuropathy (for the purpose of this section, the term acute and subacute peripheral neuropathy means temporary peripheral neuropathy that appears within weeks or months of exposure to an herbicide agent and resolves within two years of the date of onset)

- Porphyria cutanea tarda (must occur within one year of exposure to Agent Orange)
- Prostate cancer
- Respiratory cancers (cancer of the lung, bronchus, larynx, or trachea)
- Soft-tissue sarcoma (other than osteosarcoma, chondrosarcoma, Kaposi's sarcoma, or mesothelioma)

The issue is international. After all, if it is true that US soldiers exposed to Agent Orange show symptoms often associated with dioxin exposure, then there is also likely to be residual dioxin contamination in the treated areas of Vietnam. Dioxin is highly persistent.

Scientists from Vietnam, the United States, and 11 other countries have discussed the state of the science of research in the health effects of dioxin. The Vietnamese and US have agreed to a plan that addresses the need for direct research on human health outcomes from exposure to dioxin and research on the environmental and ecological effects of dioxin and Agent Orange.

Chlorinated dioxins have 75 different forms and there are 135 different chlorinated furans, depending on the number and arrangement of chlorine atoms in the molecules. The compounds can be separated into groups that have the same number of chlorine atoms attached to the furan or to the dioxin ring. Each form varies in its chemical, physical, and toxicological characteristics (Figure 2.5) The primary

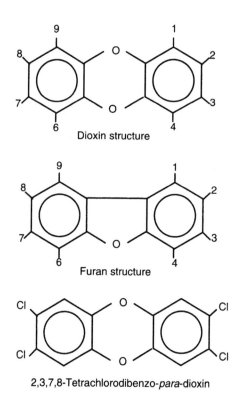

Dioxin structure

Furan structure

2,3,7,8-Tetrachlorodibenzo-*para*-dioxin

Figure 2.5   Molecular structures of dioxins and furans. Bottom structure is of the most toxic dioxin congener, tetrachlorodibenzo-*para*-dioxin (TCDD), formed by the substitution of chlorine for hydrogen atoms at positions 2, 3, 7, and 8 in the molecule.

concerns in Vietnam from prolonged exposure to dioxin are reproductive and developmental disorders that may be occurring in the general population.

Dioxins are created only unintentionally during chemical reactions, especially combustion processes; that is, they have never been synthesized for any other reason than for scientific investigation, for example, to make analytical standards for testing. The most toxic form is the 2,3,7,8-tetrachlorodibenzo-*para*-dioxin (TCDD) isomer, which is a byproduct when certain pesticides, such as those used in the Vietnam defoliants, are produced. Other isomers, which may have been present in the formulations along with the 2,3,7,8 configurations, are also considered to have higher toxicity than the dioxins and furans with different chlorine atom arrangements.

Dioxin contaminants of Agent Orange have persisted in the environment in Vietnam for over thirty years. In addition to a better understanding of the outcomes of exposure, an improved understanding of residue levels and rates of migration of dioxin and other chemicals into the environment is needed. "Hot spots" containing high levels of dioxin in soil have been identified and others are presumed to exist but have yet to be located.

Dioxin has migrated through soil and has been transported through natural processes such as wind-blown dust and erosion into the aquatic environment. Contamination of soil and sediments provides a reservoir source of dioxin for direct and indirect exposure pathways for humans and wildlife. Movement of dioxin through the food web results in bioconcentration and biomagnification with potential ecological impacts and continuing human exposure. Research is needed to develop approaches for more rapid and less expensive screening of dioxin residue levels in soil, sediment, and biological samples, which can be applied in Vietnam.

Actually, a number of defoliating agents were used in Vietnam, including those listed in Table 2.2. Most of the formulations included the two herbicides 2,4-D and 2,4,5-T. The combined product was mixed with kerosene or diesel fuel and dispersed by aircraft, vehicle, and hand spraying. An estimated 80 million liters of the formulation was applied in South Vietnam during the war.[30]

The Agent Orange problem illustrates the problem of uncertainty in characterizing and enumerating effects. There is little consensus on whether the symptoms and disorders suggested to be linked to Agent Orange are sufficiently strong and well documented; that is, provide weight of evidence, to support

Table 2.2
Formulations of defoliants used in the Vietnam War

Agent	Formulation
Purple	2,4,-D and 2,4,5,-T used between 1962 and 1964
Green	Contained 2,4,5-T and was used during 1962–1964
Pink	Contained 2,4,5-T and was used during 1962–1964
Orange	A formulation of 2,4,-D and 2,4,5-T used between 1965 and 1970
White	A formulation of picloram and 2,4,-D
Blue	Contained cacodylic acid
Orange II	2,4,-D and 2,4,5-T used in 1968 and 1969 (also sometimes referred to as "Super Orange")
Dinoxol	2,4,-D and 2,4,5-T. Small quantities were tested in Vietnam between 1962 and 1964
Trinoxol	2,4,5-T. Small quantities tested in Vietnam during 1962–1964

Source: Agent Orange website: http://www.lewispublishing.com/orange.htm; accessed on 22 April 2005.

cause and effect. This complicates bioethical decision making since the factual premises are fraught with uncertainties.

JAPANESE METAL INDUSTRIES

As in military decisions, industrial decisions also have significant long-term health considerations.

Minamata Mercury Case

One of the most telling cases of improper bioethical decision making was that of Minamata, a small factory town on Japan's Shiranui Sea. Minamata means "nitrogen," emblematic of the town's production of commercial fertilizer by the Chisso Corporation for decades, beginning in 1907.[31] Beginning in 1932, the company produced pharmaceutical products, perfumes, and plastics and processed petrochemicals. Chisso became highly profitable, notably because it became the only Japanese source of a high-demand primary chemical, diotyl phthalate (DOP), a plasticizing agent. These processes needed the reactive organic compound, acetaldehyde, which is produced using mercury. The residents of Minamata played a huge price for this industrial heritage. Records indicate that from 1932 to 1968, the company released approximately 27 tons of mercury compounds into the adjacent Minamata Bay. This directly affected the dietary intake of toxic mercury by fisherman, farmers, and their families in Kumamoto, a small village about 900 kilometers from Tokyo. The consumed fish contained extremely elevated concentrations of a number of mercury compounds, including the highly toxic methylated forms (i.e., monomethyl mercury and dimethyl mercury), leading to classic symptoms of methyl mercury poisoning. In fact, the symptoms were so pronounced that the syndrome of these effects came to be known as the "Minamata disease."

In the middle of the 1950s, residents began to report what they called the "strange disease," including the classic form of mercury toxicity; that is, disorders of the central nervous system (CNS) and peripheral nervous systems (PNS). Diagnoses included numbness in lips and limbs, slurred speech, and constricted vision. A number of people engaged in uncontrollable shouting. Pets and domestic animals also demonstrated mercury toxicity, including cat suicides and birds dying in flight. These events were met with panic by the townspeople.

Physician, Hajime Hosokawa from the Chisso Corporation Hospital, reported in 1956 that "an unclarified disease of the central nervous system has broken out." Hosokawa correctly associated the fish dietary exposure with the health effects. Soon after this initial public health declaration, government investigators linked the dietary exposures to the bay water. Chisso denied the linkages and continued the chemical production, but within two years, they moved their chemical releases upstream from Minamata Bay to the Minamata River, with the intent of reducing the public outcry. The mercury pollution became more widespread. For example, towns along the Minamata River were also contaminated. Hachimon residents also showed symptoms of the "strange disease" within a few months. This led to a partial ban by the Kumamoto Prefecture government, which responded by allowing fisherman to catch, but not to sell, fish from Minamata Bay. The ban did not reduce the local people's primary exposure, since they depended on the bay's fish for sustenance. However, the ban did acquit the government from further liability.

Some three years after the initial public health declaration, in 1959, Kumamoto University researchers determined that the organic forms of mercury were the cause of the "Minamata disease." A number of panels and committees, which included Chisso Corporation membership, studied the problem. They rejected the scientific findings and any direct linkages between the symptoms and the mercury-tainted water. After Hosokawa performed cat experiments that dramatically demonstrated the effects of mercury poisoning, Chisso managers no longer allowed him to conduct such research and his findings were concealed from the public.[32] Realizing that the links were true, the Chisso Corporation began to settle with the victims. The desperate and relatively illiterate residents signed agreements with the company for payment, which released the company from any responsibility. The agreement included the exclusion: "... if Chisso Corporation were later proven guilty, the company would not be liable for further compensation." Notwithstanding these setbacks, Minamata also represents one of the first cases of environmental activism. Residents began protests in 1959, demanding monetary compensation. However, these protests led to threats and intimation by Chisso; so the victims settled for fear of losing even the limited compensation.

Chisso installed a mercury removal device at the outfall, known as a "cyclator," but it omitted a key production phase so the removal was not effective. Finally, in 1968, the company stopped releasing mercury compounds into the Minamata River and Bay. Ironically, the decision was neither an environmental one, nor an engineering solution. The decision was made because the old mercury production method had become antiquated. Subsequently, the courts found that the Chisso Corporation repeatedly and persistently contaminated Minamata Bay from 1932 to 1968.

Victim compensation has been slow. About 4000 people have either been officially recognized as having "Minamata Disease" or are in the queue for verification from the board of physicians in Kumamoto Prefecture. Fish consumption from the bay has never stopped, but mercury levels appear to have dropped, since cases of severe poisoning are no longer reported.

Cadmium and *Itai Itai* Disease

The mines of central Japan located near the Toyama Prefecture have been removing metals from the surrounding mountains since as early as 710 A.D. Gold was the first metal to be mined from the area, followed by silver in 1589, and shortly thereafter lead, copper, and zinc. At the start of the twentieth century, the Mitsui Mining and Smelting Co. Ltd controlled the production of these mines. As a result of the Russo-Japanese War, World War I, and World War II, a surge in the demand for metals in the use of weapons and military equipment caused massive increases in the mines' production that was aided with the advent of new European technologies in mining.

Along with the huge increase in mining production came a significant increase in pollution produced from the mines. Liquid and solid wastes were dumped into the surrounding waters, including the Jinzu River that flows into the Sea of Japan, and the five major tributaries that flow into the Jinzu River. The Jinzu River water system supplies water to the surrounding city of Toyama, 30 kilometers downstream from the main mining operations.[33] This water was primarily used by the surrounding areas for irrigation of the rice paddies. In addition, water provided a source for drinking, washing, and fishing.

Large amounts of cadmium were released into the Jinzu River Basin from 1910 to 1945. Cadmium was extracted from the earth's crust during the production of other metals like zinc, lead, and copper that were being mined near the Toyama area. Cadmium is a naturally occurring element that does not corrode easily, enters the air during mining, can travel long distances, and then falls into the ground or water only to be taken up by fish, plants, animals, or humans from the environment.[34] The cadmium from the mines deposited in the river and land of the Jinzu River Basin was absorbed by the surrounding plants and animals causing fish to die and the rice to grow poorly. Furthermore, humans living in that area consumed poisoned water and rice.

As a result of the ingestion of cadmium, a previously undiagnosed disease specific to the Toyama Prefecture appeared in 1912. Initially, the symptoms were not well understood and suspected to stem from either a regional or a bacterial disease or the result of lead poisoning. However, in 1955, cadmium was linked to the disease which came to be known as *itai itai*. In 1961, the Kamioka Mining Station of Mitsui Mining and Smelting Co. Ltd was linked as the direct source of cadmium poisoning, and the Toyama Prefecture, 30 kilometers downstream, was designated the worst cadmium-contaminated area (Figure 2.6). The concentrations of cadmium in this prefecture were orders of magnitude higher than that found in background levels and were well above other industrialized locations (Table 2.3).

Cadmium exposure results in major health-related problems of *itai itai* disease, irreversible kidney damage, and bone disease (*itai itai* is Japanese for "Ouch Ouch"). After exposure to high levels of cadmium, the kidneys have decreased ability to remove acids from the blood due to proximal tubular dysfunction resulting in hypophosphatemia (low-phosphate blood levels), gout (arthritic disease), hyperuricemia (elevated uric acid levels in the blood), hyperchloremia (elevated chloride blood levels), and kidney atrophy (as much as 30%). Following kidney dysfunction, victims of the disease develop

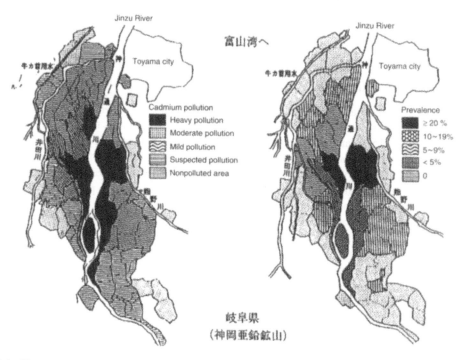

Figure 2.6 The co-occurrence of cadmium contamination with the prevalence of itai itai disease in woman over fifty years of age in Toyama Prefecture (c. 1961).
Adapted from: Kanazawa Medicine, 1998.

Table 2.3
Estimates of average daily dietary intake of cadmium based on food analysis in various countries

Country	Estimates (μg Cd per day)	Reference
Areas of normal exposure		
Belgium	15	Buchet et al. (1983)
Finland	13	Koivistoinen (1980)
Japan	31	Yamagata & Iwashima (1975)
Japan	48	Suzuki & Lu (1976)
Japan	49	Ushio & Doguchi (1977)
Japan	35	Iwao (1977)
Japan	49	Ohmomo & Sumiya (1981)
Japan	59	Iwao et al. (1981a)
Japan	43.9 (males), 37.0 (females)	Watanabe et al. (1985)
New Zealand	21	Guthrie & Robinson (1977)
Sweden	10	Wester (1974)
Sweden	17	Kjellström (1977)
United Kingdom	10–20	Walters & Sherlock (1981)
USA	41	Mahaffey et al. (1975)
Areas of elevated exposure		
Japan	211–245	Japan Public Health Association (1970)
Japan	180–391	
Japan	136	Iwao et al. (1981a)
United Kingdom	36	Sherlock et al. (1983)
United Kingdom	29	Sherlock et al. (1983)
USA	33	Spencer et al. (1979)

Source: International Programme on Chemical Safety, 1992, Environmental Health Criteria: 134 (Cadmium), World Health Organization, Geneva, Switzerland.

osteomalacia (soft bones), loss of bone mass, and osteoporosis leading to severe joint and back pains and increased risk of fractures ATSDR (1999).

SCALE IS MORE THAN SIZE

As the previous cases indicate, the timing of exposure, not just the amount of time, is of the essence. Corporations and governments have changed the health baseline of the world. The background and baseline of contaminants in the human body has increased dramatically in a relatively short time (see Table 2.3). Theo Colburn, who is known for her publications on the growth and effects of environmental endocrine disrupting compounds, sums up the problem of synthetic chemicals and the moral responsibility of corporations that gained much attention in the 1960s and 1970s:

> Every one of you sitting here today is carrying at least 500 measurable chemicals in your body that were never in anybody's body before the 1920s. . . . We have dusted the globe with man-made chemicals that can

undermine the development of the brain and behavior, and the endocrine, immune and reproductive systems, vital systems that assure perpetuity.... Everyone is exposed. You are not exposed to one chemical at a time, but a complex mixture of chemicals that changes day by day, hour by hour, depending on where you are and the environment you are in.... In the United States alone it is estimated that over 72,000 different chemicals are used regularly. Two thousand five hundred new chemicals are introduced annually—and of these, only 15 are partially tested for their safety. Not one of the chemicals in use today has been adequately tested for these intergenerational effects that are initiated in the womb.[35]

The corporate track record in the twentieth century was not good. Toxic substances sprang up at Love Canal in New York, Times Beach in Missouri, and the Valley of the Drum in Kentucky. Soon numerous other sites throughout the United States were found to be contaminated, leading to a progression of environmental laws, especially the Comprehensive Environmental Response, Compensation and Liability Act, better known as "Superfund," beginning in 1980. Much of the previous legal precedence for environmental jurisprudence had been more on the order of nuisance laws. With the greater recognition of public health risks associated with toxic substances like those found in these hazardous waste cases, the public and the environmental professionals called for a more aggressive and scientifically based approach. This changed the bioethical framework for engineering.

LOVE CANAL

The seminal and arguably the most infamous case is the contamination in and around Love Canal, New York. The beneficent beginnings of the case belie its infamy. In the nineteenth century, William T. Love had an opportunity to generate electricity from Niagara Falls and the potential for industrial development. To achieve this, Love planned to build a canal that would also allow ships to pass around the Niagara falls and travel between the two Great Lakes, Erie and Ontario. The project started in the 1890s, but soon floundered due to inadequate financing and also due to the development of alternating electrical current, which made it unnecessary for industries to locate near a source of power production. Hooker Chemical Company purchased the land adjacent to the canal in the early 1990s and constructed a production facility. In 1942, Hooker Chemical began disposing its industrial waste in the canal. This was wartime in the United States, and there was little concern for possible environmental consequences. Hooker Chemical (which later became Occidental Chemical Corporation) disposed of over 21 000 tons of chemical wastes including halogenated pesticides, chlorobenzenes, and other hazardous materials into the old Love Canal. The disposal continued until 1952 at which time the company covered the site with soil and deeded it to the City of Niagara Falls, which wanted to use it for a public park. In the transfer of the deed, Hooker specifically stated that the site was used for the burial of hazardous materials, and the company warned the city that this fact should govern future decisions on the use of the land. Everything Hooker Chemical did during those years was seemingly legal.

About this time, the Niagara Falls Board of Education was looking for a place to construct a new elementary school, and the old Love Canal seemed like a perfect spot. This area was a growing suburb, with densely packed single-family residences on streets paralleling the old canal. A school on this site seemed like a perfect solution and so it was built.

In the 1960s, the first complaints began, and intensified during the early 1970s. The groundwater table rose during those years and brought some of the buried chemicals to the surface. Children in the school playground were seen playing with strange 55-gallon drums that popped out of the ground.

The contaminated liquids started to ooze into the basements of the nearby residents, causing odor and complaints of various health problems. More importantly, perhaps, the contaminated liquid was found to have entered the storm sewers and was being discharged upstream of the water intake for the Niagara Falls water treatment plant.

The situation reached a crisis point and President Jimmy Carter declared an environmental emergency in 1978, resulting in the evacuation of 950 families in an area of 10 square blocks around the canal. But the solution presented a difficult engineering problem. Excavating the waste would have been a dangerous undertaking, and would probably have caused the death of some of the workers. Digging up the waste would also have exposed it to the atmosphere resulting in uncontrolled toxic air emissions. Finally, there was the question as to what would be done with the waste. Since it was mixed, no single solution such as incineration would have been appropriate. The US Environmental Protection Agency (EPA) finally decided that the only thing to do with this dump was to isolate it and continue to monitor and to treat the groundwater. The contaminated soil on the school site was excavated, detoxified, and stabilized and the building itself was razed. All the sewers were cleaned, removing 62 000 tons of sediment that had to be treated and moved to a remote site. At the present time, the groundwater is still being pumped and treated, thus preventing further contamination.

The cost is staggering, and a final accounting is still not available. Occidental Chemical paid $129 million and continues to pay for oversight and monitoring. The rest of the funds are from the Federal Emergency Management Agency and from the US Army, which was found to have contributed waste to the canal.

The Love Canal story had the effect of galvanizing the American public into understanding the problems of hazardous waste, and was the impetus for the passage of several significant pieces of legislation such as the Resource Conservation and Recovery Act, the Comprehensive Environmental Response, Compensation, and Liability Act, and the Toxic Substances Control Act. It also ushered in a new bioethical perspective: "toxics."

TIMES BEACH

The Time's Beach story is an example of a confluence of events that can lead to difficult bioethical decisions. It had inauspicious beginnings. Times Beach was a popular resort community along the Meramec River, about 17 miles west of St Louis. With few resources, the roads in the town were not paved and dust on the roads was controlled by spraying oil. For two years, 1972 and 1973, the contract for road spraying went to a waste oil hauler, Russell Bliss. The roads were paved in 1973 and the spraying ceased.

Bliss obtained his waste oil from the Northeastern Pharmaceutical and Chemical Company in Verona, Missouri, which manufactured hexachlorophene, a bactericidal chemical. In the production of hexachlorophene, considerable quantities of dioxin-laden waste had to be removed and disposed. A significant amount of the dioxin was contained in the "still bottoms" of chemical reactors. Incineration of the wastes was very costly. The company was taken over by Syntex Agribusiness in 1972, and the new company decided to contract with Bliss to haul away the still bottom waste without telling Bliss what was in the oily substance. Bliss mixed it with other waste oils and he used it to oil the roads in Times Beach, unaware that the oil contained high concentrations of dioxin (greater than 2000 ppm), including the most toxic congener, 2,3,7,8-dibenzo-*para*-dioxin (TCDD).

Bliss oiled roads and sprayed oil to control dust, especially in horse arenas. He used the dioxin-laden oil to spray the roads and horse runs in nearby farms. In fact, it was the death of horses in these farms that first alerted the Center for Disease Control to sample the soil at the farms. They found dioxin, but did not make the connection with Bliss. Finally in 1979, the US EPA became aware of the problem when a former employee of the company told them about the sloppy practices in handling the dioxin-laden waste. The EPA converged on Times Beach in "moon suits" and panic set in among the populace. The situation was not helped by the message from the EPA to the residence of the town. "If you are in town it is advisable for you to leave and if you are out of town do not go back." In February 1983, on the basis of an advisory from the Centers for Disease Control, the EPA permanently relocated all the residents and businesses at a cost of $33 million. Times Beach was by no means the only problem stemming from the contaminated waste oil. Twenty-seven other sites in Missouri were also contaminated with dioxins. Most of the dioxin contamination has since been cleaned up.

The contamination of Times Beach, Missouri, while affecting much of the town, was not the key reason for the national attention. The event occurred shortly after the Love Canal hazardous waste problem was identified and people were wondering just how extensively dioxin and other persistent organic compounds were going to be found in the environment. Times Beach also occurred at the time when scientists and engineers were beginning to get a handle on how to measure and even how to treat (i.e., by incineration) contaminated soil and water.

Other events, such as the worries about DDT and its effect on eagles and other wildlife, cryptosporidium outbreaks, and Legionnaire's Disease, also seem to have received greater attention due to their timing. This illustrate the importance of timing in bioethical decision making. One would be hard-pressed to identify a single event that caused the public concern about the metal lead. In fact, numerous incremental steps brought the world to appreciate lead toxicity and risk. For example, studies following lead reductions in gasoline and paint showed marked improvements in lead levels in blood in many children. Meanwhile, scientific and medical research was linking lead to numerous neurotoxic effects in PNS and CNS, especially of children. Similarly, stepwise progressions of the knowledge of environmental risk occurred for polychlorinated biphenyls (PCBs), numerous organochlorine, organophosphate, and other pesticides, depletion of the stratospheric ozone layer by halogenated (especially chlorinated) compounds, and even the effect of releases of carbon dioxide, methane, and other "greenhouse gases" on global warming (though more properly called global climate change).

Teachable Moment: The Whole Is Greater than the Sum of Its Parts

Aristotle (384 BC–322 BC) is generally credited with the famous philosophical principle, "the whole is greater than the sum of its parts." To engineers and scientists, this is synergy. We also know that in some instances the whole is less than the sum of its parts, which is antagonism. Such principles also hold for ethics. These cases have demonstrated that the combination of a few unethical decisions can lead to dramatic consequences.

The engineering profession is one of creative problems solving and optimization. Decisions lead to events that lead to consequences, which in turn, lead to other decisions, events, and consequences. At each decision point, the engineer is presented options. The design is successful when it meets the needs of the client, and it provides for the optimal outcome in terms of the public's health, safety, and welfare. Thus, every design decision is to some degree an ethical decision. And, every engineering project is an ethics-laden project. The design must lie between timidity and recklessness.

> The most exciting phrase to hear in science, the one that heralds new discoveries is not "Eureka," (I found it!) but "That's funny . . . ."
>
> Isaac Asimov (1920–1992)

One hard and fast rule is that engineers are purveyors of science; so we do not have the prerogative of messing with the facts and scientific laws. We must respect factual information, yet we must be careful not to label something prematurely as "correct," especially in light of possible contradicting information. Just because we are comfortable with the *status quo* is not a sufficient reason to hold to wrong and invalid arguments. In fact, such stubbornness is quite unscientific. Ironically, scientists are considered to be searchers of truth no matter where the journey takes us, yet like most people we have a comfort zone. Such xenophobia translates into scientific and ethical myopia. We do not like change and resist it reflexively, particularly if it means undoing some of our cherished tenets.

Surprise, even when unwelcome, as Asimov seemed to say, is a necessary part of the discovery process in science and in ethics. Women and men of science, both researchers and practitioners, must simultaneously keep an open mind and an eye toward better ways of doing things and must maintain scientific rigor. Thus, engineering is an intellectually and morally active pursuit.

Questions

1. What do the Agent Orange, Japanese metal, Love Canal and Times Beach cases have in common?
2. How has ethical responsibility changed since the early 1900s?

ACTIVE ENGINEERING

An important tenet of ethics and engineering communications is that we be clear in what we say and mean. The first step in bioethical analysis is to reach an understanding about the facts of the matter. So, then, what is meant when we say that engineering is an "active" process? The first definition in the dictionary[36] is "characterized by action, rather than by contemplation or speculation." This connotation is interesting and valuable from a number of perspectives. Obviously, the noun "action" drives the adjective "active." But among the definitions of the noun that best fit the adjective's definition, one seems to stand out; that is, "the bringing about of an alteration by force or natural agency." Alteration means that something has changed. Engineers hope and expect the change to be better than what existed before. However, if this is not the case, it is a type of failure. Failure in itself is not unethical. Only some failures are rooted in ethical breaches.

Another interesting aspect of the first dictionary definition of "active" is the contrast between an action and contemplation or speculation. Contemplation should precede any action. Prudence dictates that "you should look before you leap." And, the proper sequence of ethical or any decision making is "ready, aim, shoot," although many of us frequently "ready, shoot, aim." In other words, action-oriented people can have a natural proclivity to act, even before much or any thinking. This can lead to addressing symptoms of problems, but not solving the problems themselves. The definition is not a value judgment, but simply recognizes that the two steps, contemplation and action, are unique and sequential.

The other contrast between action and speculation, however, is value-laden. There are times, deservedly or not, when decision makers appear to suffer from "analysis paralysis." Decisions are always made with some degree of uncertainty. The key dilemma for the designer is to know when a sufficient amount of information about the risks and the benefits of a design has been ascertained. The sufficiency is a function of the severity of the outcome (costs of being wrong) and the loss of a benefit (opportunity risk). In some cases, a "50/50" flip of the coin decision-making approach is sufficient, such as whether a bridge between Chapel Hill and Durham should be painted Carolina or Duke blue (although such decisions carry great local import in the North Carolina Piedmont). Conversely, a decision as to whether to set a drinking health standard at 10 parts per million (ppm) or 15 ppm for a pollutant can account for a margin of waterborne diseases and even allow for increased mortality if the higher concentration is applied. Speculation in the former case may lead to some unhappy fans, but speculation in the latter case can translate into increased morbidity and mortality.

Another way to think about activity is to consider what one means by the opposite of active. At least two very different antonyms must be considered. If something or someone is "inactive" they are idle. Another antonym of active is passive. An active solution is one that requires added energy, whereas a passive solution is one that occurs as a matter of course. Sometimes, the passive solution is preferable, as when a noninvasive, homeopathic treatment is used instead of an invasive, pharmaceutical approach to treat a similar malady. And, at the other time, professional judgment dictates the need to "do something."

Yes, engineers do things, and what we do should give us great pride. Engineers are, by nature, a thoughtful and outwardly directed lot. We are sympathetic to the needs of others, beginning with our clients. In fact, designers, in general, are optimistic and future-oriented. We see things long before they take on physical reality. Thus, engineers are highly suited to take a long view of things. Some faith traditions may characterize us as having much "faith." Our code calls us to be "faithful agents." While this is a statement about the trust that our clients place in us, it is also a statement about our confidence and faith in ourselves as professionals. This is the engineering perspective. We "expect" the medical device, or building, or water supply to become real, to become "substance."[37] Even those engineers who are called in to fix things, such as biomedical engineers finding ways to ameliorate human suffering, environmental engineers who remediate hazardous waste sites or failure analysts who look for ways to prevent future disasters, always expect things to improve with action.

Engineers are willing to stand behind our work (e.g., we build in feedback in the operation and maintenance (O&M) process, we check progress continuously on biomedical devices, install monitoring wells to ensure waste cleanup is going as planned, and we work closely with inspectors to incorporate lessons learned from failure analyses). Hence, in addition to being faithful, we adapt. In fact, adapting is a big part of design. The dynamism of biomedicine has been a fertile ground for engineering for many centuries. A classic example was the 1592 visit of Galileo to one of the first medical institutions in Padua, Italy.[38] Galileo, acting as an engineer, took the opportunity to lecture the future medical practitioners and researchers on mathematical principles including his own theories, as well as their applications (notably the pendulum, thermoscope, and telescope). Such applications led to an enhanced understanding of physiology, such as studies by Galileo's student, Sanctorius, of human body temperature and pulse rates. One student at Padua during Galileo's visit was William Harvey, renowned for characterizing the circulatory system, which he based on the mechanical and motion laws expounded by Galileo. Such interplay between medicine and engineering, while often subtle and indirect, has evolved into the more formal collaborations we see today. Duke, Johns Hopkins, and other world class biomedical engineering programs are co-located and intellectually intertwined with leading medical schools. But, the dynamic design processes of the contemporary engineers make for complicated ethical challenges and none more so than those related to biomedicine.

The stakes are very high in biomedicine. It is truly and literally a matter of life and death. Not advancing the state of biomedical science is simply not an ethical option. Consequently, the typical precautions that may hold for other areas of engineering need special considerations when addressing biomedical challenges. For example, the "no action" alternative is seldom satisfying to biomedical practitioners. Not looking for a better device or system that can improve the quality of a patient's life is simply not acceptable. Risk assessors refer to this as "opportunity risk." That is, if we simply seek the refuge of no added risk, we may lose an opportunity to really improve things. For example, nanotechnologies (using systems that are only a few hundred molecules in size) are fraught with risks, such as the chance that working at this level may cause changes to self-replication and other cellular signals, which could cause unknown damage. However, if we do not seek the uses of nanotechnologies, such as the application of highly efficient, biologically inspired processes for drug delivery, we may lose the opportunity to make the drugs more efficacious. If we can use biologically inspired nanomanufacturing to synthesize and deliver tumor-reducing drugs, it may be possible to treat cancers that have heretofore been untreatable. So, we must be bold in applying nascent sciences and simultaneously take great care to ensure that we are not opening some awful "Pandora's box."[39] Finding this balance of acceptable and reasonable risk goes beyond science, engineering, and technology and enters the realms of ethics.

In a way, the advancement of science and its applications must follow a type of ethical index. Indices have weighted values for each of the important variables. Some more sophisticated indices have operators that can shut the index down, such as a water quality index that automatically goes to zero if certain levels of dissolved oxygen are not available (even if all the other values are fine). Thus, engineers need some type of working index for biomedical ethics. They must be able to tell when something is going awry. This is a tall order, since most of us are well trained in the physical sciences and mathematics, but few in the intricacies of biology associated with health care. The exception, of course, is the biomedical engineer, but even their perspective of biomedicine varies from that of others in the health care profession.

Choosing the right "ethics index" requires some knowledge of the common ethical models used to evaluate engineering decisions and actions.

ETHICAL THEORIES: A PRIMER

Engineers are more familiar with the works of Newton, Boyle, Einstein, and Bohr than they are with those of Hammarabi, Socrates, Hippocrates, Aristotle, Hobbes, Mill, Kant and Rawls. However, it was only four centuries ago that philosophy was central to all studies. In fact, physical science was considered to be within the realm of "natural philosophy." The Ancient and Renaissance scientists did not distinguish ethics and other aspects of philosophy from physics and other underpinning sciences of engineering. Later Descartes, Humes, and Mill tried to turn things around by attempting to place ethics under the scientific method. Although we cannot return to such thinking completely, it may be useful to consider our careers as engineers a bit more comprehensively as a starting point for understanding bioethics.

The comprehensive view allows us to consider some topics not often covered in engineering texts; concepts like good *versus* evil, moral *versus* immoral acts, and obligations *versus* prohibitions are understood by most professionals at some intuitive level, but unlike our colleagues in the humanities and social sciences, we are more likely to avoid considering them theoretically. However, reading the classical works of Aristotle, Aquinas, Kant, et al. makes the case for life being a mix of virtues and vices available to humans. In fact, our previous discussion about active engineering sets the stage for ethical engineering. Ethicists sometimes refer to goodness to be the result of "right action." Virtue can be defined as the power to do good or a habit of doing good. In fact, one of Aristotle's most memorable lines is "Excellence is habit." So, if we do good, we are more likely, according to Aristotle, to keep doing good. Conversely, vice is the power and the habit of doing evil.

TRUTH

The subjectivity or relational nature of good and evil, however, leads to some discomfort in scientific and engineering circles, where meanings of certainty and consistency of definition are crucial to problem solving. Actually, this is consistent with the perspective of philosophers, especially ethicists. Scientific facts are a crucial part of any bioethical case analysis, and so are other factors. To wit, philosophers tell us that we can determine whether a moral argument is valid (not necessarily correct) by parsing the argument into a "syllogism," which consists of four parts:

1. The factual premise
2. The connecting premise (i.e., factual to evaluative)
3. The evaluative premise
4. The moral conclusion

For example, the facts may show that exposing people to a chemical at a certain dosage (e.g., one part per million or ppm) leads to a specific form of cancer in one in every ten thousand people. We also know that, from a public health perspective, allowing people to contract cancer as a result of some human activity is morally wrong. Thus, the syllogism would be the following:

1. Factual premise: exposure to chemical X at 1 ppm leads to cancer
2. Connecting premise: A company is releasing 10 kg of chemical X per day which leads to 1 ppm exposure to people living near an industrial plant
3. Evaluative premise: decisions that allow industrial releases that lead to cancer are morally wrong
4. Moral conclusion: therefore, corporate executives who decide to release 10 or more kilograms of chemical X from their plants are morally wrong.

Science provides us with the factual premise and part of the connecting premise, but social mores and norms provide us with the evaluative premise and drive the moral conclusion. However, if we are uncertain about the facts, the validity of the argument is disrupted. Scientific uncertainties are brought about both by variability and error.[40] Variability is ever present in space and time. Every case has a unique set of factors, dependent variables, situations, and scenarios, so that what occurred will never be completely repeated again. Every cubic centimeter of soil is different from every other cubic centimeter. The same goes for a sample of water, sediment, air, and organic tissue. And, these all change with time. Taking a sample in the winter is different from that in the summer. Conditions in 1977 are different in so many ways from conditions in 2007. And, of course, there are errors. Some are random in that the conditions that led to the cases in this book are partially explained by chance and things that are neither predictable nor correctable, although we can explain (or at least try to explain) them statistically, (e.g., with normal distributions).

Other error is systematic, such as those of my own bias. I see things through a prism different from anyone else's. This prism, like yours, is the result of my own experience and expertise. This prism is my perception of what is real and what is important. My bias is heavily weighted in sound science, or at least what I believe to be sound science (as opposed to "junk science").[41] Sound science requires sufficient precision and accuracy in presenting the facts. "Precision" describes how refined and repeatable an operation can be performed, such as the exactness in the instruments and the methods used to obtain a result. It is an indication of the uniformity or reproducibility of a result. This can be likened to shooting arrows,[42] with each arrow representing a data point. Targets A and B in Figure 2.7 show equal precision. Assuming that the center of the target; that is, the bull's-eye, is the "true value," data set B is more accurate than A. If we are consistently missing the bull's-eye in the same direction at the same distance, this is an example of bias or systematic error. The good news is that if we are aware that we are missing the bull's-eye (e.g., by comparing our results with those of known standards when using our analytical equipment), we can calibrate and adjust the equipment. To stay with our archery analogy, the archer would move her sight up and to the right.

Thus, "accuracy" is an expression of how well a study conforms to some defined standard (the true value). So, accuracy expresses the quality of what we find, while precision expresses the quality of the operation by which we obtained our finding. So, the other two scenarios of data quality are shown in targets C and D. Thus, the four possibilities are that our data is precise but inaccurate (target A), precise and accurate (target B), imprecise and inaccurate (target C), and imprecise and accurate (target D).

At first blush, target D may seem unlikely, but it is really not all that uncommon. The difference between targets B and D is simply that D has more "spread" in the data. For example, the variance and

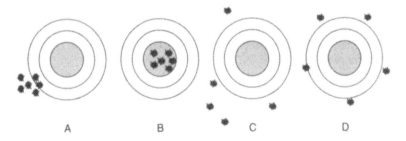

Figure 2.7 Precision and accuracy. The bull's-eye represents the true value. Targets A and B demonstrate data sets that are precise, targets B and D demonstrate data sets that are accurate, and targets C and D demonstrate data sets that are imprecise. Target B is the ideal data set, which is precise and accurate.

standard deviation of D is much larger than that of B. However, their measures of central tendency, i.e., the means, are nearly the same. So, both data sets are giving us the right answer, but almost all of the data points in B are near the true value. None of the data points in D is near the true value, but the mean (average location) is near the center of the bull's-eye; so, it has the same accuracy as target B, but with much less precision. The key is that precision and accuracy of the facts surrounding a case must be known.

Even if we agree on the facts, we all will not agree on which of the virtues and vices are best or even whether something is a virtue or a vice (e.g., loyalty). But one concept does seem to come to the fore in most major religions and moral philosophies: empathy. Putting oneself in another's situation is a good metric for virtuous acts. The "Golden Rule" is at the heart of Immanuel Kant's "categorical imperative." With apologies to Kant, here is a simplified way to describe the categorical imperative: When deciding whether to act in a certain way, ask if your action (or inaction) will make for a better world if all others in your situation acted in the same way. An individual action's virtue or vice is seen in a comprehensive manner. It is not whether one should cut corners and use deficient materials when designing a device, it is whether everyone in the same situation should do likewise. A corollary to this concept is what Elizabeth Kiss, formerly of Duke's Kenan Center for Ethics and now president of Agnes Scott College calls the "Six O'clock News" imperative. That is, when deciding whether an action is ethical or not, consider how my friends and family would feel if they heard about all of its details on tonight's TV news. That may cause me to consider more fully the possible externalities and consequences of my decision (and maybe even tempt me to "overdesign").

PSYCHOLOGICAL ASPECTS OF ETHICS

Engineers are familiar with conditions and constraints. For example, we know from chaos theory the importance of initial conditions on outcomes, but also know these can be changed by boundary conditions and constraints imposed by the engineer (i.e., interventions). These concepts are analogous to moral development. At the most basic level, a person's path toward moral decision making is a continuum. Since much of an engineer's ethics is influenced by his or her own "conscience," a basic understanding of the theory of moral development is in order.[43]

"Conscience" is something that develops rather rapidly in human development, but is honed constantly throughout life. As a child learns to communicate, especially with the first caregivers, the conscience is controlled and informed by external sources (i.e., we learn to do what we are told by an authority figure, or we learn by watching and imitating others). However, as we mature, the conscience to a greater extent is controlled internally. A mature conscience is the result of internalization. In other words, we start to depend completely on outside sources, grow to become independent, but eventually behave as interdependent members of a community.

Another doctrine repugnant to Civil Society is, that whatsoever a man does against his Conscience is Sin; For a man's Conscience, and his Judgment is the same thing; and as the Judgment, so also the Conscience may be erroneous; once a man lives in a commonwealth, the Law is the public Conscience, by which he hath already undertaken to be guided. Otherwise in such diversity, as there is of private Consciences, which are but private opinions, the Commonwealth must needs be distracted, and no man dare to obey the Sovereign Power, farther than it shall seem good in his own eyes.

Thomas Hobbes (1660)[44]

Individual, personal conscience is the building block for what the seventeenth century natural philosopher Thomas Hobbes considered to be needed in a "social contract." Hobbes argued that humans are egoistic and self-serving by nature, so an efficient society first instructs its individual members to behave in ways that support the community. Hobbes thought that society had to overcome the "state of nature" where selfish, brutish individual desires would lead to anarchy. This argument sees the role of conscience as twofold: to benefit the individual member (private conscience) and to benefit the society as a whole (the commonwealth).

Character: As mentioned in Chapter 1, Jean Piaget, Lawrence Kohlberg, and other educational psychologists have argued that moral development takes a predictable and stepwise progression. The development is the result of social interactions over time. For example, Kohlberg[45] identified six stages in three levels (see Table 1.2).

Kohlberg insisted that these steps are progressive. Every person must pass through the preceding step before advancing to the next. Thus, a person first behaves according to authority (stages 1 and 2), then according to approval (stages 3 and 4), before finally maturing to the point where they are genuinely interested in the welfare of others. My experience has been gratifying in that most of my colleagues and the majority of engineering students in my courses have indicated moral development well within the postconventional level.

We can apply the Kohlberg model directly to the engineering profession (Figure 2.8). The most basic (bottom tier) actions are preconditional; that is, engineering decisions are made solely to stay out of trouble. While proscriptions against unethical behavior at this level are effective, the training, mentorship, and other opportunities for professional growth push the engineer to higher ethical expectations. This is the normative aspect of professionalism. In other words, with experience as guided by observing and emulating ethical role models, the engineer moves to conventional stages. The engineering practice is the convention, as articulated in our codes of ethics.

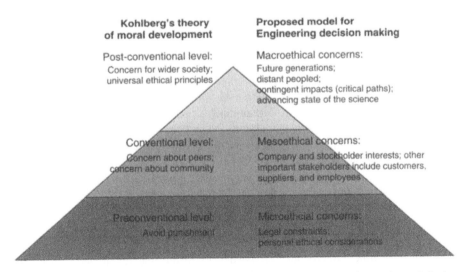

Figure 2.8 Adaptation of Kohlberg's stages of moral development to the ethical expectations and growth in the engineering profession.

Above the conventional stages, the truly ethical engineer makes decisions based on the greater good of the society, even at personal costs. In fact, the "payoff" for the engineer in these cases is usually for people he or she will never meet and may occur in future he or she will not share personally. The payoff does provide benefits to the profession as a whole, notably that we as a profession can be trusted. This top-down benefit has incremental value for every engineer. Two common sayings come to mind about top-down benefits. Financial analysts often say about the effect of a growing economy on individual companies: "A rising tide lifts all ships." Likewise, environmentalists ask us: "To think globally, but to act locally." In a like manner, the individual engineer is an emissary of the whole profession.

Bioethical considerations introduce a number of challenges that must be approached at all three ethical levels. At the most basic, microethical level, laws, rules, regulations, and policies dictate certain behaviors. For example, cloning and blastocyst research, especially that which receives federal funding, is controlled by rules overseen by federal and state agencies. Such rules are often proscriptive; that is, they tell what "not to do," but are less clear on what actually "to do."

At the next level, beyond legal considerations, the engineer is charged with being a loyal and faithful agent to the clients. Researchers are beholding to their respective universities and institutions. Engineers working in companies and agencies are required to follow mandates to employees (although never in conflict with their obligations to the engineering profession). Thus, engineers must stay within budget, use appropriate materials, and follow best practices as they concern their respective designs. For example, if an engineer is engaged in work that would benefit from a collaboration with another company working with similar genetic material, the engineer must take precautionary steps to avoid breaches in confidentiality, such as trade secrets and intellectual property.

The highest level, the macroethical perspective, has a number of bioethical aspects. Many of the research and development projects are in areas that could greatly benefit society but may lead to unforeseen costs. The engineer is called to consider possible contingencies. For example, if an engineer is designing "nanomachinery" at the subcellular level, is there a possibility that self-replication mechanisms in the cell could be modified to lead to potential adverse effects, such as generating mutant pathological cells, toxic byproducts, or changes in genetic structure not previously expected? Thus, this highest level of professional development is often where "risk tradeoffs" must be considered. In the case of our example, the risk of adverse genetic outcomes must be weighed against the loss of advancing the state of medical science (e.g., finding nanomachines that manufacture and deliver tumor-destroying drugs efficiently).

Ongoing cutting-edge research (such as the efficient manufacturing of chemicals at the cellular scale, the development of cybernetic storage, and data transfer systems using biological or biologically inspired processes, etc.) will create new solutions to perennial human problems by designing more effective devices and improving computational methodologies. Nonetheless, in our zeal to push the envelopes of science, we must not ignore some of the larger, societal repercussions of our research; that is, we must employ new paradigms of "macroethics."

William A. Wulf, President of the National Academy of Engineering, introduced the term macroethics, defining it as a societal behavior that increases the intellectual pressure "to do the right thing" for the long-term improvement of society. Balancing the potential benefits of the advances in nanotechnology to society while also avoiding negative societal consequences is a type of macroethical dilemma.[46] Macroethics asks us to consider the broad societal impact of science in shaping research agendas and priorities. At the same time, "microethics" is needed to ensure that researchers and practitioners act in accordance with scientific and professional norms, as dictated by standards of practice, community standards of excellence, and codes of ethics.[47] The engineering profession and engineering education standards require attention to both the macro- and microdimensions of ethics. Criterion 3, "Program Outcomes and Assessment," of the Accreditation Board for Engineering and Technology (ABET), Inc. includes a basic microethical requirement for engineering education programs, identified as "(f) an understanding of professional and ethical responsibility," along with the macroethical requirements that graduates of these programs should have "(h) the broad education necessary to understand the impact of engineering solutions in a global and societal context" and "(j) a knowledge of contemporary issues."[48]

Teachable Moment: The Physiome Project: The Macroethics of Engineering toward Health

In *The Bridge*, the Journal of the National Academy of Engineering, James B. Bassingthwaighte Vol. 32, No. 3 – (Fall 2002), identified and described some of the most important *engineering tools and techniques that can be used to advance health care.*

According to Bassingthwaighte, the new tools fall into four categories:

1. **Informatics and Information Flow**

   The problem in medicine and biology is that much relevant information is either irretrievable or undiscovered. Even a complete human genome cannot define human function. In fact, it is only a guide to the possible ingredients. The genetically derived aspects of the genome (i.e., proteins) are much more numerous than the genes. To get an idea of the magnitude of the problem, consider that yeast has about three proteins per gene, and humans have about 10 proteins per gene. Pretranslational selection from different parts of the DNA sequence, the posttranslational slicing out of parts of the protein, the splicing of two or more proteins together, and the combining of groups of proteins into functional, assembly-line-like complexes, all contribute to the variety of the products of gene expression. Even a completely identified proteome, which is still beyond the scientific horizon, will be like a list of the types of parts of a jumbo jet with no indication of how many should be used or where they go. The concentration of proteins, the balance between synthesis and decay in each cell type, is governed by environment, by behavior, and by the dynamic relationships among proteins, substrates, ionic composition, energy balance, and so on and, thus, cannot be predicted on the basis of the genome.

2. **The Combinatorial Dilemma**

   Sorting out the genome will leave us with a huge number of proteins to think about. The estimates of the number of genes have come down by about half from earlier estimates of 60 000 to 100 000; because new ones are also being found, 50 000 is a reasonable estimate. The level of complexity in mammalian protein expression far exceeds that of *C. elegans*, which has 19 536 genes and 952 cells. Humans might only have two or three times as many genes, but probably have a much higher ratio of proteins per gene. Assuming 10 proteins per gene, we have on the order of a half million proteins in widely varied abundance, and each protein has several possible states. If a protein in a given state interacts with only five other proteins (e.g., exchanging substrates with neighbors in a pathway or modifying the kinetics of others in a signaling sequence), then it may "connect" to any other protein through only a few links, a kind of "six degrees of separation" from any other protein. Moreover, cells contain not just proteins, but also substrates and metabolites, and they are influenced by their environments. Given the possible permutations and combinations of linkages and the many multiples further in the dynamics of their interactions, the combinatorial explosion would appear to preclude predictions.

3. **Managing Complexity**

The complexity . . . briefly described provides a basis for functionality that cannot be predicted from knowledge about each of the components. "Emergent" behavior is the result of interactions among proteins, subcellular systems, and aggregates of cells, tissues, organs, and systems within an organism. Physiological systems are highly nonlinear, higher order systems, and dynamics are often chaotic. Chaotic systems are only predictable over the short term; but they have a limited operating range. Even when Bernard (1927) defined the stability of the "milieu interieure," he meant a mildly fluctuating state rather than a stagnant "homeostasis." Biological systems are "homeodynamic"; they fluctuate, but under control, and they are neither "static" nor "randomly varying."

4. **Bioinformatics**

This vast array of information must be linked into a consistent whole. The databases must be well curated and easily accessible, and they must provide a substrate for behaviorally realistic models of physiological systems. The arguments for building large databases to capture biological data are fairly new. Federal funds support genomic and proteomic databases, but not databases of higher level physiological information. Organ and systems data acquired over the past century have not been collected in databases and are poorly indexed in the print literature. Providing searchable texts of articles online will help but will not be a substitute for organized databases. The Visible Human Project, the National Library of Medicine's effort to preserve anatomic information is a part of the morphome (which we define as providing anatomic and morphometric information), analogous to the genome and the proteome.

Bassingthwaighte continues:

All of this information must then be captured in a comprehensive, consistent, conceptual framework; that is, a model of the system that conveys understanding, and for this we will need to use engineering approaches. Understanding complicated systems, modeling them, and learning the tricks for reducing their complexity to attain computability, are in the engineering domain, and bioengineering-trained investigators will be the integrators of the future.

Of course, all models are incomplete. They come in a variety of forms, such as sketches of concepts, diagrams of relationships, schemas of interactions, mathematical models defined by sets of equations, and computational models (from analytical mathematical solutions or from numerical solutions to differential or algebraic equations). The behavior of a well developed, well documented computer model can give us some insight into the behavior of the real system.

We must do our utmost to predict well, not just the direct results of a proposed intervention, but also the secondary and long-term effects. Thus, databasing, the development, archiving, and dissemination of simple and complex systems models, and the evaluation (and rejection or improvement) of data and of models – are all part of the moral imperative. They are the tools necessary to thinking in depth about the problems that accompany, or are created by, interventions in human systems or ecosystems.

As a step toward providing these tools, Bassingthwaighte has initiated the Physiome and the Physiome Project:

A physiome can be defined as the quantitative description of the functional state of an organism. A quantitative model is a way of removing contradictions among observations and concepts and creating a consistent, reproducible representation of a system. Like the genome, the physiome can be defined for each species and for each individual within the species. The composite and integrated system behavior of the living organism is described quantitatively in hierarchical sets of mathematical models defining the behavior of the system. The models will be linked to databases of information from a multitude of studies. Without data, there is nothing to model; and without models, there is no source of deep predictive understanding.

The Physiome Project provides one response to the macroethical imperative to minimize risk while advancing medical science and therapy. The project is an effort to define the physiome, through databasing and modeling, of individual species, from bacteria to man. The project began with collaborations among groups of scientists in a few fields and is developing spontaneously as a multinational collaborative effort. Investigators first defined goals and then proceeded to put pieces together into impressive edifices. Via iteration with new experimentation, models can remove contradictions and demonstrate emergent properties. These models are part of the tool kit for the "reverse engineering" of biology. The scale of the models, like the scale of models for weather prediction, presents computational grand challenges.

The Physiome Project is not likely to result in a virtual human being as a single computational entity. Instead, small models linked together will form large integrative systems for analyzing data. There is a growing appreciation of the importance, indeed the necessity, of modeling for analysis and for prediction in biological systems as much as in physical and chemical systems.

The hierarchical nature of biological systems is being used as a guide to the development of hierarchies of models. Models at the molecular level can be based on biophysics, chemistry, energetics, and molecular dynamics, but it is obviously not practical to use molecular dynamics in describing the fluxes through sets of biochemical pathways, just as it is not practical to use the full set of biochemical reactions when describing force–velocity relationships in muscle, or to use the details of myofilament crossbridge reactions when describing limb movement and athletic performance. One cannot build a truck out of quarks.

Biological models can be defined at many hierarchical levels from gene to protein to cell to organ to intact organism. Practical models comprised of sets of linked component models, each somewhat simplified, represent one level of the hierarchy. The strategy is to avoid computing the details of underlying events and to capture, at the higher level, the essence of their dynamic behavior. But monohierarchical models are not necessarily built to adapt to changes in conditions. Handling transients is like using adjustable time steps in systems of stiff equations, but more complicated; the lower level model must be used to correct the higher level representation. Once we have very good models that extend from gene regulation to the functions of the organism, they can be used to predict the short-term and long-term efficacy and side effects of various therapies.

Questions:

1. Like genomic information, should biological information and models be put in the public domain?
2. What types of risks and benefits can be expected from this project?
3. Is this a model for medical and engineering collaboration? Why or why not?

Personal and organizational ethics and morality are affected by psychology. Attitudes are complex mental processes that influence how a person processes information and that motivate behavior.[49] They have been explored in depth in the psychological literature, where "attitude" has been defined as:

> [A] psychological tendency that is expressed by evaluating a particular entity {the object of the attitude} with some degree of favour or disfavour.[50]

Attitudes are inferred by observing an individual's response to a situation (a stimulus); but they cannot be measured directly (Figure 2.1)[51]. For example, a doctor holding a pro-life attitude, when confronted by a patient requesting a termination of pregnancy, might respond by refusing to act as the patient wishes, or by explaining his or her beliefs to the patient, or both. But if, in spite of the doctor's beliefs, he or she agrees to arrange the termination of pregnancy, then the patient might not be able to infer that the doctor has a pro-life attitude, and a knowledgeable observer might conclude that the doctor's pro-life attitude is not strongly held compared with competing pressures to act in a counter-attitude manner.

Attitude affects empathy, which is central to justice. Justice is the virtue that enables us to give others what is due to them as our fellow human beings. This means we must not only avoid hurting others by our actions, but we ought to safeguard the rights of others in what we do and what we leave undone.

The categorical imperative is emblematic of empathy. Kant uses this maxim to underpin duty ethics (so-called deontology) with empathetic scrutiny. However, empathy is not the exclusive domain of duty ethics. In consequentialism, also known as teleological ethics, empathy is one of the palliative approaches to deal with the problem of "ends justifying the means." Other philosophers also incorporated the empathic viewpoint into their frameworks. In fact, Mill's utilitarianism axiom of "greatest good for the greatest number of people" is moderated by his "harm principle," which, at its heart, is empathetic. That is, even though an act can be good for the majority, it may still be unethical if it causes undue harm to even one person. Empathy also comes into play in contractarianism, as articulated by Hobbes' social contract theory. For example, John Rawls has moderated the social contract with the "veil of ignorance" as a way to consider the perspective of the weakest, one might say "most disenfranchised," members of society. Finally, the rationalist frameworks incorporate empathy into all ethical decisions when they ask the guiding question of "what is going on here?" In other words, what benefit or harm, based on reason, can I expect from actions brought about by the decision I am about to make? One calculus of this harm or benefit is to be empathetic to all others, particularly the weakest members of society, those with little or no "voice."

The word "empathy" has an interesting beginning. It originally comes from the German word *einfühlung*, which means the ability to project oneself into a work of art, like a painting. Psychologists at the beginning of the 1900s searched for a word that meant the projection of oneself into another person, and chose the German word, translated into English as "empathy." The concept itself was known, such as the Native Americans' admonition to walk in another's moccasins, but it needed a construction. The earlier meaning of empathy was thus the ability to project oneself into another person, to imitate the emotions of that person by physical actions. For example, watching someone prick a finger would result in a visible winching on the part of the observer because the observer would know how it feels.

From that notion of empathy, it was natural to move to more cognitive role taking, imagining the other person's thoughts and motives. From here, empathy began to be thought of as the response that a person has for another's situation. Psychologists and educators, especially Jean Piaget,[52] began to believe that empathy develops throughout childhood, beginning with the child's first notion of others who might be suffering personal stress. The child's growing cognitive sense eventually allows him or her to experience the stress in others. Because people are social animals, this understanding of the stress in others, according to the psychologists, eventually leads to true compassion for others.

A problem with this notion of empathy development is that some experiments have shown that the state of mind of a person is very important in that person's ability to empathize. Apparently, small gifts or compliments significantly increase the likelihood that a person will show empathy toward third parties. A person in a good mood tends to be more understanding of others. If this is true, then empathy is (at least partly) independent of the object of the empathy, and empathy becomes a characteristic of the individual.[53]

Charles Morris defines empathy as:[54]

The arousal of an emotion in an observer that is a vicarious response to the other person's situation . . . Empathy depends not only on one's ability to identify someone else's emotions but also on one's capacity to put oneself in the other person' place and to experience an appropriate emotional response. Just as sensitivity to non-verbal cues increases with age, so does empathy: The cognitive and perceptual abilities required for empathy develop on as a child matures.

Such definitions of empathy seem to be widely accepted in the moral psychology field. But there are some serious problems with such a definition.

First, we have no way of knowing if the emotion triggered in the observer is an accurate representation of the stress in the subject. We presume that a pin prick would be felt in a similar way because we have had this done to us and we know what it feels like. But what about the stress caused by a broken promise? How can an observer know that he or she is on the same wavelength as the subject when the stress is emotional?[55]

If a subject says that she is sad, the observer would know what it is like to be sad, and would share in the sadness; that is, the observer would empathize with the subject's sadness and be able to tell the subject what is being felt. But is the observer really feeling what the subject is feeling? There is no way to define or measure "sadness," and thus there is no way to prove that the observer is actually feeling the same sadness that the subject is feeling.[56] An existentialist and empiricist might say that this is true for everything, even physical realities, but that is beyond the scope of this discussion.

The second problem relates to nonhuman animals. Psychologists have studied empathy exclusively as a human–human interaction, and yet many nonhuman animals can exhibit empathy. Witness the actions of a dog when its master is sick. You can read the caring and sympathy and hopefulness in the dog's eyes.[57] Since sentience and pain management are important to biomedical engineering, these uncertainties are no small matter.

Humans also have strong emotional feelings toward nonhuman animals. The easiest to understand in these terms is the empathy we feel when animals are in pain. We do not know for sure that they are in pain of course, since they cannot tell us, but they act in ways similar to the way humans behave when they are in pain and there is every reason to believe that they feel pain in the same way. Anatomical studies on animals confirm that many of their nervous systems do not differ substantially from those of humans, indicating that they feel pain. Indirect measures, such as tomography, also support the contention that animals feel pain in ways similar to humans.

More problematic are the lower animals and plants. There is some evidence that trees respond physiologically when they are damaged, but this is far from certain. The response may not be pain at all, but some other sensation (if we can even suggest that trees have sensations). And yet many of us are loathe to cut down a tree, believing that the tree ought to be respected for what it is, a center of life. This idea was best articulated by Albert Schweitzer in his discussions on the "reverence for life," or the idea that all life is sacred.

Empathy toward the non-human world cannot be based solely on sentience. Something else is going on. When a person does not want to cut down a tree because of caring for the tree, this is certainly some form of empathy, but it does not come close to the definitions used by the psychologists.

The third problem with this definition of empathy is that there is a huge disconnect between empathy and sympathy. If an observer watches a subject getting a finger pricked, the observer may know exactly what it feels like, having had a similar experience in the past. So there is great empathy. But there might be little sympathy for the subject. The observer might actually be glad that the subject is being hurt, or it might be funny to the observer to watch the subject suffer.

Years ago on the popular television show "Saturday Night Live," there was an occasional bit where a clay figure, Mr Bill, suffered all manner of horrible disasters and ended up being cut, mangled, crumbled, and squashed. Watching this may have elicited some empathy on the part of the observers, but certainly there was no sympathy for the destruction of the little clay man. His destruction was meant to be funny.

We could argue that a lack of sympathy might indicate that there must be a lack of empathy also. How is it possible for someone to empathize with another person getting a finger pricked, but think it to be humorous? Perhaps there has been no empathy at all. Or perhaps we have conditioned ourselves to laugh at others when they get hurt as a defense mechanism (e.g., "whistling in the dark") to somehow separate the violence from our own experience. Or we have learned from and have become desensitized by mass media and video games to destroy others without regret.

FAIRNESS

Empathy is not a moral value in the same way that loyalty, truthfulness, and honesty are moral values. Each of us can choose to tell the truth or to lie in any particular circumstance, and a moral person will tell the truth (unless there is an overwhelming reason not to, such as to save a life). But it is not possible to choose to have or not to have empathy. One either has empathy or one does not. One either cares for those in need or one does not.

Because we believe that empathy is worthwhile, and respect and admire people who have empathy, we tend to assign moral worth to this characteristic, and we believe that people with empathy are virtuous. On the other hand, we do not condemn those who do not have empathy. For example, people who contribute to various relief organizations such as CARE and Catholic Charities do so because they have empathy for those in need, but many people choose not to contribute. They lack some measure of empathy for others in need, but this does not make them bad people. They simply choose not to contribute.

Can engineers not have empathy and still do good engineering? That is, is empathy necessary for good engineering? Certainly on a personal level, engineers read the same newspapers and hear the same TV news as everyone else, and thus their lack of empathy ought not to be any more or less criticized than the lack of empathy by anyone else. But the truth is that responsibility of professional engineers is supererogatory to everyday ethics. Engineering ethics is a different layer on top of everyday common morality, and engineers share many responsibilities not required of nonengineers. By virtue of their training and skills engineers serve others and have certain responsibilities that relate to their place in society. The oft-quoted first canon in many codes of engineering ethics

> The engineer shall hold paramount the health, safety, and welfare of the public

is very clear. It states that the engineer has responsibility to the "public," not to a segment of the public that fits the design paradigm, or that segment that employs the engineer, or that segment that has power and money. The engineer is responsible to the public. Full stop. And in so doing, the engineer must help that segment of the public least able to look out for themselves. There is a *noblesse oblige* in engineering, the responsibility of the "nobles" to care for the less fortunate.

Thus, to be an effective and "good" engineer requires that we be able to put ourselves in the place of those who have given us their trust. The implications for justice are that it is much easier to export "canned" answers and solutions to problems from our vested viewpoints. This view must span time and space. What will the product performance look like in ten years if the project is implemented? What happens if some of the optimistic assumptions are realized? The users will be left with the consequences. It is much better, but much more difficult, to see the problem from the perspective of those with the least power to change things. We are empowered as professionals to be agents of change. So, as agents of change and justice, engineers must strive to hold paramount the health, safety, and welfare of all the public, we must be competent, and we must be fair.

Value as a Bioethical and Engineering Concept

In engineering, one might consider "value" through the idea of value engineering. This concept was created at General Electric Co. during World War II. As the war caused shortages of labor and materials, the company was forced to look for more accessible substitutes. Through this process, they saw that the substitutes often reduced costs or improved a product.[58] Consequently, they turned the process into a systematic procedure called "value analysis."

Value engineering consists of assessing the value of goods in terms of "function." Value, as a ratio of function to cost, can be improved in various ways. Oftentimes, value engineering is done systematically through the four basic steps:[59] information gathering, alternative generation, evaluation, and presentation. In the information gathering step, engineers consider what the requirements for the object are. Part of this step includes function analysis, which attempts to determine what functions or performance characteristics are important. In the next step, i.e., alternative generation, value engineers consider the possible alternative ways of meeting the requirements. Next, in evaluation, the engineers assess the alternatives in terms of functionality and cost-effectiveness. Finally, in the presentation stage, the best alternative is chosen and presented to the client for the final decision.[60]

In the realm of economics, value is considered to be the worth of one commodity in terms of other commodities (or currency). There are three main value theories in economics. The first, an intrinsic theory of value, holds that the value of an object, good, or service is contained in the item itself. These theories tend to consider the costs associated with the process of producing an item when assigning the item value. For example, the labor theory of value, a model developed by David Ricardo, holds that the value of a good is derived from the effort of its production, reduced to the two inputs in the production frontier, labor, and capital.[61] In this model, if a lamp is produced in five hours by 3 people, then the lamp is worth $3 \times 5 = 15$ man-hours. On the other hand, the subjective theory of value holds that goods have no "intrinsic" value, outside the desire of individuals to have the items. Here, value becomes a function of how much an individual is willing to give up in order to have that item.[62]

Similarly, the marginal theory of value accounts for both the scarcity and the desirability of a good, holding that the utility rendered by the last unit consumed determines the total value of a good. The main difference between the labor theory of value (and the concept of intrinsic value) and marginal theory of value is that the former accomodates a form of value derived from utility – from satisfying human desire.[63]

Furthermore, the common understanding of the term "value" generally means that the item is of importance for one reason or the other. At the same time, to say something is "valuable" would mean that it costs a lot of money, meaning that a high demand for the item in society has driven the prices up. Therefore, person A might value her beaded necklace because she derives pleasure from it, but her diamond necklace would be considered valuable, because of the large number of other members of the society that also feel they would derive pleasure from it (which causes a high demand that increases the price).

The so-called diamond – water paradox, is a noteworthy example of the role of scarcity in economic value theory. Diamonds, which have relatively little use, i.e., only aesthetic value, have an extremely high price when compared to water, which is essential to life itself. The example illustrates the importance of scarcity in the economic value. Here, as diamonds are far scarcer than is water, they have the higher price. However, this is situational. In the middle of the desert, where water is extremely scarce, someone would almost certainly be willing to pay more money for water than for diamonds. Thus, that person would value water more highly.

Since utility is a common metric for value in biomedicine, the differing definitions should give us pause. Often, the design's value as perceived by decision makers is what drives ethics. However, we may be defining value in substantially different ways.

TECHNICAL OPTIMISM VERSUS DISMAL SCIENCE

The National Academy of Engineering has declared:

> Engineers and their inventions and innovations have helped shape the changes that have made our lives more productive and fruitful.[64]

But what does it mean to become more fruitful? One way to determine such success is to consider the economic implications.

In economics, technology can be considered a tool of "empowerment," in that it empowers producers to generate more output from given levels of the two inputs, labor, and capital. A catalyst in chemistry is a substance that increases the rate of reaction; in this context, one might say that technology is a "catalyst" for the production of output – as it increases the amount of output we get from given inputs. Technology allows for the use of more advanced capital, which results in better and faster ways to create output. Producers are rendered more efficient, as they are able to produce more output, given the same amount of input. This results in greater profit, which results in economic growth.

Technology improves utility as it allows for better use of scarce resources, which are then allocated through the economic system, facilitating the achievement of Pareto optimality. That is, given a decision of whether to take an action, the selected alternative must improve the lot of at least one member of the affected group and cannot worsen the plight of any other member (known as a Pareto improvement). Thus, economics would drive the decision toward improved utility, at least for that one person.

Therefore, technology broadens the horizon through which economics operates. When firms invest, they increase capital; increasing our input means more output and more economic growth. And when capital is depreciating, it is less productive, yielding less output. Technology also allows higher levels of sustainability for capital.

As discussed in Chapter 1 the economist, Malthus did not realize that technology could increase food supply. And his modern day disciple Paul Ehrlich, gives an exceedingly grim prognosis for the future: "Each year food production in underdeveloped countries falls a bit further behind burgeoning population growth, and people go to bed a little hungrier. While there are temporary or local reversals of this trend, it now seems inevitable that it will continue to its logical conclusion: mass starvation."[65] Not only does Ehrlich state that the world is headed toward calamity, he is convinced that there is nothing anyone can really do that will provide anything more than temporary abatement. To focus on Ehrlich's attitude towards technology as part of the solution to the impending problem is to see Ehrlich's "technological pessimism," so to speak. Ehrlich's lack of confidence in technology to deal with the problems plaguing the future is seen in his statement: "But, you say, surely Science (with a capital "S") will find a way for us to occupy the other planets of our solar system and eventually of other stars before we get all that crowded."[66] Ehrlich was sure that "the battle to feed humanity is over." He insisted that India would be unable to provide sustenance for the 200 million person influx in its population by 1980. He was wrong – Ehrlich did not count on the "green revolution."

As this predictive framework earned the title of "dismal science" for economics, engineers look upon the same problems with more of a technical optimism. Engineers bump up the Malthusian curve by finding ways to improve conditions. In *The Engineer of 2020*, one finds descriptions of the various ways engineers in the future will help to solve the very same problems about which Ehrlich (and the Malthusian model in general) is concerned. Where Ehrlich considered technology's role in solving the problem would only be seen through how "improved technology has greatly increased the potential of war as a population control device,"[67] engineers look towards technology not as Ehrlich's "means for self-extermination"[68] but rather they opt to use it to support and improve life in the future.

According to National Academy of Engineering, the world's population will approach 8 billion people; much of this increase will be seen in groups that are today considered underdeveloped countries, mainly in Asia and Africa. Apparently, "by 2015, and for the first time in history, the majority of people, mostly poor will reside in urban centers, mostly in countries that lack the economic, social, and physical infrastructures to support a burgeoning population."[69] However, engineers see in the challenge posed by the highly crowded and densely population world of 2020 as an opportunity for "the application of thoughtfully constructed solutions through the work of engineers."[70] Likewise, engineers look upon the necessity for improved health care delivery in the world of future with confidence. The key word is confidence, and not "arrogance." Engineers must make advanced technologies accessible to this ever-growing global population base. In the next twenty years, positive implications on human health, will be possible due to improved air quality and the control and clean up of hazardous waste sites, and focused efforts to treat diseases like malaria and AIDS.[71]

Engineers believe that they can solve the problems posed by the future, as opposed to views like the one posed by Ehrlich who sees a future where "small pockets of *Homo sapiens* hold on for a while in the Southern Hemisphere, but slowly die out as social systems break down, radiation poisoning takes effect, climatic changes kill crops, livestock dies off, and various man-made plagues spread. The most intelligent creatures ultimately surviving this period are cockroaches."[72] Indeed, this dramatic example serves to illustrate the differences between engineer's technical optimism and doomsayer's technical pessimism, so to speak. But what does the future call for – the proverbial idealist or the modern-day skeptic? Human kind can either resign itself to failure and deem the problems it will come to face unsolvable, or it can press forward, attempting to solve the problems it faces and overcome tomorrow's challenges. While the question is subjective, it becomes clear that in order to have progress, most engineers choose creativity and action.

NOTES AND COMMENTARY

[1] National Academy of Engineering, *The Engineer of 2020: Visions of Engineering in the New Century* (Washington, DC: National Academy Press, 2004), 49.

[2] V.R. Potter II, "What Does Bioethics Mean?" *The Ag Bioethics Forum* 8, no. 1 (1996): 2–3.

[3] A.S. Daar, H. Thorsteinsdóttir, D.K. Martin, A.C. Smith, S. Nast, and P.A. Singer, "Top Ten Biotechnologies for Improving Health in Developing Countries," *Nature Genetics* 32 (2002): 229–32.

[4] *Oxford Dictionary of Biochemistry and Molecular Biology* (New York, NY: Oxford University Press, 1997).

[5] Ibid.

[6] Ibid.

[7] The President's Council on Bioethics, "Working Paper 1, Session 4: Human Cloning 1: Human Procreation and Biotechnology," (17 January 2002).

[8] L.R. Kass, The President's Council on Bioethics, Transmittal Memo to "Human Cloning and Human Dignity: An Ethical Inquiry," (Washington, DC: 2002).

[9] L.R. Kass, 2005, Letter of Transmittal to President George W. Bush, *Alternative Sources of Human Pluripotent Stem Cells*, a White Paper of the President's Council on Bioethics (Washington, DC, 10 May 2005).

[10] The President's Council on Bioethics, Executive Summary, "Human Cloning and Human Dignity: An Ethical Inquiry," (Washington, DC: 2002), xxvii.

[11] P. Singer, *Rethinking Life and Death: The Collapse of Our Traditional Ethics* (New York, NY: St. Martin's Griffin, 1994).

[12] M. Schooyans, *Bioethics and Population: The Choice of Life* (St Louis, MO: Central Bureau, Community Center for Vital Aging, 1996).

[13] P. De Coppi, G. Bartsch Jr., M.M. Siddiqui, T. Xu, C.C. Santos, L. Perin, G. Mostoslavsky, A.C. Serre, E.Y. Synder, J.J. Yoo, M.E. Furth, S. Soker, and A. Atala "Isolation of Amniotic Stem Cell Lines with Potential for Therapy" *Nature Biotechnology* 25(1) (2007): 100–6.

[14] Church of Scotland, 2006, Society, Religion and Technology Project, Patenting Life: An Introduction to the Issues, http://www.srtp.org.uk/scsunpat.shtml (accessed 17 September 2006).

[15] K.D. Warner, "Are Life Patents Ethical? Conflict between Catholic Social Teaching and Agricultural Biotechnology's Patent Regime," *Journal of Agricultural and Environmental Ethics* 14, no. 3 (2002): 301–19.

[16] D. Rees and S. Rose, *The New Brain Sciences: Perils and Prospects* (Cambridge, UK: Cambridge University Press, 2004).

[17] President's Council on Bioethics, Staff Background Paper: Organ Transplantation: Ethical Dilemmas and Policy Choices (Washington, DC, 2003).

[18] Ibid.

[19] US Department of Health, Education and Welfare, National Commission for the Protection of Human Subjects of Biomedical and Behavioral Research, *The Belmont Report: Ethical Principles and Guidelines for the Protection of Human Subjects of Research*, 18 April 1979.

[20] R.V. Orzeck shared this case that will appear in his book, *So Now You'll Know*, to be published in 2007.

[21] One of the common themes of this book, along with a systematic approach and the need for professional trust, is that of empathy. Schweitzer's reverence for life has been characterized as a "bioempathetic" viewpoint. See: A. Sweitzer, *Out of My Life and Thought* (translated by A.B. Lemke) (Henry Holt & Co., New York, NY, 1990), 157.

[22] This discussion draws upon the ideas of Diana Chang, who conducted undergraduate research in an independent study course that I facilitated at Duke.

[23] F.B. Orlans, T.L. Beauchamp, R. Dresser, D.B. Morton, and J.P. Gluck, *The Human Use of Animals: Case Studies in Ethical Choice* (New York: Oxford University Press, 1998).

[24] Ibid.

[25] T.F. Budinger and M.D. Budinger refer to this approach as the "four As:" (1) acquire facts; (2) alternatives; (3) assessment; and (4) action (T.F. Budinger and M.D. Budinger, *Ethics of Emerging Technologies: Scientific Facts and Moral Challenges*) (Hoboken, NJ: John Wiley & Sons, 2006).

[26] This discussion draws upon the ideas of Zach Abrams, who conducted undergraduate research on GMOs in my Ethis in professions course of Duke.

[27] G. Tulloch, *Euthanasia: Choice and Death* (Edinburgh, UK: Edinburgh University Press, 2005).

[28] S. Ewen and A. Pusztai, "Effect of Diets Containing Genetically Modified Potatoes Expressing *Galanthus nivalis* Lectin on Rat Small Intestine," *The Lancet*, 354 (1999), 9187.

[29] M.L. King Jr., *Strength to Love*, Fortress Edition (May 1981) (Minneapolis, MN: Augsburg Fortress Publishers, 1963).

[30] Agent Orange website: http://www.lewispublishing.com/orange.htm (accessed 22 April 2005).

[31] A principal source for the Minamata case is the Trade & Environment Database, developed by James R. Lee, American University, The School of International Service, http://www.american.edu/TED/ (accessed 19 April 2005).

[32] This is an all too common professional ethics problem, i.e., lack of full disclosure. It is often, in retrospect, a very costly decision to withhold information about a product, even if the consequences of releasing the information would adversely affect the "bottom line." Ultimately, as has been seen in numerous ethical case studies, the costs of not disclosing are severe, such as bankruptcy and massive class action lawsuits, let alone the fact that a company's decision may have led to the death and disease of the very people they claim to be serving, their customers and workers!

[33] International Programme on Chemical Safety, United Nations Environmental Programme, "Cadmium." Environmental Health Criteria (EHC134), Geneva, Switzerland, 1992.

[34] Agency for Toxic Substances and Disease Registry, US Department of Health and Human Services "Toxicological Profile for Cadmium," Washington, DC, 1999.

[35] T. Colburn, Speech at *the State of the World Forum* (San Francisco, CA: 1996).

[36] *Webster's Ninth New Collegiate Dictionary* (Springfield, MA: Merriam-Webster, 1990).

[37] Christian Scripture has defined faith as "the substance of things hoped for, the evidence of things not seen" (Hebrews 11:1). This is an extension of the Judaic outlook for the "promised land" (a metaphor to the engineer's expectation to see the designs reach the build phase, to the medical researcher's search for cures of obdurate maladies, or to the city planner's long-range plan, envisioning green spaces and copious public amenities). This optimistic view lends a temporal dimension to faithfulness. For example, the concept of sustainability in terms of public health and environmental quality is a requirement to take (or to avoid) actions based on their impact on future generations. This means that engineers must be faithful agents to distant and future people. An action that is most expedient for the present may not be the best if it has severe effects in the long run.

[38] This account can be found in J. Enderle, S. Blanchard, and J. Bronzino, ed., *Introduction to Biomedical Engineering* (Burlington, MA: Elsevier Academic Press, 2005).

[39] The Pandora's box is an interesting and useful ethical metaphor that conveys the potential of unintended, negative outcomes from a seemingly innocuous or beneficial act. The term comes from the Greek myth about a box left by Mercury with Epimetheus and Pandora for safekeeping. Epimetheus warned Pandora not to open the box, but eventually upon hearing voices asking to be freed, Pandora's curiosity got the best of her. She opened the lid and out came diseases, vices, and other ills to humanity. Hence, the myth has been used as a warning not to unadvisedly or prematurely rush into the unknown (e.g., viral research, neurotechnologies, nanotechnologies, and new drug therapies). However, an often forgotten part of the story is that Pandora opened the box a second time and "hope" was released. Biomedicine and engineering are modern manifestations of this hope.

[40] Another way to look at uncertainty is that it is a function of variability and ignorance. This has been well articulated by L. Ginzburg in his review of ecological case studies in US Environmental Protection Agency, 1994, Peer Review Workshop Report on Ecological Risk Assessment Issue Papers, Report Number EPA/630/R-94/008. According to Ginzburg, "variability includes stochasticity arising from temporal and spatial heterogeneity in environmental factors and among exposed individuals. Ignorance includes measurement error, indecision about the form of the mathematical model, or appropriate level of abstraction." Thus, variability can be lessened by increased attention, e.g., empirical evidence, and "translated into risk (i.e., probability) by the application of a probabilistic model," but ignorance cannot. Ignorance simply translates into confidence intervals, or "error bounds" on any statement of risk.

[41] See, for example, Physical Principles of Unworkable Devices, http://www.lhup.edu/~dsimanek/museum/physgal.htm. Donald E. Simanek's humorous but informative site on why perpetual motion machines cannot work, i.e., their inventors assumed erroneous "principles." This site is instructive to biomedical decision makers to beware of "junk science." Sometimes a good way to learn why something works the way it does is to consider all the reasons that it fails to work.

[42] My apologies to the originator of this analogy, who deserves much credit for this teaching device. The target is a widely used way to describe precision and accuracy.

[43] It is probably safe to say that most engineers lack a substantial amount of formal training in psychology and the behavioral sciences, so no previous background in these areas is needed this text. However, I have observed many engineers who are gifted in "people skills." Also, most have gained knowledge and have read extensively in these areas after their baccalaureate education. An interesting and valuable change in engineering education has been greater emphasis on the student's grasp of the social sciences and the humanities.

My own educational interests were a bit ahead of this trend (by three decades), not due to prescience or an unusual sense of what I would need to prepare for the challenges of a very rewarding career. No, my second major in psychology was because I met a beautiful young woman in a sophomore course. She happened to be a psychology major, so I found myself taking an inordinate amount of social science courses, which eventually translated into a major. Oh, by the way, my insistence paid off. I married her and we recently celebrated our thirtieth anniversary. I have never drawn a critical path or fault tree for contingent probabilities of this outcome, and the benefits derived (at least by me) but perhaps, I shall.

Analyze that!

[44] T. Hobbes, 1660, *Leviathan*.

[45] Lawrence Kohlberg, *Child Psychology and Childhood Education: A Cognitive-Developmental View* (New York, NY: Longman Press, 1987).

[46] J.B. Bassingthwaighte, "The Physiome Project: The Macroethics of Engineering toward Health," *The Bridge* 32, no. 3 (2002): 24–9.

[47] J.R. Herkert, "Microethics, Macroethics, and Professional Engineering Societies," in *Emerging Technologies and Ethical Issues in Engineering* (Washington, DC: The National Academies Press, 2004), 107–14.

[48] Accreditation Board for Engineering and Technology (ABET), Inc., *Criteria for Accrediting Engineering Programs: Effective for Evaluations during the 2004–2005 Accreditation Cycle* (Baltimore, MD: ABET, 2003).

[49] I. Ajzen, *Attitudes, Personality, and Behaviour* (Milton Keynes: Open University Press, 1988).

[50] A.H. Eagly and S. Chaiken, *The Psychology of Attitudes* (Orlando, FL: Harcourt Brace & Company, 1993).

[51] Ibid.

[52] J. Piaget, *The Moral Judgment of the Child* (New York, NY: The Free Press, 1965).

[53] S. Vaknin, *Malignant Self Love: Narcissism Revisited* (Macedonia: Lidija Rangelovska Narcissus Publications, 2005).

[54] C.G. Morris, *Psychology – An Introduction*, 9th ed. (Englewood Cliffs, NJ: Prentice-Hall, 1996).

[55] This is one of the problems with B.F. Skinner's brand of behaviorism, as articulated in *Beyond Freedom and Dignity* (Hackett Publishing, 1971). Certainly, persons act out on what they have learned and that learning is an aggregate of their responses to stimuli. However, human emotions and empathy are much more than this. Empathy is a very high form of social and personal development. So, although one might be able to "train" an ant or a bee to respond to light stimuli, or a pigeon to "play ping-pong" (as Skinner did), even these lower animals have overriding social complexities. At the heart of humanity are freedom and dignity, in spite of what some behaviorists tell us.

[56] Vaknin, *Malignant Self Love*.

[57] The concept may be innate and extended to other animals, such as elephants sensing "awe" for their ancestral graveyards.

[58] "Value Engineering," 24 February 2006. DOD Value Engineering Program, http://ve.ida.org/ve/ve.html (accessed 16 March 2006).

[59] "What Is the Value Method?", 5 May 2006. Systemic Analytic Methods and Innovations, http://www.value-engineering.com (accessed 5 May 2006).

[60] "The Value Engineering (VE) Process," 11 March 2005, US Department of Transportation Federal Highway Administration, http://www.fhwa.dot.gov/ve/veproc.htm (accessed 16 March 2006).

[61] "Value in Economics" The Columbia Encyclopedia, 6th ed. 2001–2005, http://www.bartleby.com/65/va/value2.html (accessed 2 May 2006).

[62] David Ricardo, *On the Principles of Political Economy and Taxation* (John Murray, London: 1821), http://www.econlib.org/library/Ricardo/ricP.html (accessed 15 March 2006).

[63] Value in Economics.

[64] National Academy of Engineering, *The Engineer of 2020*: *Visions of Engineering in the New Century* (Washington, DC: National Academy Press, 2004), 48.

[65] P. Ehrlich, *The Population Bomb* (New York, NY: Ballantine Books, 1968).

[66] Ibid., 20.

[67] Ibid., 69.

[68] Ibid.

[69] National Academy of Engineering, 27–8.

[70] Ibid.

[71] Ibid., 28–9.

[72] Ehrlich, *The Population Bomb*, 78.

Chapter 3

An Engineered Future: Human Enhancement

Since engineering is the application of the sciences to improve the human condition, we should consider how much and what types of enhancements are ethical. Thus, the chapter addresses one overriding bioethics question:

Bioethics Question: When is human enhancement ethically right and when is it ethically wrong?

In Chapter 1, the term "enhancement" appears as one of the goals of engineering. Chapter 2 stresses the importance of "active" engineering. The combination of these qualities, active enhancement, is encouraged by the engineering profession. But enhancement of the human species is far from a morally neutral enterprise. To demonstrate, here is another thought experiment:

Thought Experiment

Dave is a biomedical engineer working on an aerosolizer that will allow a steroid to be administered easily and delivered to the target organ (in this case, the lung's alveolar region) with nearly 100% efficiency. This will be a monumental advancement in drug therapies for asthmatics where much of a particular drug does not find its way to the target alveoli and is lost to other inefficacious regions. However, in the process, since the drug is delivered with such efficiency, metabolism mimics natural corticosteroids so well that most drug testing protocols will not be able to distinguish the pharmaceutical steroids from those endogenously and naturally produced by the body. In other words the device causes steroid tests to be false negatives (i.e., show no steroid use even when artificial steroids are being used).

Your company forecasts a large legal profit, but it also predicts possible illegal activities that may lead to a 25% increase in steroid use by at-risk populations, such as high school athletes.

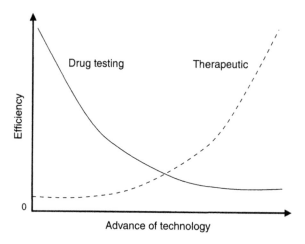

Figure 3.1 Hypothetical relationship between competing values. In this instance, as a device's technology improves therapeutically, the ability to monitor illegal and dangerous drug use decreases.

Is it ethical to produce such a device?

Clearly, there are pros and cons in almost any technological advancement. Figure 3.1 shows the hypothetical relationship between two competing outcomes. In real situations, numerous competing values are involved (Figure 3.2). Often, the difference between engineering success and failure is

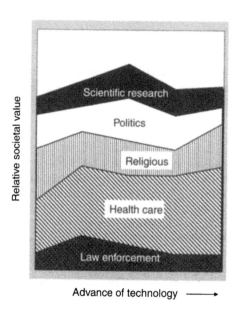

Figure 3.2 Hypothetical relationship among values of a technology perceived by society.

measured by change, in particular, the difference between the *status quo* and the conditions given by a designed intervention. And, the value that society places on each aspect of this change determines whether something is moral or immoral. These relationships can also be likened to a force field (discussed in Chapter 7), where a decision is influenced by factors of varying weights.

Each engineer changes the future of the client and, in turn, incrementally changes the future. Some changes are profound, such as the future invoked by the designs of Copernicus on space travel, Orville and Wilber Wright and Henry Ford on transportation, Paracelsus and Galen on medicine, and Newton and Galileo on monitoring devices (optics and remote sensing, respectively). But, as the controversial Edward O. Wilson's quote reminds us, enhancement evokes negative futures as well:

> *Homo sapiens*, the first truly free species, is about to decommission natural selection, the force that made us . . . . Soon we must look deep within ourselves and decide what we wish to become.[1]

Wilson may not have intended it to mean this, but the quote does serve as a warning for unchecked biomedical enhancements.

So, then, is enhancement good or evil? The answer, as usual, is "it depends."

Bioethics operates between the realm of the clearly wrong and the clearly right. This is evident by the words we use to compliment or to condemn the work of a professional. Consider the following adjectives:

- Although painful, the therapy was worthwhile.
- He exposed his patient to unnecessary pain.
- The stent's failure rate is unacceptable.
- The defibrillator's reliability is tolerable.

Sound judgment is a balance between boldness and care. We are apt to be cautions.

A common human failing is that we are frequently xenophobic. We fear and resist change. We tend to keep doing things the way we have always done things. Maybe this is not so much a "failing" as an adaptive skill of the survival of the species. If we suffered no ill effects after eating that berry, we will eat more of that particular species. But, even a similar-looking species may be toxic, so we are careful to eat only the species that did not kill us, as best as we can differentiate.

So, we have a paradox as professionals. We do not want to expose our clients or ourselves to unreasonable risks, but we must to some extent "push the envelope" to find better ways of doing this. This requires that we often suppress new ways of looking at problems. Sometimes, the facts and theories may be so overwhelmingly convincing that we must change our worldview. This is consistent with Thomas S. Kuhn's concept of paradigm shift[2] introduced in Chapter 1. Scientists are often very reluctant to accept these new ways of thinking (Kuhn said such resistance can be violent). In fact, they are often not dramatic reversals (revolutions), but modifications of existing designs (evolutions). Some say that the bicycle was merely a mechanically re-rendering of the horse (e.g., the saddle seat, the linear and bilaterally symmetrical structure, and the handle bars), as was the automobile. Often, as the great philosophers tell us, "we may look but we do not see!"[3]

PROFESSIONAL *ZEITGEIST*: HOW ENGINEERS THINK

We may need to refine our question, given the nature of professional judgment:

Bioethics Question: How can engineering problem-solving skills be applied to bioethical issues like human enhancement?

To understand bioethics from an engineering perspective, we should briefly distinguish engineering thinking from that of other scientific disciplines. Beyond a unique and shared vocabulary, different professions use selective problem-solving tools. These are manifestations of their particular place and time, their *zeitgeist*. For example, mathematicians and engineers share many commonalities, such as the application of mathematical reasoning to solve problems. I have had some interesting discussions with Duke mathematician John Trangenstein about the major differences between the two professions. After all, both groups speak the same language: mathematics. Trangenstein is a Professor of both Mathematics and Civil and Environmental Engineering, so he is directly aware of the differences. One principal difference seems to be the way that practitioners of the two disciplines see problems and the methods that they use to solve them.

The first difference is between the practitioner and the inquirer. Differences in vernacular can point to dissimilarities in the way various groups go about solving problems. Practitioners frequently perceive mathematics as a means to an end; that is, a tool to help to solve the problem, whereas the goal of mathematics is to explain shape, quantity, and dependence. This perspective characterizes mathematicians as not really "creating" anything or solving any problem (unless it is applied mathematics, the branch of mathematics that applies math to other domains).

Fermat's Last Theorem, for which the proof was published in 1986, provides an example of inquiry. However, the theorem itself was published by seventeenth-century mathematician Pierre de Fermat 357 years ago. In other words, when we set aside all the applications of discoveries in mathematics, development of the subject in and of itself becomes much less urgent, as "urgency" here is used to refer to the degree to which discoveries in a science directly impact human life.[4]

The concepts of efficiency, efficaciousness, and effectiveness also distinguish the practitioner within this framework. Often, mathematics requires the examination of relationships that already exist. Therefore, the concept of effectiveness to mathematicians involves proving or explaining these already existing phenomena. Success and failure become terms to describe the ability or inability to access a system that already exists. For engineers and design professionals, the terms apply somewhat differently. Designing something to improve a process or to solve a problem requires that the terms efficiency and effectiveness refer to whether the design is conducive to the purpose for which it was created, and whether the design function performs some task, respectively. Mathematics draws from deductive reasoning to observe patterns and relationships that already exist.

Engineering, on the other hand, is not an exclusively deductive science, as engineers also base knowledge from experience and observation. They first generate rules based on observations (i.e., the laws of nature: chemistry, physics, biology) of the world around them, the way things work. Once engineers have this understanding, they may apply it by using the rule to create something, such as a new technology, designed to reach some end. History is filled with such examples, ranging from keeping food cold to delivering nutrition to comatose patients to finding better means of delivering drugs to specific tumor sites. According to the National Academy of Engineering, "technology is the outcome of engineering; it is rare that science translates directly to technology, just as it is not true that engineering is just applied science."[5] Innovations in engineering occur when a need or opportunity arises (hence the adage "necessity is the mother of invention"). Therefore, engineers may first develop an understanding of the thermodynamics behind a phase change heat pump, and then apply this knowledge when society experiences a need to keep food cold. Likewise, engineering researchers must adapt and change their understanding of electromagnetic forces at the nanoscale when developing new drug delivery nanotechnologies.

In his groundbreaking book *Pasteur's Quadrant*, Donald Stokes details the evolution of the scientific process as a dynamic form of the post-World War II paradigm. First, it is useful to consider the historical progression of the ideas of "basic" and "applied" research.

In 1944, Vannevar Bush, Franklin D. Roosevelt's director of the wartime Office of Scientific Research and Development, was asked to consider the role of science in peacetime. He did this through his work "Science, the Endless Frontier," through two aphorisms. The first aphorism was that "basic research is performed without thought of practical ends." According to Bush, basic research is to contribute to "general knowledge and an understanding of nature and its laws." Seeing an inevitable conflict between research to increase understanding and research geared toward use, he held that "applied research invariably drives out pure."[6]

Today, Bush's "rugged individual approach" has been largely displaced by the paradigm of teamwork, and a combination of pure and applied science. Here the emphasis is on the cooperative effort. This is conceptualized by Frank LeFasto and Carl Larson who in their book *When Teams Work Best* hold that teams are "groups of people who design new products, stage dramatic productions, climb mountains, fight epidemics, raid crack houses, fight fires"[7] or pursue an unlimited list of present and future objectives. The paradigm recognizes that to be effective, we need not only groups of people who are technically competent but also groups of people who are good at collaborating with one another in order to realize a common objective.[8] Much of biomedical and biosystem engineering is teamwork. If we are to succeed by the new paradigm, we have to act synergistically.

In dealing with these paradigms through time, attempts to classify, separate, and define the ideas of basic and applied research have seemingly led to some irresolution. According to Stokes, "the differing goals of basic and applied research make these types of research conceptually distinct."[9] Basic research is defined by the fact that it seeks to widen the understanding of the phenomena of a scientific field – it is guided by the quest to advance knowledge. While Bush felt that basic and applied research were at least to some degree in discord, Stokes points out that "the belief that the goals of understanding and use are inherently in conflict, and that the categories of basic and applied research are necessarily separate, is itself in tension with the actual experience of science."[10] To support this claim, many influential works of research are in fact driven by both of these goals. A prime example is the work of Louis Pasteur, who not only sought to understand the microbiological processes he discovered, but also sought to apply this understanding to the practical objective of preventing the spoilage of vinegar, beer, wine, and milk.[11]

Similarly, these goals of understanding and use are very closely related as Stokes brings to attention that the traditional fear of earthquakes, storms, droughts, and floods brought about the scientific fields of seismology, oceanic science, and atmospheric science. However, the idea that there is disparity between basic and applied research is captured in the "linear model" of the dynamic form of the postwar paradigm (Figure 3.3). It is important to keep in mind though that in the dynamic flow model each of the successive stages depends upon the stage before it.

Figure 3.3 Linear model of flow from basic research to useful, engineered designs.
Adapted from: D.E. Stokes, 1997, *Pasteur's Quadrant*, The Brookings Institution, Washington, DC.

This belief that advances in science, including biomedicine, are made applicable through a dynamic flow from science to technology is widely accepted among research and development managers. The process illustrated in Figure 3.3 has come to be called "technology transfer" as it describes the movement from basic science to technology. The first step in this process is basic research, which charts the course for practical application, eliminates dead ends, and enables the applied scientist and engineer to reach their goal quickly, and economically. Then, applied research involves the elaboration and the application of the known. Here, scientists convert the possible into the actual. The final stage in the technological sequence, development, is the stage where scientists systematically adapt research findings into useful materials, devices, systems, methods, and processes.[12]

However, this postwar paradigm has been called into question for several reasons. It has been criticized for being too simple an account of the flow from science to technology. This is understandable, as the oversimplification may be due to the effort of the scientific spokesman in the postwar era to be able to communicate these concepts to the public. Less defensible is the model's assumption that there is a one-way flow from scientific discovery to technological innovation. In other words, the problem lies in the supposition that science exists entirely outside of technology. On the contrary, throughout history there has been a reverse flow, a flow from technology to the advancement of science. Examples date as far back as Johannes Kepler, who helped invent the calculus of variations by studying the structure of wine casks in order to optimize their design. Therefore, history illustrates that science has progressively become more technology derived.[13]

Returning to the distinctness of basic and applied research, critics believe that "the terms basic and applied are, in another sense, not opposites. Work directed toward applied goals can be highly fundamental in character in that it has an important impact on the conceptual structure or outlook of a field. Moreover, the fact that research is of such a nature that it can be applied does not mean that it is not also basic."[14] So then, it becomes apparent that a model that recognizes the synthesis of the goals of understanding and use, such as that which is seen in the work of Louis Pasteur, is needed. In the one-dimensional model, which consists of a line with "basic research" on one end and "applied research" on the other (as though the two were polar opposites), Stokes initially placed Pasteur in the center. Pasteur had equal (and strong) commitments to understanding the microbiological process and to controlling the effects of these processes. Thus, Pasteur might also be represented by two points: one at the "basic research" end of the spectrum and the other at the "applied research" end of the spectrum. This placement leads Stokes to suggest a different model that reconciles the shortcomings of this one-dimensional model (Figure 3.4). The two-dimensional model also better fits the profession of engineering, which straddles the basic and applied sciences.

In the seemingly elegant model, the degree to which a given body of research seeks to expand understanding is represented on the vertical axis and the degree to which the research is driven by considerations of use is represented on the horizontal axis. Here, unlike in the one-dimensional model, it is possible to portray a body of research that is equally committed to potential use and the furthering of understanding, like Pasteur's research or that of any biomedical engineer.[15] Pasteur's work challenged

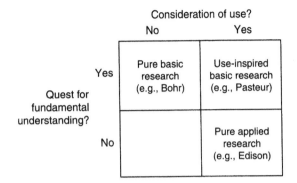

Figure 3.4 Typology of research according to advancing the state of the science (understanding) and providing practical applications (use).
Adapted from: D.E. Stokes, 1997, *Pasteur's Quadrant*, The Brookings Institution, Washington, DC.

the existing science paradigm in that it showed that there need not be a strict divide between basic and applied research; in this way it brought about a paradigm shift.

Thomas Kuhn, actually changed the meaning of the word "paradigm," which had been almost the exclusive province of grammar (a fable or parable). Kuhn extended the term to mean an accepted, specific set of scientific practices. The scientific paradigm is made up of what is to be observed and analyzed, the questions that arise pertaining to this scientific subject matter, to whom such questions are to be asked, and how the results of the investigations into this subject matter will be interpreted. The paradigm can be harmful if it allows incorrect theories and information to be accepted by the scientific and engineering communities. Such erroneous adherences can result from "groupthink," a term coined by Irving Janus, a University of California, Berkeley psychologist. Groupthink is a collective set of systematic errors (biases) held by and perpetuated by a group.[16]

With this background in engineering *zeitgeist*, we can begin to address the engineer's stake in human enhancement.

IMPROVEMENT *VERSUS* ENHANCEMENT

> In the year 3535
> Ain't gonna need to tell the truth, tell no lies.
> Everything you think, do and say, is in the pill you took today . . . .
>
> In the year 5555
> Your arms hanging limp at your sides.
> Your legs got nothing to do.
> Some machine is doing that for you.
>
> Denny Zager and Rick Evans, *In the Year 2525* (1964)

> Gentlemen, we can rebuild him. We have the technology. We have the capability to build the world's first bionic man. Steve Austin will be that man. Better than he was before. Better, stronger, faster.
>
> Opening Narration, *The Six Million Dollar Man*, Television Series (1974)

Which is it? Does our technological presence and future foretell us as victims or as beneficiaries of manipulation? In fact, we have seen both. Eugenics has been an excuse for genocide and bigotry. Medical advances have provided a quality of life to people who a generation ago would not survived.

Engineers find themselves in the middle of this debate. In fact, engineers will be agents who determine whether the future is the one feared by Zager and Evans in their hit tune, *In the Year 2525*, or the one envisioned by the screenwriters of *The Six Million Dollar Man*. Humans have never been satisfied with leaving "well enough alone." We found the synergy of hunting in teams preferable to that of the lone wolf. We began to use fire as a protective device and culinary aid. We learned to abide by the social contracts articulated in rules and dogma, finding these more desirable than constantly competing with our fellow humans as "brute beasts."[17] We harnessed botanical and zoological principles, changing agriculture and farming, thus allowing us to live communally. We learned the laws of mathematics and applied them in ways to change our environment, such as the Roman's application of Archimedes' principles of trajectories to develop catapults and calculate missile efficiencies. We continue to fight communicable diseases with medical treatment and engineering advances (Figure 3.5). Likewise, we are making medical and scientific progress against previously intractable, chronic diseases like cancer.

It can be argued that our species has made progress, notwithstanding the wars, injustices, immoralities, and other attendant problems of this advance. The challenge of making this call is complicated by the subjective nature of progress. Take the word "improvement." Engineers are familiar with the term as a euphemism. For example, the wooded area near your neighborhood was likely considered by real estate developers, engineers, and planners as "unimproved." This simply meant that it had not yet had the "benefit" of anthropogenic improvement. But, as a child, most of us recognized the benefit of an

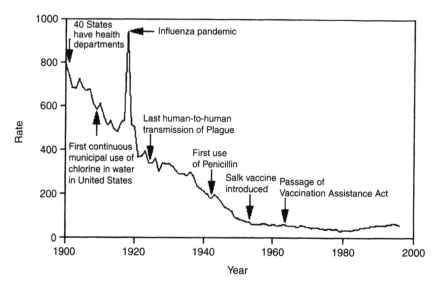

Figure 3.5 Crude infectious disease death rate (per 100 000 populations) in the United States from 1990 through 1996. Adapted from: Center for Disease Control and Prevention, 1999, Achievements in Public Health, 1900–1999: Control of Infectious Diseases, *Morbidity and Mortality Weekly Report*, 48 (29), pp. 621–9.

area devoid of streets, curbs, houses, parking lots, and other "amenities." At an early age our intuition informed us of the inherent value of this property.

Likewise, intuition comes into play when we think about human enhancement. For example, even if we agree that the next generation will have to be smarter than this one as the problems and opportunities become more complex, we intuitively know that we must not manipulate our species in certain ways to make us smarter. For example, if we are able to increase cranium size *in utero* (i.e., within the womb) and this increase is found to make the person smarter, would it be ethical to carry out such manipulation? Many of us would be opposed to this. Some opposition owes to a historical track record of similar "advances" being accompanied by unintended consequences later (such as the effects of the *in utero*-administered drug diethylstilbestrol (DES) in mothers whose daughters later developed cervical cancer after adolescence). Other opposition is on moral grounds.[18] We do not have the moral perspective or authority to make this decision for the unborn. Some are opposed because of aesthetics. Will the child's larger head be perceived as unappealing and lead to social challenges?

The subtle intuition is quite interesting from a moral perspective. For one thing, human enhancement cannot be completely based on utility. As in the case of an increased cranium size, it could be argued that having a third arm is a benefit. How many times have you been trying to do something that would have been so much easier if you had another arm? Or if you had eyes behind your head? Both of these would in fact add utility to numerous tasks. But few of us would opt for either. Why not?

Engineering Intuition

> Engineering problems are under-defined, there are many solutions, good, bad and indifferent. The art is to arrive at a good solution. This is a creative activity, involving imagination, intuition and deliberate choice.
>
> Sir Ove Arup (1895–1988), Danish-English Engineer-Architect

It can be argued that engineering is a blend of information, creativity, and ethics.[19] The engineer operates within a range of constraints and opportunity (Figure 3.6). Too far in the direction of risk means the engineer is not dutifully protecting and ensuring the safety of the client and the pubic at large. Too far in the direction of caution means the engineer is not taking advantage of the personal and public investment in education and experience that should have engendered creativity in solving society's problems. The "sweet spot" in the center of the diagrams is determined by knowledge, with a hefty dose of intuition.

Engineers are risk managers. Moreover, the discussion also begs the question as to how far should an engineer push the envelope before becoming unethical. The stakes are very high. For example, modifying the design of a medical device, such as an implanted defibrillator, could well increase its efficiency. However, it could also introduce new risks. Unlike paintings, sculpture, and other arts, the "art" of engineering is a balance of mathematics, physical science, and creativity. As Sir Ove aptly stated, creativity embodies "imagination, intuition and deliberate choice." These attributes set engineers apart from many other professions.

Engineers *versus* Economists

As discussed in the previous chapter both engineers and social scientists, especially economists, certainly seek maximum utility and usefulness of their projects, but their approach in these matters differs. For the economist, the task can be viewed from two perspectives. First, economics considers the aspect that deals with the distribution of goods and services and management of economic systems. Economists see utility, at least in part, as a means of finding the best way to allocate scarce resources. When they find the best solution to any problem of allocation or distribution, they have reached what is known as Pareto optimality,[20] the point at which no one else can be made any better off without making someone else any worse off.

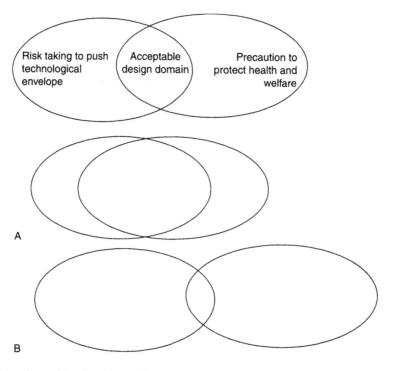

Figure 3.6 Span of control for the engineer. Diagram A represents an engineering problem with large tolerances. The domain for acceptable design is large. Diagram B represents a problem with tight tolerances, with the domain for design acceptability much smaller. Diagram B problems do not allow as much risk taking, so creative solutions are more difficult, but arguably even more necessary than those represented by Diagram A.

Second, economists' decisions are focused primarily on a specific business and often consider the perspective of a firm or the individual making economic decisions. How do we optimize our profits, or lower our individual costs? Here, the goal is to make ourselves or our individual firms as well off as we can, given the constraints. In this view, the principal concern is not the degree to which others become better or worse off. In other words, economists must choose whether they are seeking maximum profitability or Pareto optimality as the measure of success in a given project.

In the consumer model, for example, the goal is to maximize the level of utility that a consumer achieves, while subject to the constraints of the budget (essentially the total amount of money available in a given project). This utility maximization is the endpoint. Some argue that utility maximization is an ethical construct. This is a "greed is good" argument, since the overall benefit of a large sector of society maximizing its own utility means overall growth. Certainly, there is some truth in this. Capitalism does indeed provide many benefits. And, the failed models of socialism are often due to the reduced incentives for individuals. In the producer model, firms optimize by maximizing profits, given the limitations they face due to the wage rate (the price of labor) and the rental rate (the price of capital).

None of these models in practice, however, produce the ideals in every situation. For example, if something prized by society does not lend itself well to a monetary value, such as open space, health care, or sensitive environmental habitats, the consumer and producer models begin to collapse. Garrett Hardin's "Tragedy of the Commons" addresses this deficiency (see discussion in Chapter 9).

Essentially, then, engineers must be creative and intuitive in modeling success. The bioengineering aspects of this blend are well articulated by James B. Bassingthwaighte, Professor of Bioengineering, Biomathematics, and Radiology at the University of Washington:

> Current risk/benefit analyses of potential new drug therapies are not sufficient to the task. In general, current analyses are based on the inference that a drug acts on a single protein, usually an enzyme or a transporter; efficacy and side effects are determined later by observation. We need a great leap forward that will enable us to make "knowledgeable" calculations of risks and benefits. We need information on which decisions can be based. In the United States, new technologies, such as gene-insertion, stem-cell infusion, and new pharmaceuticals are within the purview of regulatory agencies, such as the Food and Drug Administration, rather than scientific funding agencies. The mission of regulatory agencies is to protect us against speculative or risky advances. These agencies depend heavily on large, expensive clinical trials in which some human subjects are put at risk in the interest of protecting others. For novel interventions, which offer great possibilities but little evidence for predicting success and a high risk of failure, harm, or damage, we must find another way to move ahead, but with minimal risk. In other words, we must find ways that enable us to follow our intuition and insight by maximizing our ability to predict risks and benefits.[21]

Thus, human enhancement is an optimization between risk and opportunity.

Intuiting Value

The foregoing discussion provides some important guidelines for human enhancement.

Optimization for engineers involves creativity and intuition. Essentially, engineering involves design under constraint.[22] Engineers strive to optimize the level of elegance of a given structure while subject to the constraints of fixed physical and mental resources, as well as the demands of the marketplace. At the same time, engineers additionally face the greatest constraint of all: safety. This constraint becomes more important than the aesthetic and economic aspects because, as articulated by Henry Petroski:

[T]he loss of a single life due to structural collapse can turn the most economically promising structure into the most costly and can make the most beautiful one ugly.

Thus, the engineered project can be thought of as an index with minimum and maximum operators; that is, certain criteria must be met or the whole project fails. If the electrical system of a medical device fails when it should have worked, the whole device is a failure.

The main issues here are measures of success, so to speak. Success is the right amount and type of human enhancement. For instance, we can quantitatively measure the level of success of an engineering design in terms of effectiveness. For instance, let's assume that some type-y pump has been designed to administer x milligrams of a specific drug to a sick patient's body. We can consider the person's actual physical condition, and then assess whether the condition is due in part to the administration of x drug by this device. Thus, efficiency is relatively easy to measure in engineering. Effectiveness, however, may not be. Suppose that this drug is being administered to this patient with the immediate goal of his recovery. Now, let us assume that our patient is, for the sake of the model, a convicted killer who cannot work and will continuously be supported by the state. Is it effective for society at large to keep him alive, "artificially" considering what is good for society at large? Or, to complicate it further, what if the only person we know who would benefit from this design is a person serving a life sentence for murder?[24] Would the device be worth the investment? Would it be a "success" from an economic perspective to allocate these resources, which may have been used elsewhere, to this patient? Thus, the effectiveness is determined not only by how efficiently our device works, but on how we place a value on the outcome of the design.

Intuition comes into play when we perceive and assign value. How do we classify a life as a success or a failure? One common (even legal) approach is to assign a sort of expected net worth to every person over the age of 25 (two to three years after most college undergraduates graduate) based on their material wealth. This method has major shortcomings. Then, what if we found a way to measure a person's success taking into account net worth, education, accomplishments, interpersonal skills, vocational skills, life trajectory, progeny, and charitable donations – in some sort of comprehensive model? However all-encompassing our algorithms, however accurate our calculations, there is one thing that assures our model will be flawed: contrasting and often opposing cultural views of value. Simply put, the factors that comprise the ideas of success for the majority of the members of one culture are often entirely different than that of a different culture. To provide an extreme example, we compare life goals in the Buddhist monk community to those of the college junior on the fast track to Wall Street. As the monk strives for enlightenment, denouncing material gains, our ideas of "success" often involve material compensation. So to many the college student is more likely to be a "success"!

Reasoning is a very important element of ethical decision making. People go about it in myriad ways, but most engineers have a few things in common.

Deductive and Inductive Reasoning: Precursors to Intuition

Intuition has always been an asset for engineers, and its value is increasing. The term "intuition" is widely used in a number of ways, so it needs to be defined here so that we are clear about what we mean by intuition, and, more importantly, what engineering intuition is not. One of the things that set engineers apart from most other scientists is the way that engineers process information.

There are two ways of looking at data to derive information and to gain knowledge. These are deductive and inductive reasoning. When we "deduce," we use a general principle or fact to give us information about a more specific situation. This is the nature of scientific inquiry. We use general theories, laws, and experiential information to provide accurate information about the problem or the situation we are addressing. In scientific investigations, a cause is often deduced from an observed effect. For example, low dissolved oxygen levels in a stream will not support certain fish species, so we reason that the fish kill is the result of low oxygen. A damaged liver produces certain unique enzymes or normal enzymes at a much higher rate than does a healthy liver. From the measurements, we deduce the lack of a certain fish species to be caused by the lack of oxygen or the spike in enzymes to be caused by the damaged liver. This demonstrates a product of deductive reasoning; this is known as synthesis.

Engineers and other technical professionals also engage in inductive reasoning or analysis. When we induce, we move from the specific to the general and from the effect to the cause. We attribute the fish kill to the low dissolved oxygen levels in a stream that results from the presence of certain substances that feed microbes that, in turn, use up the oxygen. We conduct experiments in microcosms that allow us to understand certain well-defined and well-controlled aspects of a system. We induce from these observations larger principles beyond our specific study.

Inductive reasoning is also called "abstraction," because it starts with something concrete and forms a more abstract ideal. Philosophers have argued for centuries regarding the value of inductive reasoning. Since induction is the process that takes specific facts, findings, or cases and then generally applies them to construct new concepts and ideas, abstraction leaves out specific details, unifying them into a whole based on a defined principle. For example, a brown-feathered chicken, a white-feathered chicken, and a polka dot-feathered chicken can all be integrated because each is a chicken, albeit with differences. The feather color, then, can be eliminated under the principle or criterion of being a chicken (i.e., chickenness), i.e., color is not "relevant." A brown chicken, brown bear, and brown paper bag can be integrated under the criterion of having brown color. The other aspects (besides brownness) of each item's characteristics are not relevant in this case, so they are omitted.

The peril of induction is that any conclusion must be limited. For example, our experiment may show a direct relationship between an independent and a dependent variable, but one does not know just how far to extend the relationship beyond the controlled environment of the laboratory. We may show that increasing X results in growth of Y, but what happens in the presence of A, B, C, and Z? Engineers realize this and must be arbiters of what is useful and what will happen in real-world settings.

So, like other professionals, engineers build up a body of information and knowledge from deductive and inductive reasoning. They must rigorously apply scientific theory (deduction) and extend specific laboratory and field results (induction). Over time, the engineer's comfort level increases. To observe the decision making of a seasoned engineer might well lead to the conclusion that the engineer is using a lot of "intuition." Engineers learn about how their designs and plans will work in two ways:

1. Their formal and continuing education, i.e., what others tell them.
2. What they have experienced personally.

The engineer learns both subject matter (i.e., content) and processes (i.e., rules). The scientific and practical content is what each engineer has learned about the world. Physical facts and information about matter and energy and the relationships between them are the content of engineering. Rules are the sets of instructions that the collective profession of engineering has written (literally and figuratively) over time that tells us how to do things.[25]

Bioethics Question: Can intuition be a reliable guide to ethical decision making?

The content and rules accumulated from the individual engineer's academic experience and professional practice leads to intuition. Thus, intuition can be explained as the lack of awareness of why or how professional judgments have come to be. Kenneth Hammond, a psychologist who has investigated intuitive processes, says that intuition is in fact:

> [A] cognitive process that somehow produces an answer, solution, or idea without the use of a conscious, logically defensible step-by-step process.[26]

Therefore, intuition is an example of something that we know occurs, and probably quite frequently, but it is not deliberative. Nor can intuition be explained explicitly after it occurs. It can be argued that intuition, then, is really a collective memory of the many deductive and inductive lessons learned (content), using a system to pull these together, sort out differences, synthesize, analyze, and come to conclusions (rules). The more one practices, the more content that is gathered and the more refined and tested the rules become.

Creativity

The right solution in one instance may be downright dangerous in another. Or as the National Academy of Engineering puts it:

> Engineering is a profoundly creative process.[27]

However, engineers must always design solutions to problems adhering to the relevant constraints and within the tolerances called for by the problem at hand. When it comes to human enhancement and other bioethical issues, this may be a balance between natural and artificial systems. This balance depends on data from many sources. Good data make for reliable information. Reliable information adds to scientific and societal knowledge. Knowledge, with time and experience, leads to wisdom.

Building a structure such as a hazardous waste treatment facility or a device such as a drug delivery mechanism is part of the solution to an engineering problem. In other situations, the engineering solution calls for a process that may or may not require the design and construction of a structure or device. Certainly, whenever an engineered system is called for it must be accompanied by thoughtfulness about the operation and maintenance (O&M) and the life cycle analysis (LCA) needed. Such a process represents the entire solution to the engineering problem, such as instituting recycling or pollution prevention besides adding the new structure or changes in diet and exercise in addition to the new device. Such intuitive thinking has gained currency in that it is a vital part of sustainable design that applies to all engineering disciplines, not just environmental engineering. Standard practice in civil and mechanical engineering now embodies sustainable design; for example, we now expect engineers to design for the environment (DFE), design for recycling (DFR), and design for disassembly (DFD), as well as to consider ways to reduce the need for toxic chemicals and substances and to minimize the generation of waste when they conceive of new products and processes.[28] Engineering seldom, if ever, can rely exclusively on a single scientific solution, but is always a choice among many possible solutions dictated by the particular conditions.

All engineers apply physical principles. Most also apply ample amounts of chemistry to their respective engineering disciplines. But, biosystem and biomedical engineers must also account for biology. The concern for biology ranges across all kingdoms, phyla, and species. Engineers use biological principles and concepts to solve problems (e.g., cellular and nanoscale processes to deliver drugs and nutrients, bacteria and fungi adapted to treat wastes, macrophytic flora to extract contaminants (i.e., "phytoremediation") and to restore wetlands, and higher animals to test substances for metabolic and other physiological effects). Engineers use microbes as indicators of levels of contamination (e.g., algal blooms, species diversity, and abundance of top predators and other so-called sentry species) and act as our "canaries in the coal mine" to give us early warning about stresses to ecosystems and public health problems. And, arguably most important, we study organisms as endpoints in themselves. We care principally about human health.

Scientists strive to understand and to add to the knowledge of nature. This entails making decisions about what needs to be studied. In this way, science is a social enterprise. For example, the reason we know more about many aspects of the environment today is that science has decided or been forced to decide to give attention to these matters.[29] Engineers have devoted entire lifetimes to ascertaining how a specific scientific or mathematical principle should be applied to a given event (e.g., why compound X evaporates more quickly while compound Z under the same conditions remains on a particle surface). Such research is more than academic. For example, once we know why something does or does not occur, we can use it to prevent disasters (e.g., choosing the right materials and designing a ship hull correctly or determining whether these compounds may readily pass through a cell's membrane). In turn, the new knowledge is applied to real-life problems. For instance, in a disaster situation, compound X may not be as problematic in a spill as compound Z if the latter does not evaporate in a reasonable time, but compound X may be very dangerous if it is toxic and people nearby are breathing air that it has contaminated. Also, these factors affect what the Coast Guard, fire departments, and other first responders should do when they encounter these compounds. The release of volatile compound X may call for an immediate evacuation of human beings whereas a spill of compound Z may be a bigger problem for fish and wildlife (it stays in the ocean or lake and makes contact with plants and animals). Thus, when deconvoluting a failure to determine responsibility and to hold the right people accountable, one must look at several compartments.

Arguably, the majority of engineers and scientists are most comfortable with the physical compartment. This is the one we know the most about. We know how to measure things. We can even use models to extrapolate from what we find. We can also fill in the blanks between the places where we take measurements (i.e., interpolations). In turn, engineering intuition allows us to assign values of important scientific features and extend the meaning of what we find in space and time. For example, if we use sound methods and apply statistics correctly, measuring the amount of crude oil on a few ducks can tell us a lot about the extent of an oil spill's impact on waterfowl in general. And, good models can even give us an idea of how the environment will change with time (e.g., is the oil likely to be broken down by microbes and, if so, how fast?). This is not to say that the physical compartment is easy to deal with. It is often very complex and fraught with uncertainty. But it is our domain. Missions of government agencies, such as the Office of Homeland Security, the US Environmental Protection Agency, the Agency for Toxic Substances and Disease Registry, the National Institutes of Health, the Food and Drug Administration, and the US Public Health Service, devote considerable effort in essentially getting the science right. Universities and research institutes are collectively adding to the knowledge base to improve the science and engineering that underpins the physical principles that underpin public health and environmental consequences from contaminants, whether these be intentional or by happenstance.

Another important aspect of engineering intuition is determining the anthropogenic aspects of a problem. Scientists use this term to represent the human component of an event ("anthropo" denotes human and "genic" denotes origin). This includes the *gestalt* of humanity, taking into account all of the factors that society imposes down to the things that drive an individual or group. For example, the anthropogenic compartment would include the factors that led to a ship captain's failure to stay awake. However, it must also include why the fail-safe mechanisms did not kick in. Likewise, in designing a medical device, the human factors (or ergonomics), such as ease of use and maintenance, influence the failure rate. These failures certainly have physical factors that drive them; for example, a release valve may have rusted shut or the timer's quartz mechanism failed because of a power surge, but there is also an arguably more important human failure in each. For example, one common theme in many disasters is that the safety procedures are often adequate in and of themselves, but the implementation of these procedures was insufficient. Often, failures have shown that the safety manuals and data sheets were properly written and available and contingency plans were adequate, but the workforce was not properly trained and inspectors failed in at least some crucial aspects of their jobs, leading to horrible consequences.

This brings up the controversial topic of "cause and effect" and the credible science needed to connect exposure to a risk and a negative outcome. Scientists frequently "punt" on this issue. We have learned from introductory statistics courses that association and causation are not synonymous. We are taught, for example, to look for the "third variable." Something other than what we are studying may be the reason for the relationship. In statistics classes, we are given simple examples of such occurrences:

Studies show that, in the summer, people who wear shorts in Illinois eat more ice cream.

Therefore, wearing shorts induces people to eat more ice cream.

The first statement is simply a measurement. It is stated correctly as an association. However, the second statement contains a causal link that is clearly wrong for most occurrences.[30] Something else is actually causing both variables, i.e., the wearing of shorts and the eating of ice cream. For example, if one were to plot ambient average temperature and compare it to either the wearing of shorts or the eating of ice cream, one would see a direct relationship between the variables; that is, as temperatures increase so does short wearing and so does ice cream eating.

I said that we scientists often punt on causality. Punting is not a bad thing (Ask the football coach who decides to go for the first down on fourth and inches and whose team comes up a half-inch short. He would have likely wished he had asked for a punt!). It is only troublesome when we use the association argument invariably (the football coach who always punts on fourth and short might be considered to lack courage). People want to know what our findings mean. Again, the medical science community may help us deal with the causality challenge. The best that science can usually do in this regard is to provide enough weight of evidence to support or reject a suspicion that a substance causes a disease. The medical research and epidemiological communities use a number of criteria to determine the strength of an argument for causality, but the first well-articulated criteria were Hill's Causal Criteria.[31] These criteria (Table 3.1) are a set of guidelines[32] that enhance intuition in bridging observed effects with possible causes, such as coherence of facts, biological gradient, and the requirement that a cause precede the effect.

Table 3.1
Hill's criteria for causality

Factors to be considered in determining whether an event is caused by an effect:

Criterion 1: Strength of Association. For something to cause an effect, it must be associated with that effect. Strong associations provide more certain evidence of causality than is provided by weak associations. Common epidemiological metrics are used in association include risk ratio, odds ratio, and standardized mortality ratio.

Criterion 2: Consistency. If a cause is consistently associated with an effect under different studies using diverse methods of study of assorted populations under varying circumstances by different investigators, the link to causality is stronger. For example, if the carcinogenic effects of Chemical X is found in mutagenicity studies, mouse and Rhesus monkey experiments, and human epidemiological studies, there is greater consistency between Chemical X and cancer than if only one of these studies showed the effect.

Criterion 3: Specificity. The specificity criterion holds that the cause should lead to only one disease and that the disease should result from only this single cause. This criterion appears to be based in the germ theory of microbiology, where a specific strain of bacteria and viruses elicits a specific disease. This is rarely the case in studying most chronic diseases, since a chemical can be associated with cancers in numerous organs, and the same chemical may elicit cancer, hormonal, immunological, and neural dysfunctions.

Criterion 4: Temporality. Timing of exposure to a causative agent is critical. This criterion requires that cause must precede the effect. For example, in a retrospective study, the researcher must be certain that the manifestation of a disease was not already present before the exposure to the chemical. If the disease was present prior to the exposure, it may not mean that the chemical in question is not a cause, but it does mean that it is not the sole cause of the disease (see "Specificity" above).

Criterion 5: Biologic Gradient. This is another essential criterion for pharmacological and chemical risks. In fact, this is known as the "dose-response" step in risk assessment. If the level, intensity, duration, or total level of pharmaceutical or chemical exposure is increased, a concomitant, progressive increase should occur in the efficacious or toxic effect.

(Continued)

Table 3.1
Hill's criteria for causality—cont'd

Factors to be considered in determining whether exposure to a chemical elicits an effect:

Criterion 6: Plausibility. Generally, an association needs to follow a well-defined explanation based on known biological system. However, "paradigm shifts" in the understanding of key scientific concepts do change. A noteworthy example is the change in the latter part of the twentieth century of the understanding of how the endocrine, immune, neural systems function, from the view that these are exclusive systems to today's perspective that in many ways they constitute an integrated chemical and electrical set of signals in an organism.[a]

Criterion 7: Coherence. The criterion of coherence suggests that all available evidence concerning the natural history and biology of the disease should "stick together" (cohere) to form a cohesive whole. By that, the proposed causal relationship should not conflict or contradict information from experimental, laboratory, epidemiologic, theoretical, or other knowledge sources.

Criterion 8: Experimentation. Experimental evidence in support of a causal hypothesis may come in the form of community and clinical trials, *in vitro* laboratory experiments, animal models, and natural experiments.

Criterion 9: Analogy. The term analogy implies a similarity in some respects among things that are otherwise different. It is thus considered one of the weaker forms of evidence.

[a] Candace Pert, a pioneer in endorphin research, has responded the concept of mind/body, with all the systems interconnected, rather than separate and independent systems.

MORAL COHERENCE

The most exciting phrase to hear in science, the one that heralds new discoveries is not "Eureka," (I found it!) but "That's funny . . . ."

Isaac Asimov (1920–1992)

Hill's guidelines for causality can be difficult to grasp from a medical perspective. In fact, one of those most challenging for undergraduate students (at least those in engineering) is that of coherence. To be coherent, any causal conclusion must not contradict present substantive knowledge. This condition is well aligned with another of Hill's criteria, consistency, which requires that the same cause–effect relationship be observed repeatedly. So, to be coherent and consistent the relationship of factual information must not be in opposition with absolute standards and conventions, and such a relationship must be observed consistently in all venues. An engineer, for example, finds trouble when a design does not adhere to best practices and the norms of the profession. This is a challenge to engineering innovation. Aristotle advised the search for the "golden mean" (although his writings never call it such). A wise engineer must look for the optimal design, which lies between carelessness and timidity.

Scientific method requires such coherence and consistency. In fact, part of the scientific review process is to ensure that any proposed findings from a study meet these two criteria before the study is allowed to be published. Unfortunately, this is more difficult, or at least not nearly so acceptable in ethical decision making, including those confronted in bioethics. But to be valid and systematic, moral arguments must strive for moral coherence. Some would argue that moral decision making is different in kind from scientific decision making. This is indicated by such axioms as that of Ralph Waldo Emerson in his essay "Self-Reliance":

Foolish consistency is the hobgoblin of small minds.

This quote is popularly used without the modifier "foolish," which is quite telling. I doubt that Emerson or even most transcendentalist thinkers would have had difficulty in thoughtful or wise consistency, as dictated by the scientific method. In fact, foolish consistency is not consistency at all. To use a colloquialism, many moralists are "one-trick ponies." They have become comfortable with a template for moral, intellectual, or ideological correctness, and have long since quit thinking about it. We see this in the sciences as well, where we fall in love with a construct and resist even slight improvements in even in the face of contravening evidence. So, perhaps Emerson was warning us about mindless thinking (which out of this context seems to be oxymoronic). We should not simply adopt someone else's template, but we should refine our own.

In considering how best to use engineering problem-solving skills to address human enhancement, there is a corollary question: How do we decide what is important?

It is interesting to hear certain arguments for moral incoherence, such as that of policy makers who in their own lives would find an act to immoral, but who have no problem advising in favor the act in the policy arena. They say something to the effect of "I believe that X is wrong, but I would never presuppose to vote that everyone had to also believe this." It boils down to two factors. Is the decision about morality? And, is the moral obligation so strong that it should be universal (i.e., Kant's categorical imperative). For example, if I think squares are better than circles; most of us would agree that my preference for a shape should be mandated to others; that is, I should not be allowed to legislate against your preference for a shape. This is an amoral decision. It has nothing to do with right or wrong, only preference.

This does not mean that shape preference is always amoral. For example, the swastika is a shape. Before Hitler, if one were asked whether they liked the shape or not, the question was probably an amoral one. However, after the heinous acts associated with that symbol, the question becomes a moral one. Emerson's concept of foolish consistency helps us to draw this distinction. Thus, the question of whether one likes dogs more than cats or Mexican food more than Italian food is really an amoral one. However, if someone asks if research on animals or genetic modification is right or wrong, there has been a sufficient historical track record on each to make them moral questions.

The second factor, importance, can be tricky. The concept of importance is value-laden. What you believe to be important, even essential, can vary substantially from what I believe. We may look at the same evidence and come to a very different conclusion about its meaning. For example, in the animal research question, you may hold that the cure for cancer is sufficiently important to justify animal experiments, whereas the pain and suffering of the animals may be the most important factor for me. In the former, you are basing your value on the end, while in the latter, I am basing my value on the means. So, even our methods of arriving at importance vary.

One of the major criticisms, and often rightly deserved, is that the way we go about ascribing moral importance is incoherent. Sometimes, whole ideologies are subject to this criticism, such as when a political party's platform strongly supports the death penalty but opposes abortion, while the opposing party's platform strongly favors abortion as a right, but opposes public executions even for the most heinous crimes. Both issues appear to be rooted in an underlying value on human life, but the moral positions seem incoherent. That is, they do not seem to grow from the absolute standard and from the body of substantive knowledge (Hill's coherence criterion). And, they seem inconsistent. The application of the moral imperative appears to vary from issue to issue.

Perhaps, this is because our assumptions about these positions are misplaced. Our assumption that it is the value of human life may be correct, but the definition of "human" and "life" differ between the two camps. This is one of the problems of ethics, at least philosophical ethics, which troubles many engineers and scientists. We are most comfortable with facts. But here, even the factual premises can be contrived. Does a person who commits a sufficiently horrible act cease to be human? Does the ontogenic stage determine whether someone is human? Such questions, from a purely scientific and factual perspective have to be answered, "No." Further, no moral issue is so important that there is some "greater truth" that allows for fact-fudging, as some have argued.

CREATIVITY AND BIOETHICS[33]

Bioethics Question: When does pushing the envelope become unethical and, conversely, when does not pushing the envelope become unethical?

Since, creativity embodies "imagination, intuition and deliberate choice," engineering shares much with medicine. In both professions, human factors, judgment, and intuition must be underpinned by sound science. This is a sort of "uncommon sense." In many engineering and medical situations, the simplest and most rationale approach is the best. The wise physician advises: "When you hear hooves, think horses, not zebras." However, if we only "think horses" what do we do when a "zebra" shows up? Thus, we need a modicum of attention to exceptional situations.

Intuition allows the engineer to push the envelope without introducing unacceptable risks. Intuition enables the engineer to think outside of the box within the realm of necessary precaution. Intuition builds upon what the engineer knows, allowing imagination to lead. Intuition provides warning, even when convenient and available facts seem to suggest a certain approach. Human enhancements may seem to be beneficial, for example, when merely deductive reasoning is applied. However, the same enhancements may well raise "red flags" when the engineer's intuition becomes active.

One criterion for assessing the morality of human enhancement is how an approach differs from what is natural (e.g., a natural childhood). However, what is the difference between a "natural" future and an "engineered" future? This is the stuff of science fiction, but it is also a reality of contemporary society. Humans refuse to quietly succumb to the natural processes of disease, aging, loss of memory, limits to intelligence, physiological restrictions, and even seemingly superficial characteristics like "beauty."

Society has demanded enhancements throughout human history, with external adornments (e.g., tattoos, markings, piercings) and defenses (e.g., pharmaceuticals, hormones, and nutritional supplements). However, as most baseball fans can attest, enhancements can also be seen as something to be avoided. Many have argued that Barry Bonds' imminent phenomenal feat of surpassing the legendary Henry Aaron's lifetime homerun record should be asterisked. Bonds, in a number of baseball fans' and Congressional inquisitors' minds, somehow cheated by using physiological enhancements (allegedly using anabolic steroids or other hormonally active agents). Others see little difference between pharmaceutical enhancements and the out-of-the-ordinary training and nutritional regimens that separate professional athletes from the rest of us. For example, I am asthmatic and as such would have nowhere near the "normal" lung capacity and disease resistance allowed by my daily use of corticosteroids, which is not all that different biochemically from the drugs allegedly used by Bonds. So the act does not differ much, but the situations do, at least in the minds of many.

One could argue that all engineering is really about enhancement. In fact, one of the distinguishing aspects of our species is our ability to change things, usually for something we perceive to be "better." Julian Savulescu, founder and director of the Uehiro Centre for Practical Ethics at Oxford University and head of the Melbourne-Oxford Stem Cell Collaboration, has stated:

> What distinguishes us is our minds and our capacity to make decisions based on reason and for what we judge to be better. . . . Recently, medicine has sought to change evolution. It tries to enable us to have healthier lives. I think we are about to take a much bigger step, and that is to choose our evolution not just to be healthier beings but to be better beings, to be beings who can have a greater opportunity of having a better life.[34]

Engineering is a human endeavor. This begs the question as to when *Homo sapiens* started acting "human." Noted bioethicist Arthur Caplan from the Department of Medical Ethics at the University of Pennsylvania wonders when this happened and what makes us human:

> Is it fire? Is it metal? Is it speaking? Is it standing up? So we get into these battles about the properties that define us. Even if we go back, though, very simply, to say 10 000–12 000 years ago to our agriculturally using ancestors, if they were to see us today, living seventy-eighty years of age, riding around in cars, flying in the air, computing, talking through the airwaves like we're doing right now, they would think we were divine.[35]

While we show glimpses of divinity, we are not divine, at least not consistently.[36] Unfortunately, the same species that gave us Albert Schweitzer and Mother Teresa gave us Adolph Hitler and Joseph Stalin.[37] And, the latter two dictators notoriously supported "enhancements" (genocide and eugenics) for those humans they deemed worthy.

Genetics is a new frontier for human enhancement. Caplan argues:

> [W]e have just seen the mapping of the human genome. That's a marvelous accomplishment. We have figured out the location and placement of all of our genes that sit on our chromosomes. The next mission is, however, to figure out what they correlate with, what they control, what they contribute to, what they cause. That's an explosion of new understanding of the genetic or hereditary contribution to all of our traits and properties. We are going to see that available both to test ourselves, test our embryos, test our children, and say: are these traits or these behaviors going to develop or not? And, ultimately, we will be able to engineer them to actually try and genetically change them. So that is an unprecedented explosion of knowledge.[38]

But what types of checks and balances are needed against human enhancements? How far can research and practice go in advancing human enhancement systems and technologies before there is a sufficiently dire threat to present and future generations? The answers to these questions are at the heart of the current debates among scientists, policy makers, and the public.

We are familiar with anatomical enhancements, especially those enabled by plastic surgery. Engineers have played a major role in developing devices and systems to support such surgical procedures, albeit indirectly. Now, however, enhancement is becoming much more efficient, approaching what many conceive to be eugenics. For example, successful cloning could provide eugenic enhancement, by allowing parents to avoid the genetic defects that may arise when human reproduction is left to chance. Genetic manipulation could also preserve and perpetuate those traits that society holds dear (e.g., athletic prowess, intelligence quotients, and creativity). This could eventually include cloning to perpetuate genetically engineered enhancements; that is, we are moving toward a truly engineered future. Is society ready and are engineers specifically prepared for this brave new world?[39]

THE ETHICAL QUANDARY OF ENHANCEMENT

Are we going to engineer hard-working children?

Julian Savulescu

Here's another thought experiment:

You have a disease of the connecting tissue in your brain. You have two years to live. The good news is that right up to the end you will be able to maintain your physical and mental capabilities as they now exist. You will have no depreciation in your quality of life. However, your physician tells you a new therapy exists that would completely cure your disease. Your life expectancy would increase 30 years. There is, however, one complication. You would become a completely different person; no better or worse, just different. You would no longer be you and you would have no memory of the old "you."

Would you accept the therapy?

What is "personhood?" There has been some debate, for example, about personality effects resulting from treatment of neurological diseases, such as Parkinson's disease. Pharmaceuticals, especially psychotropic drugs, have been known to change drastically the thinking and personality characteristics of persons being treated. Thus, it is quite plausible that tissue implants, neurotechnologies, and other emergent techniques could have similar, possibly more dramatic, results.

Bioethics Question: Is it ethical to enhance future people?

In a way, parents are making decisions about the personhood, or at least the physical and mental characteristics, of their children. Does society have a place in encouraging an enhanced population in the future?

Consider the following excerpts from an exchange between two previously mentioned, leading bioethicists, Arthur Caplan and Julian Savulescu, moderated by Leonard Lopate:

CAPLAN: I am not, here in favor of *carte blanche* when it comes to where biotechnology/bioengineering might take us . . . . But let's not say, in principle, "our nature is what it is; we shouldn't touch it.

SAVULESCU: I think it's a difficult decision whether a drug or an enhancement is actually in somebody's interests. There is a recent study that showed that you can take monkeys that are naturally lazy and introduce the gene from the reward center into their brain and they become very hardworking.

LOPATE: It sounds like Aldous Huxley.[40]

CAPLAN: It sounds like Microsoft. (Laughter.)

SAVULESCU: Well, the question was: Are we going to engineer hardworking children? It's not clear to me that that would be in the child's interest. But there are others.

LOPATE: But some parents might want that, the parents who push their kids to get only As.

SAVULESCU: Well, that's right. But the question is whether it's in the child's interest. We were talking about principles, and I think the fundamental principle has to be: is this modification in the individual's interest?

There are other cases like impulse control. Impulse control, our ability to control our temper and our impulses, is strongly correlated with socioeconomic success and staying out of prison. We all know people with a hot temper, but even to milder degrees the ability to control your impulses will determine whether you stay in or out of prison to some degree. Now, if somebody has a very hot temper, I don't see the problem with offering that person or their parents the opportunity to calm that down. After all, we laud parents who try to educate their children to control their tempers. Why not do it biologically?

LOPATE: But isn't there also a trade-off usually, and the trade-offs are side effects, sometimes loss of other abilities? There may be impulse control on the one hand, but then this person is also a writer who finds out that he can't write anymore because something seems to have been lost.

SAVULESCU: We have to guard against hubris or excessive pride and arrogance in this area and a belief that we know what is best, and we ought to be very admitting of various different ways of living our lives, and I think we should be reluctant to use these things. But in the end we have to make a judgment based on a balance of the costs and the benefits.

The point that Arthur is making is exactly correct; we can't just say because there are risks, there are side effects, we should never consider this. We have to as rational beings make a judgment about what is best for us and our children.

Here are three very intelligent and thoughtful people trying to answer a deceptively simple question: "Should we engineer hard-working children?"

Note that question is not "Can we engineer hard-working children?" Obviously, we continue to get better at that. The question is "Ought we?" One way to start the argument is to address the risk trade-off mentioned by Lopate. Engineers often work within ranges of acceptability, such as when ample margins of safety, ample cost savings, and ample product utility are achieved in a single product. So, a first step would be to list the pros and cons. Let us name a few of each:

Potential Benefits of Enhancement:

1. Better performance translates into acceptance into better schools, which translates into better careers and better quality of life for the student.
2. Higher collective intelligence improves the pool of scientists, physicians, engineers, policy makers and other contributors to society.
3. The resulting improvements to the technical pool enhance the chances to solve some of society's biggest problems (e.g., diseases, war, pestilence, famine).
4. Parents will feel proud of their children's achievements.
5. Children will feel proud of their own achievements.

Potential Drawbacks of Enhancement:

1. Kids lose autonomy over their own lives.
2. Presently unknown side effects (e.g., neurological, psychiatric, personality, social) could show up later.

3. Presently unforeseen societal impacts can rise (e.g., what are ramifications of collective increase in intelligence quotient (IQ) and other metrics, including "IQ inflation?").
4. In addition to medical side effects, there are possible societal side effects (combination of increased intelligence and personality changes could translate into threats, e.g., increased incidence of sociopathologies).
5. Possible feeling that achievement is external to child (no "ownership" since the success was enhanced by a drug, device, or genetics).

Even if we could name every cost and risk, we are still left with the task of evaluating their importance. How, for example, do we put relative weight on social *versus* individual costs? This is a challenge for utilitarianism. Even if we subscribe to the belief that the best approach is to find the "greatest good" for society, as John Stuart Mill advises, is this really the best approach? Recall that even Mill, the utilitarian, advises this only if it does not render undue harm to any member of society.

Enhancement is complex. Few would argue that disease treatment and prevention are absolutely necessary and are worth many of the risks in advancing the state of science to achieve these enhancements. If there were an *in utero* device or treatment that can prevent a child's debilitating physical or psychological malady, for example, we would likely recommend it. But, does a child's sex or the color of a child's eyes warrant a similar interest? Does finding in an unborn child a trait considered undesirable by society (e.g., possible lower intelligence) invite the abortion of that child ("culling the herd" so to speak)? The central peril of "designer children" is well articulated by the President's Council on Bioethics[41]:

> Ironically, as we advance technologically, in some ways, we seem to become a more barbaric society.
>
> The problem has to do with the control of the entire genotype and the production of children to selected specifications.

The benefits of human enhancement are found in human health and well-being, whereas the actual and potential downsides involve individual risks, fair and equitable distribution of benefits, and the effects of changes to DNA on future generations. In fact, all types of human enhancement have a profound effect on the individual person and on society (Table 3.2). This can be likened to an engineering design problem. The "public" wants to continue to enhance performance in several selective areas, but such enhancement must be constrained by numerous conditions, including the respect for persons, justice, and beneficence. It is not simply a matter of choosing between outlawing and having free reign on human enhancement. For example, many enhancements are absolutely necessary, such as those that provide healthier bodies and minds. Thus, there are risks of not enhancing the present and next generation. These are opportunity risks. In other words, if we do not act soon to deal with some of these challenges, there is a cost to certain members of society.

SCIENTIFIC DISSENT

We like to think that since we are driven by the scientific method, scientists are as a group more rational and open-minded than most. However, history has shown that scientists can be quite closed-minded once they adopt a paradigm, as evidenced by the resistance against Copernicus, Galileo, Newton, and Einstein.

Table 3.2

Actual and perceived risks associated with selected human enhancements

Enhancement	Genetic (DNA) manipulation issues	Risk to individuals	Fairness issues	Opportunity risks	Macroethical issues
Preplantation genetic diagnosis (PGD)	Yes. Preselection of certain traits.	Yes. Implantation procedure carries risks to mother and child. Preselection of certain traits has also been known to cause birth defects and mishaps. "Slippery slope" wherein PGD leads to devaluing certain traits and increasing abortions of classes of people in an effort to give birth to the perfect, "designer" baby and nearly certain risk of countless embryos dying during scientific trials and utilization by the public.	Yes. Loss of human diversity if the practice becomes widespread; even if the practice does not become widespread, some sense of individuality is lost. Many argue that geneticists are playing God by manipulating a natural code of genes.	Yes. If disease is observed, the embryo could be treated (although a decision not to implant and to destroy the embryo is more likely). Some also argue that the preselection could make populations more intelligent, stronger, and more successful (although same arguments were made by Nazis for their "eugenics" programs].	Yes. Beginning of life (BOL) issues, for example, female embryos and fetuses have been destroyed in an attempt to have a male child. Tampering with genetics implies value judgment about perceived worth of people with specific traits, which translates into judgments about differential worth of individuals. Opportunity risk of losing population advantages of diversity (e.g., disease resistance, collective intelligence, nonmonetized benefits) akin to monoculture problems in agriculture (where the values of certain traits are not fully understood until after they are lost).
In vitro fertilization (IVF)	Yes. Reproductive efficiencies are manipulated in hopes of a very specific goal.	Yes. Change in ability to conceive may introduce new susceptible subpopulations (e.g., older women who cannot conceive naturally, but can conceive via IVF), which can increase overall risks (e.g., birth defect risks associated with mother's age).	Yes and no. Today, IVF is a widespread procedure, and many argue that children should not be solely available to those who are naturally able to conceive. The religiously motivated, the financially constrained, and those who see this as an unjust means (e.g., the overproduction of embryos) to a worthwhile end (i.e., parenthood) view IVF as unfair.	Possibly. Some may say that allowing certain individuals to reproduce could richen the gene pool. Also, couples with disease in their family history may choose this method to avoid the chance of a medical condition with the pregnancy.	There are macroethical issues at hand. Once again, many feel that IVF is playing God. By planting life, some would say that you are interfering with the natural order of reproduction.

(Continued)

Table 3.2
Actual and perceived risks associated with selected human enhancements—cont'd

Enhancement	Genetic (DNA) manipulation issues	Risk to individuals	Fairness issues	Opportunity risks	Macroethical issues
	Indirect genetic manipulation since the DNA itself is not altered, but genetic material is different if IVF were not available (unique person would not have existed).	As with other procedures dealing with genetics and the body, IVF carries a risk to the individual. There is the chance of discomfort, faulty operations, side effects, birth defects, and continued infertility.			"Slippery slope" toward exclusive implantation of "best" possible offspring (e.g., "successful" people's eggs and sperm).
Cosmetic surgery	No. Although a strict social Darwinian might argue that when individuals are more attractive (e.g., facial symmetry) they are artificially changing the gene pool since they have increased their likelihood of reproducing.	Yes. Attendant reconstructive surgical risks (e.g., mishaps, iatrogenic failures, and nosocomial infections), Body–mind identity risks.	Yes. Financial constraints on selection, e.g., most cosmetic surgery is not covered by health insurance. Some would also argue that certain cultures push cosmetic surgery upon people (such as in the media) and this causes unfair and unrealistic standards.	Yes. For example, if surgery is not performed on individuals in certain cultures, they would be outcast or even actively harmed. Also, once deemed more attractive, individuals often become more powerful and this causes certain benefits, such as financial and social successes.	There are definitely larger ethical issues at hand. By tampering with natural selection, one is prioritizing cosmetics over function, and this precedent is upsetting to many, since it would redefine the individual to fit ephemeral standards of beauty.

Performance enhancing drugs	No. Unless indirect, e.g., changes that a chemical compound (i.e., endocrine disruptor) induce in the hormonal system that lead to mutagenic responses and changes in DNA.	Yes. In addition to risk to the person taking the drugs, it may encourage their use in sensitive subpopulations, especially children. Specific risks of medication (increased heart rate, blood clotting, etc.). Performance enhancing drugs can have very harmful, synergistic reactions with other medications taken by the user.	Yes. This is the central issue of many discussions about athletic and academic performance (i.e., is it fair to others that a drug is used to enhance performance?) Safety and risk are directly related to medical monitoring, which is less available to certain disadvantaged groups. Conversely, these prescriptions are becoming more readily available, and similar chemical forms (such as energy drinks and over-the-counter nutritional supplements) are continuously developed.	Yes. When properly prescribed and used, certain drugs can enhance the quality of life (e.g., less pain, more mobility in ageing). Increased reliability, if "winning" is a measure of success. Medication can also increase sexual, physical, and mental capacity. This could result in career, financial, and personal successes (at least temporarily).	Yes. Redefining individual performance and success (e.g., athletic achievement). This issue has been heated, as professional athletes were using performing enhancing drugs to achieve unfair advantages.
Neurological and behavior altering drugs	No. Unless indirect, e.g., changes that a chemical compound (i.e., endocrine disruptor) induce in the hormonal system that lead to mutagenic responses and changes in DNA.	Yes. All drugs carry risk. Psychotropic drugs have the added risk of effects on personality. These effects may be exacerbated in children. Sometimes the "treatment" can be directed toward the personality characteristics rather than the "disease," e.g., "masculine" behavior in a classroom may be discouraged. Also neurological drugs can cause an abundance of side effects that must be treated by other costly medications.	Yes. Safety and risk are directly related to medical or educational system monitoring, which is less available to certain disadvantaged groups, e.g., auditing resources may be more limited in lower SES school populations. Also, some groups oppose medicating behavior with drugs, arguing that this alters a natural code of behavior predetermined by higher powers.	Yes. When properly prescribed and used, certain drugs can enhance the quality of life (e.g., properly diagnosed cases of attention disorders and hyperactivity). Certain drugs not only enhance quality of life, but also in diseases such as multiple sclerosis and asthma, a drug can lengthen and improve the quality of life.	Yes. Redefining individual mental performance (e.g., scholastic achievement). "Raising the bar" on all due to the enhancement of a few. Introducing a new "default" standard of performance, wherein one's natural way shifts to what society deems to be "better".

(Continued)

Table 3.2
Actual and perceived risks associated with selected human enhancements—cont'd

Enhancement	Genetic (DNA) manipulation issues	Risk to individuals	Fairness issues	Opportunity risks	Macroethical issues
Gene therapy	Yes. There is the risk of misclassification, e.g., classifying a characteristic as a "disease" that needs to be eliminated in a society (e.g., genes associated with aggressive behavior).	Yes. Treatment, whether *in utero* or subsequent, carries medical risks.	Yes. For example, certain SES growths may be less likely to receive the therapy. Certain racial or ethical characteristics could be "treated". Potentially threatens individuality. Potentially reduces human diversity.	Yes. Certain debilitating diseases can be or have the potential to be eliminated (e.g., cystic fibrosis and hemophilia). Quality and length of life can be improved.	Yes. The extent of gene therapy will provide both improved quality of life and loss of diversity. This means that the society must take care in assigning values and in deciding what is truly a disease or trait that must be treated. Also society must decide who should be treated, and to what extent makes someone eligible for gene therapy.
Extended longevity programs	Many extended longevity programs are not genetically altering. However, over time if the patients are receptive to the drugs, it is possible that indirect effect will be selection for a longer lifespan. While the mode of action is unknown, one may make use of genetic manipulation of the cell's telomere production to prevent apoptosis. This requires alteration of the cell's DNA.	Still premature to determine precise risk of extended longevity programs. As with any manipulation programs, there are risks that the patient will experience counter-indications and side effects (such as a higher blood pressure and organ damage). Possibly will result in medical failures in test subjects, including embryonic deaths, as a side effect of research to improve efficacy.	This is undoubtedly a drug that will be most accessible to a selected group of people. Could increase the gap that already exists between lower and higher socioeconomic strata.	Increasing life span could increase productivity and quality of life.	Questions about the "right" amount of time to live will arise. Ethics of tampering with the natural life cycle needs to be addressed.

Growth hormone therapy	Similar to medication for altering one's lifespan, can indirectly manipulate gene populations with time.	Associated with numerous adverse effects, e.g., heart and liver problems. In addition, undesired physical traits (e.g., increased body hair and masculinization in females) can cause physical and emotional discomfort.	Disproportionate availability to certain populations. However, some may argue that certain drugs prescribed to correct congenital problems (e.g., thyroid disorders) are fair as they restore quality of life.	One can become stronger, and restricting ailments can be corrected with this therapy. If increased height is a desired outcome, not using the therapy could present obstacles.	Growth hormone therapy has several ethical questions at hand. For example, many people will risk their health (or the health of their children) for materialistic benefits, such as increased sports performance. Other cases are less controversial, such as people suffering from thyroid problems.
Artificial intelligence	Does not directly manipulate genes. However, some believe that with time, people rely less on their own cognitive processes (already observed with "spell check" and search applications), with the concomitant negative effect on future populations.	Minimal physical risk to individual health. However, individuals can become dependent on artificial intelligence and this can affect their mental capabilities and well-being.	Artificial intelligence definitely has unfair advantages, as the availability of these resources is disproportionate, and directly associated with socioeconomic status.	Machines can save time, and this saves money. Artificial intelligence can aid in international communication and business successes.	Artificial intelligence can have some macroethical issues if it gives one an unfair advantage against others in major issues in life. Something such as landing a great job can enhance financial success and personal happiness.

This table is benefited from input from students of my Professional Ethics Seminar at Duke. The insights of Sebastian Larion, Madison Li, Lee Pearson, and Jamie Rudick were particularly valuable and have been incorporated into the table.
Note: SES = Socioeconomic status.
 DNA = Deoxyribonucleic acid.

As mentioned, Thomas S. Kuhn has captured this reticence in *The Structure of Scientific Revolutions* (University of Chicago Press, 1970). Basically, Kuhn argued that a scientific community needs a set of received tenets to practice its trade. These form the "educational initiation that prepares and licenses the student for professional practice" and they are necessarily "rigorous and rigid." Scientists can be very sensitive when they are challenged on how they see the world. So, "normal science" may violently oppose change and will have strong gatekeepers against new, especially opposing, ideas. Kuhn characterized the shift away from wrong ideas, even in the face of compelling evidence as "the tradition-shattering complements to the tradition-bound activity of normal science." This means that new ideas and a new ethos (Kuhn's "paradigms") call for a whole new set of assumptions and the reevaluation of prior facts. The established group does not like to admit when it is wrong, even in the face of confounding evidence.

Unfortunately, numerous tactics are used against dissenters, including those who do not hold the "consensus" view of their fellow scientists. Such tactics are often unfair and illogical, but they can be quite effective in stifling disagreements. Recall the discussions about gate keeping and Kuhn's contention that the scientific community resists change and Irving Janus' groupthink. Today, the biological research community largely favors embryonic stem cell research, for example, and this has had a chilling effect on dissent. The reasons are varied, such as seeing pro-life scientists as obstacles to the march of biomedical research advancement. The tactics may be used for even more banal reasons such as "following the money" (e.g., pushing for legislation to increase funding and competing for limited embryonic stem cell research dollars). Thus, the so-called pro-life scientists are dissenters in this debate. Here are a few of the tactics employed against dissent, with an example of now each is used against those opposed to stem cell research and in favor of personhood status of the unborn.

- *Ad homenin* attacks: Pro-life scientists are considered "far-right extremists." They are labeled as "anti-intellectuals." Even calling anti-stem cell scientists "pro-life" is code for these terms.
- Red herring: This is an intentional distraction away from the facts. A recent *Washington Post* article ran the headline, "House GOP Pushes New Abortion Limits." The article emphasizes that the House Republicans are "imposing new restrictions on abortion" and seeking to "impose incremental restrictions on abortion" while "averting a direct confrontation over women's constitutional rights to obtain the procedure." The "imposing" and "averting" actions are red herrings. The measures were being drawn against partial birth abortion. Yet people's attention is drawn away from the actual heinous acts that are being diverted toward the fact that Congress is "restricting."
- Argument by generalization: Broad conclusions about pro-life scientists are drawn from a small number of unrepresentative cases, such as the subtle innuendo that they must support abortion clinic bombings since violent extremists say the same things about the preciousness of life that antiabortion scientists do (of course this is absurd since the Roman Catholic Church, many Protestants, Moslems, Jews, and tribal religions say the same thing).

  Another example is the incest and rape argument. The implication that the reason for abortion on demand is needed overgeneralizes and exaggerates the relatively small number of actual instances. This is known to ethicists as "cherry-picking" or "argument by select reading."
- Argument from authority: Pro-choice doctors are the majority (allegedly), so they call the shots. The pro-abortionists are often perceived (at least they perceive themselves) to be the smartest people in the room.

- Two wrongs make a right: This is a charge of wrongdoing that is answered by a rationalization that others have sinned, or might have sinned. For example, I find that the cashier has given me a ten-dollar bill instead of the one-dollar bill that I should have received. I tell myself that it is okay to keep it, since most other people would also keep it. This seems to be the argument that even if a physician believes abortion to be immoral another doctor would not want to buck the system and advise a patient not to have an abortion. So, the doctor makes an immoral decision and rationalizes it, not on the basis of its morality, but on the basis that others would do the same.
- Pious fraud: This is a fraud that is undertaken for some "greater good" (utilitarianism). The biological scientists who know that they are using a deceptive definition of life, or of humans, or of a person are doing so because of some greater good (e.g., preventing overpopulation, keeping babies from being born into unhappy homes, advancing the state of medical science).
- Least plausible hypothesis: When scientist ignore all of the most reasonable explanations of what a human person indeed is, this makes their desired definition the "only" definition. Rather than simply referring to the same definition of a human, otherwise talented scientists Watson, Crick, Sagan, and others (and bioethicists like Peter Singer) do mental gymnastics to come up with some harebrained and dangerous definitions of "human." Ethicists apply the "Ockham's razor" in such arguments, which requires that the simplest is the best. Scientists also refer to "the most parsimonious," suggesting that a reasonable explanation is preferred to a more intricate and rare possibility. As mentioned, medical practitioners are counseled to expect first a horse and not a zebra when they hear the sounds of hoofs. It is only after the reasonable explanations are exhausted that the complicated ones are needed.
- Argument *ad nauseum*: If they keep making the same argument, everyone including the biologists themselves starts believing it. The pro-choice people have managed to define the fetus as something other than human so often that it is the mantra of many scientists.

These arguments, when applied to bioethics, can be highly ironic. For example, many of the same groups (e.g., American Medical Association (AMA)) who argue for applying the "precautionary principle" to risk assessment, such as in cancer prevention or global climate change, do not apply this basic principle to research related to human life. This principle tells us that "where there are threats of serious or irreversible damage, lack of full scientific certainty shall not be used as a reason for postponing" measures to prevent a problem (from Principle 15 of the Rio Declaration, United Nations, 1992). This is akin to numerous ethical constructs, including Kant's categorical imperative, Mill's harm principle (analogous to the Hippocratic oath), and Rawl's veil of ignorance. If we applied this reasoning regarding the protection to human life, is the scientific community completely certain that an embryo is not a human being (of course, many of us are strongly convinced that an embryo is a human being)? The principle is an attempt to prevent making a tragic mistake with profound consequences. Many argue that the consequences of devaluing human life are already dreadfully obvious.

Here is the tough lesson for bioethics and especially for engineers working in organizations. Gatekeeping sounds like a utilitarian approach and, in fact, it is. The benefits of the ends (scientific advances) justify even somewhat unscrupulous or even unscientific means (gatekeeping to stifle dissent). It is a beguiling and treacherous path for scientific types who are naturally drawn to conclusions, ethical and otherwise, based on numerical and quantitative information. That is how we have been trained. But this approach can be woefully flawed in logic. Steven Kelman of Harvard University[42] points out these flaws using a logical technique of *reductio ad absurdum* (from Greek, "reduction to the impossible"). An assumption is made for the sake of argument, and a result found, but it is so absurd that the original assumption must have been wrong.[43] The ethically right thing to do is not necessarily the one that we would have expected. Foremost, it may violate one of our cherished principles. For one thing, gatekeeping can work against the advice from Feynman and Snow that scientists must be honest, or at least tell the truth, at all times.

Philosophers, from Socrates up to modern times, have agreed that there are a few ethical principles held by nearly every society, and that there really are no new principles, only new ways of applying them. The moral underpinning of cultures is amazingly similar throughout space and time. This seems to support the view of moral objectivism over that of moral relativism, at least for the principles that matter. For example, it would be absurd to think of a society with principles other than those shared by most of us, such as one that believed dishonesty is preferable to honesty, cowardice is better than courage, or that injustice is better than justice. Modern philosophers are more divided than the Ancients about the source of these principles of moral obligation. Up to a century or so ago, most philosophers held that all ethical principles are objective, i.e., they are discovered. Many modern philosophers see them as socially-derived, invented and subjective. Most modern intellectuals consider the pre-modern view that principles are universal and absolute to be dogmatic and narrow-minded.[44]

We must ask, then, are ethical principles created, like the arts, and or they mentally discovered, like the sciences? Engineering stands at a crossroads between these two perspectives. After all it is both an art and a science. So, to some extent at least, engineering ethics is also both artificial (invented) and natural (discovered). But, the ethical good cannot be derived exclusively using the scientific method.

Kelman gives the example of telling a lie. Using the pure quantitatively based benefit–cost ratio, if the person telling the lie has much greater satisfaction (however that is quantified) than the dissatisfaction of the lie's victim, the benefits would outweigh the cost and the decision would be morally acceptable. At a minimum, the effect of the lie on future lie-telling would have to be factored into the ratio, as would other cultural norms. Another of Kelman's examples of flaws of utilitarianism is the story of two friends on an Arctic expedition, wherein one becomes fatally ill. Before dying, he asks that the friend return to that very spot in the Arctic ice in ten years to light a candle in remembrance. The friend promises to do so. If no one else knows of the promise and the trip would be a great inconvenience, the benefit–cost approach instructs him not to go (i.e., the costs of inconvenience outweigh the benefit of the promise because no one else knows of the promise).

These examples point the fact that quantitatively-based information is valuable in bioethical decision making, but care must be taken in choosing this information and in giving appropriate weights to it compared to other subjective and non-quantifiable information. Otherwise, the decks are stacked against dissent, even when it is a morally superior or even obligatory position.

NOTES AND COMMENTARY

[1] E.O. Wilson, *Consilience, The Unity of Knowledge* (New York, NY: Alfred A. Knopf, 1998).

[2] T.S. Kuhn, *The Structure of Scientific Revolutions* (Chicago, IL: University of Chicago Press, 1962).

[3] For example, in Judeo-Christian scripture, this admonition appears in both the New and the Old Testament." The Prophet Isaiah (6: 9–10) warns:

Also I heard the voice of the Lord, saying, whom shall I send, and who will go for us? Then said I here am I; send me and he said, go, and tell this people:

Hear ye indeed, but understand not; and see ye indeed, but perceive not; Make the heart of this people fat; and make their ears heavy, and shut their eyes; lest they see with their eyes and hear with their ears, and understand with their heart, and convert, and be healed.

Jesus Christ later quotes this passage in seeming frustration with the disciples (Mark 8: 16–21). Sensing (e.g., hearing and seeing) is a necessary, but not sufficient, component of a paradigm shift. It must translate into a more complete understanding of the new process. Sometimes this is referred to as "thinking outside of the box."

[4] I realize that I am inviting some angry letters from mathematicians. This is not meant as a criticism, but as a distinction between the focus of the practitioner and that of the inquirer. Both foci are absolutely necessary. However, engineers are seldom permitted the luxury of waiting 350 years for a "proof" and often must be empirical and adaptive. However, as soon as a new paradigm is proven, or at least evidence shows it is superior to existing approaches, the engineer must be open minded to accept the change.

[5] National Academy of Engineering, *The Engineer of 2020: Visions of Engineering in the New Century* (Washington, DC: National Academy Press, 2004).

[6] D.E. Stokes, *Pasteur's Quadrant* (Washington, DC: The Brookings Institution, 1997).

[7] J. Fernandez, "Understanding Group Dynamics," *Business Line*, 2 December 2002.

[8] Fernandez, *Business Line*.

[9] Stokes, *Pasteur's Quadrant*, 6.

[10] Ibid., 12.

[11] Ibid.

[12] Ibid., 10–11.

[13] Ibid., 18–21.

[14] Brooks, *Applied Science and Technological Progress*, p. 1706.

[15] Stokes, *Pasteur's Quadrant*, 70–3.

[16] I. Janus, *Groupthink: Psychological Studies of Policy Decisions and Fiascoes*, 2nd ed. (Boston, MA: Houghton Mifflin Company, 1982).

[17] See Thomas Hobbes' *Leviathan*.

[18] I hesitate to bring this up, but if you are an Internet user, you have likely been "spammed" with offers for sexual "enhancements," especially increasing penis size. What keeps you from responding (besides a well-warranted fear of computer viruses and worms)? Humans are naturally wary of such claims.

[19] This section is based on research and narrative by Rayhaneh Sharif-Askary, a gifted Duke undergraduate student. Her interests in how various professionals perceive the world began in my Ethics in Professions course. She wanted to continue this pursuit after the course, so she gathered a wealth of information from numerous sources, including interviews. She found many similarities, but a number of differences in how engineers' perceptions *versus* those of other professions or *versus* those of the nontechnical "lay" public.

[20] Named after Italian philosopher-economist Vilfredo Federico Damaso Pareto (1848–1923), whose interests and contributions included income distribution and personal choice. His academic training was in engineering and his work in Pareto efficiency helped form microeconomic theory.

[21] J.B. Bassingthwaighte, The Physiome Project: The Macroethics of Engineering toward Health, *The Bridge* 32, no. 3 (2002): 24–9.

[22] National Academy of Engineering.

[23] H. Petroski, To Engineer Is Human: The Role of Failure in Successful Design (New York, NY: St. Martin's Press, 1985).

[24] Incidentally, researchers are usually proficient at justifications. In this case, they would likely argue that although this person is the only one we know of who would benefit, the research would advance the state of knowledge and would likely benefit numerous others in the future.

[25] The discussion on intuition draws upon R.M. Hogarth, *Educating Intuition* (Chicago, IL: University of Chicago Press, 2001).

[26] Ibid. and K. Hammond, *Human Judgment and Social Policy: Irreducible Uncertainty, Inevitable Error, Unavoidable Injustice* (New York, NY: Oxford University Press, 1996).

[27] National Academy of Engineering.

[28] For example, see S.B. Billatos and N.A. Basaly, *Green Technology and Design for the Environment* (London, UK: Taylor & Francis Group, 1997).

[29] For example, see D.E. Stokes, *Pasteur's Quadrant* (Washington, DC: Brookings Institute Press, 1997); and H. Brooks, "Basic and Applied Research," in *Categories of Scientific Research* (Washington, DC: National Academy Press, 1979), 14–18.

[30] This is a typical way that scientists report information. In fact, there may be people who, if they put on shorts, will want to eat ice cream, even if the temperature is $-30°$. These are known as "outliers." The term outlier is derived from the prototypical graph that plots the independent and dependent variables (i.e., the variable that we have control over and the one that is the outcome of the experiment, respectively). Outliers are those points that are furthest from the line of best fit that approximates this relationship. There is no standard for what constitutes an outlier, which is often defined by the scientists who conduct the research, although statistics and decision sciences give guidance in such assignments.

[31] A. Bradford Hill, "The Environment and Disease: Association or Causation?", Proceedings of the Royal Society of Medicine, *Occupational Medicine* 58 (1965): 295.

[32] At the 2006 Joint Conference of the International Society of Exposure Analysis and the International Society of Environmental Epidemiology in Paris, France (2 September through 6 September 2006), a number of speakers stressed the importance of not using Hill's criteria as anything more than a guide. In fact, it was pointed out that Hill did not refer to them as "criteria." This is particularly important when advising policy makers on such controversial and precedent-setting issues, such as the precautionary principle.

[33] This section is based on research and narrative by Duke undergraduate student Rayhaneh Sharif-Askary.

[34] Interview on the *Leonard Lopate Show*, New York Public Radio, 18 November 2004.

[35] Ibid.

[36] The question of when humanity rises is not simply anthropological but also spiritual and religious. For example, most faith traditions expound a set a circumstances where humans become distinguished from other species. One Native American story shows that at some point in time humans were separated by a great growing gorge. Incidentally, as a dog lover this is one of my favorites since it also addresses human–canine affinity. In this story, as humankind was being separated by the chasm and just before all of the species were moved away, the dog jumped across to join mankind.

Judeo-Christian scripture (Genesis 3) recounts the beginning as when the first true humans, Adam and Eve, partake of the fruit of the tree of knowledge of good and evil. Roman Catholic tradition (*Catechism of the Catholic Church*, Paragraph 390) asserts that "the account of the fall in Genesis 3 uses figurative language, but affirms a primeval event, a deed that took place at the beginning of the history of man. Revelation gives us the certainty of faith that the whole of human history is marked by the original fault freely committed by our first parents." While this is traditionally considered to be the "fall" of mankind (e.g., in John Milton's *Paradise Lost*), it is also a distinctive demarcation of the "rise" of humanity. Humans gained a unique perspective among creatures, our awareness of what is right and wrong; that is, the essence of ethics. Almost simultaneously, this awareness made obvious the need to improve our condition. One may argue, then, that ethics and engineering were born as twins.

37 One of the great ironies and mysteries of the twentieth century is how two men, Schweitzer and Hitler, could have grown up in the same time, place, and culture, yet one was to become an icon of goodness and the other an icon of evil.

38 Ibid.

39 This is the title of Aldous Huxley's prescient 1932 book. The book posits questions about the ethics of human cloning as a means toward human enhancement. It also considers societal themes such as how human minds can be manipulated by outside sources such as the media, government, and peers and whether this is always bad, steps a government may take to keep order among its people (e.g., social contract theory) especially distinguishing between protection and personal freedom, the values of conforming *versus* diversity, and the dangers and advantages of psychological conditioning. These are all fertile ethical topics (see next endnote).

40 Lopate is referring to the novel by Huxley, *A Brave New World* (New York, NY: Harper, 1932). The novel, written just before the rise of Nazi totalitarianism in the mid-1930s, explores modern topics that are being debated today, including reproductive technologies, genetic engineering, behavioral conditioning, and psychotropic drug therapy, especially Huxley's invention, *Soma*. Natural reproduction in the womb has been completely replaced by an industrial system, where embryos are manipulated along an assembly line to produce various castes of *Homo sapiens*. These castes are ranked in a manner similar to that of a university grading system; from the highly intelligent Alpha Pluses descending to the quite dull Epsilons. However, each caste member is conditioned to be satisfied with his or her station in society (menial work for the Epsilons and academic research for the Alphas). Away from work, society is highly hedonistic, addicted to the pleasure-inducing soma. Interestingly, they are also highly materialistic and consume goods constantly. As such, Henry Ford is considered the new god; even the dating system is based on him. The novel, for instance, begins in the year 632 A.F. ("After Ford"). Away from work, people spend their lives in constant "pleasure." The hedonism includes unencumbered sex. In fact, the Brave New World social contract encourages it, but committed love, marriage, and parenthood are considered to be obscene and are forbidden by the totalitarian rules.

41 The President's Council on Bioethics, "Human Cloning and Human Dignity: An Ethical Inquiry," (Washington, DC: 2002), 105.

42 S. Kelman, "Cost–Benefit Analysis: An Ethical Critique," *Regulation* 5, no. 1 (1981), 33–40.

43 This is also known as proof by contradiction.

44 See, for example: P. Kreeft, *What Would Socrates Do? The History of Moral Thought and Ethics* (New York, NY: Barnes and Nobles, 2004).

Chapter 4

The Bioethical Engineer

Design professionals share many things in common. They have undergone rigorous academic and professional training. They share numerous technical interests. The public perceives their competence and grants them responsibilities not given to other members of society. They form a community of scholars, and they are held to strict codes of behavior.

PROFESSIONAL TRUST

Since the professional does profess, he asks that he be trusted. The client is not a true judge of the value of the service he receives; furthermore, the problems and affairs of men are such that the best of professional advice and action will not always solve them.... The client is to trust the professional; he must tell him all secrets which bear upon the affairs in hand. He must trust his judgment and skill.

Everett C. Hughes (1988)[1]

What distinguishes professions like engineering, medicine, and law from nonprofessional careers? There is an ongoing debate, sometimes useful, sometimes not, about what it means to be a professional. If 100 people were asked what it means to be a professional it is likely to lead to 100 different answers.

In an undergraduate professional ethics course at Duke, I have queried students about what it means to be a professional and how the label "professional" should be applied. To start, we begin with a brainstorming session. I stand at the whiteboard and simply record the careers that are named by the students. The class is always a mix of engineering students with arts and science students. Each student is given an opportunity to name a career. The career mentioned is not discussed, just written on the board as named. After a few minutes, the recommendations begin to dissipate. At this point, we stop recommending and begin discussing the reasons each career was listed. Interestingly, the first criterion is often competence, followed by education. It is not necessarily the importance but the difficulty of the career (getting there and staying there) that seems to be the reason competence and educational requirements are included. For example, if the career choice takes two years and an associate degree compared to specialized baccalaureate degree followed by practice, the latter is more likely to be considered a profession. When asked to expand on whether a plumber, mechanic, or technician has to

140 Biomedical Ethics for Engineers

be very competent and knowledgeable, some students begin to consider them to be professionals, but most reconsider the criteria for professions. After some give and take, a few careers are unanimously considered to be professions: physicians; attorneys; engineers; and architects. Others are considered by most of the students to be professionals: clergy; accountants; and nurses. One group that is particularly divisive is the military officers. After some discussion, however, about the Uniform Code of Military Justice,[2] and particularly the stringent requirements for gaining and keeping a military commission, a number of students consider military officers to be professionals.

Many of the qualities of professional careers are shared by other jobs (Table 4.1). There are some essential aspects, however, that do seem to set professional careers apart from others. Prominent among these is that a professional is someone who has been granted a great deal of trust from the public, owing to that person's competence, skills, and ethics, to carry out a certain set of practices. This trust-granting by society is often exclusive, e.g., only board-certified physicians may treat most illnesses professionally, only members of the bar may represent people in court, and only professional engineers (PEs) may stamp a design of a building. A professional is not necessarily smarter or even more skillful than a person in a nonprofessional field. It just so happens that some areas that are greatly needed by society are also very complicated and difficult for most of us to understand completely. I am not a medical doctor, nor am I an attorney, but from time to time, I need both of them. Because these fields are professional fields, I do not really need to know anything at all about how they work. I just need to show up and be

Table 4.1
Essential and desirable attributes of professionals and nonprofessional career specialties

Professional Attribute	A	B	C	D
Trust of those receiving service	♦			♦
Accepting complete responsibility for successful outcome to the service provider	♦			♦
Competence in area of specialization	♦		♦	
College degree	♦		?	?
Licensure/certification	♦			
Accreditation of Undergraduate Program	♦			?
Respect for fellow members	?	♦		?
Mentoring by senior members (residences, internships, minimum time advised by licensed professional)	♦			
Graduate degree		♦	?	
Continuing education requirements	♦			♦
Code of ethics and standards of professional practice	♦			♦
Self-regulation and enforcement of practice	♦			?
Review boards/panels to oversee practitioner performance and behavior, especially for complaints	♦			

♦ = Attribute is present and expected by clients.
A, essential to professional; B, desirable to professional; C, essential to nonprofessional; D, desirable to nonprofessional; ?, applies to some but not all career specialties.

a willing patient or client of their services. Thus, the buyer's caveat, i.e., *caveat emptor*, does not hold. Instead, for professionals, there is an expectation of trust, i.e., *credat emptor*, on the part of the patient or client. Engineers fall into the trust category.[3]

Bioethics plays a part in numerous professions. In particular, engineers and physicians are helping professionals, whose work is expected to enhance the quality of life. The care for the public's health and safety called for in the engineering code is an example of the need for bioethical sensibilities. This is also true for a number of nonprofessional specialties. For example, wildlife managers are not necessarily considered to be "professionals" in the same sense that physicians and engineers are, but their work results in the protection and enhancement of ecosystems that improve the quality of life of millions of people. Furthermore, the wildlife manager, forester, and ecologist engage in bioethical decisions throughout their careers, addressing concerns about biological diversity and sustainability. Therefore, the attributes in Table 4.1 each have elements directly or indirectly related to bioethics. Among these, arguably trust of the individual and the collective public is paramount.

Contemporary society demands many things from its medical, engineering, legal, and other professionals, but for those in technical fields two expectations are prominent and unviolable: trust and competence. These expectations are built into the codes of practice and ethics of each technical discipline. The work of technical professions is both the effect and the cause of modern life. When undergoing medical treatment and procedures, people expect physicians, nurses, emergency personnel, and other health care providers to be current and capable. Society's infrastructure, buildings, roads, electronic communications, and other modern necessities and conveniences are expected to perform as designed by competent engineers and planners. But how does society ensure that these expectations are met? Much of the answer to this question is that society cedes a substantial amount of trust to a relatively small group of experts, the professionals in increasingly complex and complicated disciplines that have grown out of the technological advances that began in the middle of the twentieth century and grew exponentially in its waning decades.

Professions are not neatly subdivided as they once were. A visit to the hospital shows that not only do many of the physicians specialize in particular areas of medicine (e.g., neuromedicine, oncology, and geriatrics), but all of these physicians must rely on chemists, radiologists, and tomographic experts to obtain data and information about their patients (e.g., from serum analysis, magnetic resonance and CT scans, and sonography). In fact, many of the solutions (cures?) to health problems require an intricate cacophony and harmony among doctors, biomedical engineers, and technicians. For example, a drug delivery system requires the understanding of the biochemical needs of the patient, the fluid mechanics of the pharmacology, and the actual design of the apparatus. This is a continuum among science, engineering, and technology (Figure 4.1). So, every permutation is possible:

- Science is applied as engineering, which leads to technologies.
- Engineering frontiers push the need for scientific research, which calls for new technologies.
- New technologies increase scientific understanding (e.g., better diagnostic equipment), which leads to improved engineering (e.g., new clinical engineering designs).

Bioethics operates between the realm of the clearly wrong and the clearly right. Perception can differ from reality. Technical facts may be, intentionally or unintentionally, exaggerated or hidden. Like so many engineering concepts, place and time are crucial. The right manner of saying or writing something in one situation may be wholly inappropriate in another. The words we use are crucial. The first place to find the words of our profession is in the codes of ethics.

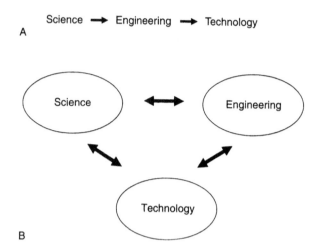

Figure 4.1 Relationships between science, engineering, and technology. The traditional view (A) is that basic research leads to applied research that in turn is adopted by engineers who develop necessary technologies. In many systems, the process can begin at any point (B). New technologies can lead to advances in scientific research and engineering. Engineering can lead to basic and applied research. In biological sciences, engineering is in the pivotal position of leading to scientific and technological advances.

CODES OF ETHICS: WORDS TO LIVE BY

> All engineering projects are communal; there would be no computers, there would be no airplanes, there would not even be civilization, if engineering were a solitary activity. What follows? It follows that we must be able to rely on other engineers; we must be able to trust their work. That is, it follows that there is a principle which binds engineering together, because without it the individual engineer would be helpless. This principle is truthfulness.
>
> Joseph Bronowski[4]

The eminent scientist Bronowski correctly captured and characterized engineering as a collaborative effort with a shared sense of purpose. Engineering practice can be likened to the invention process. We may be able to identify the inventor, but the application, improvement, and adaptations of new ideas radiate from the invention. The only way that collaborations can work effectively is that they be fact based and honest. This leads to trust. The trustworthiness of the practicing engineer within the engineering community must first be met before the engineering profession can expect to be trusted by our clients and the public. One of the key means by which the engineering profession articulates the elements of this trust is by its codes of ethics.

Professionals must codify the expectations of their members. These codes are broad statements that provide guidance. As such, they provide little in the way of details on how a practitioner or researcher should behave and even less on possible extenuating circumstances in a given situation. There have been numerous attempts to write specific codes. For example, the Biblical book of Leviticus contains a great number of specific details on proper Judaic behavior. And, approximately 4000 years ago in the kingdom of Babylon, the concept of professionalism and professional risk was extended to the physical well-being of the "engineer" himself and his family. The Code of Hammurabi, Babylon's sixth rule, lays the guidelines for the consequences of engineering failure (see Discussion Box: The Code of Hammurabi). These rules were predominantly based in an "eye for eye" reciprocation type of fairness.

Professional codes, including those for engineering disciplines, serve a number of important functions. They provide:

1. A collective recognition by members of their obligations.
2. An articulation that ethical behavior is the norm.
3. A guide or reminder in specific situations that could be hazardous to the individual practitioner's career.
4. An update on the status of the profession regarding when to develop and modify the code of ethics (most recently, sustainability was added to the American Society of Civil Engineers code, but current dialogues about macroethics are ongoing).
5. An educational tool, driving discussion in classes and professional meetings.
6. An indication to the public that the profession is seriously concerned with responsible, professional conduct.[5]

However, there are a number of ways in which simple adherence to the codes falls short in ensuring responsible conduct.

Discussion Box: The Code of Hammurabi

Hammurabi reigned over the Babylonian Empire from 1792 BC until he died in 1750 BC. He is best known as the author of the Code of Hammurabi. Whether the Code was functional during Hammurabi's time is subject to debate. It may merely have been a monument to the prince. However, the Code eventually became the subject of study under Babylonian law.

A total of 282 separate codes were enumerated. Of these, a number were precursors to modern professional codes, including medical and engineering codes of ethics. Like the modern codes, they addressed both intent and consequences although the latter were emphasized. A few are listed below:

53

If any one be too lazy to keep his dam in proper condition, and does not so keep it; if then the dam break and all the fields be flooded, then shall he in whose dam the break occurred be sold for money, and the money shall replace the corn which he has caused to be ruined.

55

If any one open his ditches to water his crop, but is careless, and the water flood the field of his neighbor, then he shall pay his neighbor corn for his loss.

The top of the stele (a stone or wooden slab, generally taller than it is wide, erected for commemorative purposes) depicting the Code of Hammurabi.

Photo credit: Fritz Milkau (1859–1934).

56

If a man let in the water, and the water overflow the plantation of his neighbor, he shall pay ten gur of corn for every ten gan of land.

215

If a physician make a large incision with an operating knife and cure it, or if he open a tumor (over the eye) with an operating knife, and saves the eye, he shall receive ten shekels in money.

216

If the patient be a freed man, he receives five shekels.

217

If he be the slave of some one, his owner shall give the physician two shekels.

218

If a physician make a large incision with the operating knife, and kill him, or open a tumor with the operating knife, and cut out the eye, his hands shall be cut off.

219

If a physician make a large incision in the slave of a freed man, and kill him, he shall replace the slave with another slave.

220

If he had opened a tumor with the operating knife, and put out his eye, he shall pay half his value.

221

If a physician heal the broken bone or diseased soft part of a man, the patient shall pay the physician five shekels in money.

222

If he were a freed man he shall pay three shekels.

223

If he were a slave his owner shall pay the physician two shekels.

224

If a veterinary surgeon perform a serious operation on an ass or an ox, and cure it, the owner shall pay the surgeon one-sixth of a shekel as a fee.

225

If he perform a serious operation on an ass or ox, and kill it, he shall pay the owner one-fourth of its value.

226

If a barber, without the knowledge of his master, cut the sign of a slave on a slave not to be sold, the hands of this barber shall be cut off.

227

If any one deceive a barber, and have him mark a slave not for sale with the sign of a slave, he shall be put to death, and buried in his house. The barber shall swear: "I did not mark him wittingly," and shall be guiltless.

228

If a builder build a house for some one and complete it, he shall give him a fee of two shekels in money for each sar of surface.

229

If a builder build a house for some one, and does not construct it properly, and the house which he built fall in and kill its owner, then that builder shall be put to death.

230

If it kill the son of the owner the son of that builder shall be put to death.

231

If it kill a slave of the owner, then he shall pay slave for slave to the owner of the house.

232

If it ruin goods, he shall make compensation for all that has been ruined, and inasmuch as he did not construct properly this house which he built and it fell, he shall re-erect the house from his own means.

233

If a builder build a house for some one, even though he has not yet completed it; if then the walls seem toppling, the builder must make the walls solid from his own means.

234

If a shipbuilder build a boat of sixty gur for a man, he shall pay him a fee of two shekels in money.

235

If a shipbuilder build a boat for some one, and do not make it tight, if during that same year that boat is sent away and suffers injury, the shipbuilder shall take the boat apart and put it together tight at his own expense. The tight boat he shall give to the boat owner.

274

If any one hire a skilled artisan, he shall pay as wages of the . . . five gerahs, as wages of the potter five gerahs, of a tailor five gerahs, of . . . gerahs, . . . of a ropemaker four gerahs, of . . . gerahs, of a mason . . . gerahs per day.

LIMITATIONS OF CODES OF ETHICS

The codes of the various engineering professions are quite similar, and all provide a first line of defense when ethical questions arise. With enough diligence and information, most ethical problems can be solved using the basic premises of the codes. At times, however, the codes are unclear about the details of circumstances, or sometimes the codes give contradictory answers.

The codes fail to define the referenced "public" to whom the engineers owe primary responsibility. For example, suppose a certain pesticide is shown to be a carcinogen and is banned in Canada and the United States. Chemical manufacturing companies often have huge financial investments in the production of new pesticides and they fear wasting this investment just because the regulating agencies do not give them permission to use the product in Canadian and American agriculture. Company decision makers may then be tempted to sell the product overseas where no such bans exist, and the sale would be perfectly legal. A company engineer working on the pesticide project must make an ethical decision. Who is the engineer's public? Decision makers often shift risks from one group of people to another (see Risk Shifting: Organochlorine Pesticides).

Risk Shifting: Organochlorine Pesticides

Often, as the old saying goes, "where you stand depends on where you sit." For example, on the whole, a larger percentage of older Americans tend to have much more positive impressions of the molecule 1,1,1-trichloro-2,2-bis-(4-chlorophenyl)-ethane, best known as DDT, than do today's younger generations. DDT, introduced during World War II, had an impressive record of reducing infectious diseases and of destroying human parasites such as lice. Yet today DDT is generally condemned as a threat to health and the environment.

So who is right? Is DDT "good" or "bad"?[6] When I ask this question in various forums, I have noticed that younger respondents are prone to categorize DDT as "bad." One of the influences they mention is Rachel Carson's seminal work, *Silent Spring*,[7] which exemplified the negative change in thinking about organochlorine pesticides in the 1960s, particularly that these synthetic molecules were threats to wildlife, especially birds (hence the "silent" Spring), as well as to human health (particularly cancer).

DDT

Conversely, most of these students are temporally removed from when Allied troops were devastated by tropical vector-borne diseases like malaria, yellow fever, and typhus. When queried, they are unlikely to know that the chemist Paul H. Müller won the 1948 Nobel Prize for Physiology or Medicine for synthesizing DDT. To many, his discovery was one of the major triumphs of the twentieth century. But, Müller, in spite of the fanfare, was circumspect about his discovery, especially as it was a harbinger of a greater reliance on chemical biocides. In his 1948 acceptance speech, he was prescient in articulating the seven criteria for an ideal pesticide:[8]

1. Great insect toxicity.
2. Rapid onset of toxic action.
3. Little or no mammalian or plant toxicity.
4. No irritant effect and no or only a faint odor (in any case not an unpleasant one).
5. The range of action should be as wide as possible, and cover as many Arthropoda as possible.
6. Long, persistent action, i.e., good chemical stability.
7. Low price (= economic application).

Thus, the combination of efficaciousness and safety was, to Müller, the key. Disputes between the pros and cons of DDT are interesting in their own light. The environmental and public health risks *versus* the commercial benefits can be hotly debated. Today's students are rightfully concerned that even though the use of a number of pesticides, including DDT, has been banned in Canada and the United States, we may still be exposed by importing food that has been grown where these pesticides are not banned. In fact, Western nations may still allow the pesticides to be formulated at home, but not allow their application and use. However, the pesticide comes back in the products we import, known as the "circle of poisons."

Arguments of risks *versus* risks are arguably even more important. In other words, it is not simply a matter of taking an action, e.g., banning worldwide use of DDT, which leads to many benefits, e.g., less eggshell thinning of endangered birds and less cases of cancer. No, what it sometimes comes down to is trading one risk for another. Since there are yet to be reliable substitutes for DDT in treating many disease-bearing insects, policy makers must decide between ecological and wildlife risks and human disease risk. Also, since DDT has been linked to some chronic effects like cancer and endocrine disruption, how can these be balanced against expected increases in deaths from malaria and other diseases where DDT is part of the strategy for reducing outbreaks? Is it appropriate for economically developed nations to push for restrictions and bans on products that can cause major problems in the health of people living in developing countries? Some have even accused Western nations of "eco-imperialism" when they attempt to foist temperate climate solutions onto tropical, developing countries. That is, we are exporting fixes based upon our values (anticancer, ecological) that are incongruent with the values of other cultures (primacy of acute diseases over chronic effects, e.g., thousands of cases of malaria compared to a few cases of cancer and to threats to the bald eagle from a global reservoir of persistent pesticides).

Sound engineering demands finding an optimal solution within specified constraints. In this instance there are at least five outcomes:

1. Acute effects from malaria and other vector-borne diseases
2. Acute health effects (e.g., poisonings) from the use of pesticides
3. Chronic health effects (e.g., cancer) from the use of pesticides

4. Acute ecological effects (e.g., loss of diversity and changes in relative abundance of certain species) from the use of pesticides

5. Chronic ecological effects (e.g., eggshell thinning of predatory birds).

Thus, to determine whether to use DDT, policy makers must decide not only whether to ban DDT but must consider what the ban would cost in terms of death and disease and whether these costs can be addressed and mitigated. For example, can suitable substitutes be found, or can the vectors be addressed without using pesticides (e.g., removing mosquito breeding areas)? Finding substitutes for chemicals that work well on target pests can be very difficult. In fact, the chemicals that have been formulated to replace DDT have been found to be either more dangerous, e.g., aldrin and dieldrin (which have also been subsequently banned) or much less effective in the developing world (e.g., pyrethroids). For example, by spraying DDT in huts in tropical and subtropical environments, fewer mosquitoes are found compared to those found in untreated huts. This likely has much to do with the staying power of DDT in mud structures compared to the higher chemical reactivity of pyrethroid pesticides.

In another example of risk shifting, the Allied Chemical Company had operated a pesticide formulation facility in Hopewell, Virginia, since 1928. The Hopewell plant had produced many different chemicals over its operational life. Reflecting the nascent growth of petrochemical revolution in the 1940s, the plant began to be used to manufacture organic[9] insecticides that had been invented recently, DDT being the first and most widely used. In 1949 the company started to manufacture chlordecone (trade name Kepone), a particularly potent herbicide that was so highly toxic and carcinogenic (Table 4.2) that Allied withdrew its application to the Department of Agriculture to sell this chemical to American farmers. It was, however, very effective and cheap to make, and so the company started to market it overseas.

Chlordecone
(each line intersection is a carbon)

In the 1970s, the US Congress amended the Federal Water Pollution Control Act to establish the national pollutant discharge elimination permit system (NPDES). One of the NPDES permit requirements was that Allied list all the chemicals it was discharging into the James River. Recognizing the problem with Kepone, Allied decided not to list it as part of their discharge, and a few years later "tolled" the manufacture of Kepone to a small company called Life Science Products Co., set up by two former Allied employees, William Moore and Virgil Hundtofte. The practice of tolling, long-standing in chemical manufacture, involves giving all the technical information to another company as well as an exclusive right to manufacture a certain chemical – for the payment of certain fees, of course. It can also add a degree of separation of corporate accountability. Life Sciences Products set up a small plant in Hopewell and started to manufacture Kepone, discharging all of its wastes into the sewerage system.

Table 4.2
Properties of chlordecone (Kepone)

Formula	Physico-chemical properties	Environmental persistence and exposure	Toxicity
1,2,3,4,5,5,6,7,9, 10,10-dodecachloroocta-hydro-1,3,4-metheno-2H-cyclobuta (cd) pentalen-2-one ($C_{10}Cl_{10}O$).	Solubility in water: 7.6 mg L^{-1} at 25 °C; vapor pressure: less than 3×10^{-5} mmHg at 25 °C; log Kow: 4.50	Estimated half-life ($T_{1/2}$) in soils is between 1 and 2 years, whereas in air it is much higher, up to 50 years. Not expected to hydrolyze, biodegrade in the environment. Also, direct photodegradation and vaporization from water and soil is not significant. General population exposure to chlordecone mainly through the consumption of contaminated fish and seafood.	Workers exposed to high levels of chlordecone over a long period (more than one year) have displayed harmful effects on the nervous system, skin, liver, and male reproductive system (likely through dermal exposure to chlordecone, although they may have inhaled or ingested some as well). Animal studies with chlordecone have shown effects similar to those seen in people, as well as harmful kidney effects, developmental effects, and effects on the ability of females to reproduce. There are no studies available on whether chlordecone is carcinogenic in people. However, studies in mice and rats have shown that ingesting chlordecone can cause liver, adrenal gland, and kidney tumors. Very highly toxic for some aquatic species such as Atlantic menhaden, sheepshead minnow, or Donaldson trout with LC_{50} between 21.4 and 56.9 mg L^{-1}.

Source: United Nations Environmental Programme, 2002, "Chemicals: North American Regional Report," Regionally Based Assessment of Persistent Toxic Substances, Global Environment Facility.

The operator of the Hopewell municipal wastewater treatment plant soon found that he had a dead anaerobic digester (i.e., the anaerobic bacteria had ceased to breakdown the organic contaminants received by the plant). He had no idea what killed his digester, and tried vainly to restart it by giving it antacids to increase the pH of the wastewater. Meanwhile, in 1975, one of the workers at the Life Sciences Products plant visited his physician, complaining of tremors, shakes, and weight loss. The physician took a sample of blood and sent it to the Center for Disease Control in Atlanta for analysis. The worker was found to have an alarmingly high $8\,mg\,L^{-1}$ of Kepone in his blood. The State of Virginia immediately closed down the plant and took everyone into a health program. Over 75 people were found to have Kepone concentrations in their blood. The Kepone that killed the digester in the wastewater treatment plant flowed into the James River, and over 100 miles of the river was closed to fishing due to the Kepone contamination. The sewers through which the waste from Life Science flowed was so contaminated that they were abandoned and new sewers had to be built. These sealed sewers are still under the streets of Hopewell, and serve as a reminder of corporate avarice.

Questions

1. How can the engineer working in such organizations help to ensure that risk shifting is ethical?
2. What are some preventive actions that an engineer can take when he or she observes untoward activities?
3. Analyze these cases from a life cycle perspective. What actions could have been taken to reduce the concentrations of Kepone in people? . . . DDT in the environment?
4. What bioethical lessons can be derived from these cases (e.g., what are some deficiencies of companies purely motivated by monetary profit, or at least by the cost–benefit calculations?)

Another environmental example of "risk shifting" is that of justice. The overall population risk may be lowered by moving contaminants to sparsely populated regions, but the risk to certain groups is in fact increased. This type of risk shifting can also occur internationally, such as when a nation decides that it does not want its population or ecosystems to be exposed to a hazardous substance, but still allows the manufacture and shipping of the substance outside of its borders.[10]

Risk shifting is also common in health care. Numerous medical technological breakthroughs are accompanied by actual and potential risks. For example, as mentioned in Chapter 2, recombinant technologies to develop vaccines against infectious diseases pit acute disease treatment and prevention against possible risks of DNA research. Other biotechnologies needed to improve health in developing countries, such as bioinformatics to identify drug targets and to examine pathogen–host interactions, can also pose societal risks, such as loss of privacy and possible misuse of personal health information.

When engineers retire or take on other responsibilities, they do not relinquish the code of ethics. Nor, does an engineer move in and out of the profession on a project-by-project basis. For example, the vice president of Morton-Thiokol, manufacturers of the space shuttle *Challenger*, was asked by the company president to "take off his engineering hat and put on his management hat," and the vice president subsequently agreed to the ill-fated launch. Obviously, an engineer does not have such a prerogative. The duties to the company are always deferential to those specified in the code of ethics.

Engineers owe a responsibility to society, much like physicians. The physician has a commitment to help if he or she possibly can. If a person becomes ill on an airplane flight, for example, the physician is ethically called to assist. It does not matter that he or she might be retired, or have a practice quite different from the immediate emergency needs, or even may have not practiced medicine for a long time because of a career change. The physician is still ethically obligated to help. Likewise, the engineer who practices in a discipline other than biomedical, environmental, or other biologically oriented subdisciplines still has a modicum of professional bioethical responsibilities. Being an engineer is a lifetime commitment to assist individuals and society whenever possible and proper.

The engineering code of ethics is not always useful as a primary source of ethical decision making in nonprofessional situations. For example, in the example above, if the vice president is not an engineer or a physician, the codes of ethics do not apply. Another problem with codes of ethics is that they can be internally inconsistent. For example, the first canon admonishes engineers to hold paramount the health and safety and welfare of the public – all three at the same time and all three as paramount. But there are times when this cannot be done. For example, a traffic engineer may decide that allowing holiday traffic the use of a bridge that is under construction enhances public welfare, posing a minor (or at least acceptable) risk in public safety. In this case, the engineer must balance and optimize within two design constraints – welfare and safety. In an optimization regime, we are seldom able to maximize more than one variable concurrently with another.

Codes can also have contradictory statements. Two canons that exist in one form or another in various engineering codes can at times be internally inconsistent and even mutually exclusive:

Canon 1: Engineers . . . shall hold paramount the safety, health, and welfare of the public.

Canon 4: Engineers . . . shall act for each employer or client as faithful agents or trustees.

Canon 1 mandates that the welfare of the public is first, but Canon 4 admonishes the engineer to be faithful to the employer. In isolation, there is nothing wrong with either canon, but adhering to both can result in problems. When in doubt, Canon 1 must dictate the engineer's decision, since it must be held paramount.

RIGHT OF PROFESSIONAL CONSCIENCE

Bioethics Question: What are the bounds on the engineer's right of conscience?

Thought Experiment

Betty is an electrical engineer who has been developing new designs for optical systems used to miniaturize various products. In fact, she has become so good at designing these systems, she was "hired away" from a small company two years ago by a large conglomerate that owns, among many ventures, a medical equipment company. Up to six months ago, her optical systems were used in diagnostic and telemetric devices, but since then she has been spending most of her time adapting them to miniaturize and to improve the efficiency of implanted pumps and regulators. Last month, however, she was asked to start work on adapting an optical system to miniaturize a wired transducer that is presently used in a triggering mechanism for a secret system that is being funded by a defense contract. Rumor has it that the mechanism is designed to improve the efficiency of a landmine. Betty is generally opposed to armed conflict, but is especially opposed to landmines.

What should she do?

Every profession consists of people from various walks of life, with numerous perspectives, and a widely ranging set of beliefs. There is much commonality in thinking amongst people practicing in the same discipline, but there are also substantial differences in perspectives. Some of this divergence lies in what a professional believes to constitute right and wrong behavior. In fact, each professional discipline allows a certain degree of latitude on the types of work that an individual must perform. A professional practitioner may refuse to perform certain tasks, represent certain clients, and take certain ideological positions that conflict with the individual's conscience. This is known as the right of professional conscience, which is:

> [T]he right of an employee to refuse to partake in unethical conduct when forced to do so by an employer. This may occur in work or non-work situations and may not necessarily involve breaking the law. Conscientious refusal may be done by either simply not participating in the activity that one sees as immoral, or it may be done with the hope of making a public protest that will draw attention to the situation that one believes is wrong.[11]

Thus, the idea of conscience-driven morality is difficult. For example, you may completely agree with Betty about landmines. However, what if you found out that the work she is doing would prevent landmines from remaining active after their military use? For example, what if the mines could be remotely destroyed using Betty's device? Is it morally permissible to argue that the mines are going to be made anyway, so why not make them safer? Or, is it Betty's "duty" to quit the project, no matter the potential value, simply because the whole warfare scenario is immoral?

To understand the right of conscience, both of the terms are important: right and conscience. Let us first consider the second term. Theologians spend a lot of time thinking about the conscience, so that might be a good place to start to find a definition. For example, Richard M. Gula explains that conscience is:

> ... often used but little understood. Trying to explain conscience is like trying to nail Jell-O to the wall, just when you think you have it pinned down, part of it begins to slip away.... We all know we have a conscience, yet our experiences of conscience are ambiguous. We struggle with conscience when facing great decisions of life, such as the choice of a career, or of conscientious objection to war, or whether to pay taxes which support defense projects. Yet we even feel the pangs of conscience over petty matters, such as jaywalking.[12]

Gula considers the quandary that even though we hear that "conscience enjoys inviolable freedom," society and its institutions have codified formal and informal rules "so absolute in character we wonder whether conscience matters at all."[13] This uncertainty is at the heart of understanding when an engineer or other professional may invoke the right of conscience.

Perhaps distinguishing conscience from other human qualities can help, or in engineering parlance, we can find the solution by difference. Conscience is not the same as personality or even the Freudian concept of superego. Sigmund Freud[14] hypothesized in his structural theory that the human personality consists of three parts: id; ego; and superego (Figure 4.2). The id is the most primitive component, seeking maximum pleasure and is oriented toward "primitive desires" (e.g., sex, rage, and hunger). The superego is the normative part of the personality, which polices behavior and enforces standards of

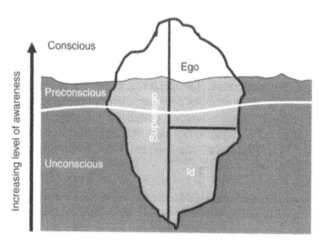

Figure 4.2 Freud's concept of the mind can be likened to an iceberg. The id is completely submerged (unconscious), whereas the ego and superego include awareness at the conscious, preconscious, and unconscious levels.

morality. The ego attempts to reconcile urges of the id with the norms of the superego. This balance gives rise to the sense of self.

The superego operates to allow the person to live in society; that is, it commands an act to gain approval, whereas the conscience is owned by the person. Thus, conscience is oriented toward a value, while the superego is oriented toward an authority. Conscience attends to a future large pattern of a desired personality archetype, whereas superego attends to past individual acts in themselves (strongly influenced by trying to avoid previously punished acts). Since conscience informs behavior on the basis of moral principles, a "good" conscience is a valuable commodity in discerning ethical choices. As such, the right of conscience of individual engineers must be respected to some extent in mediating bioethical issues.

With the distinction between conscience and superego, let us modify the earlier thought experiment.

Thought Experiment

The use of landmines in military operations is opposed by many. Even many of those who see the usefulness of the technology in certain applications agree that mines left in the ground pose a standing threat to innocent people. Jan is a mechanical engineer developing a trigger mechanism for landmines that is 25% more reliable than existing mechanisms. Jan is opposed to the use of landmines, but works for the defense contractor because she needs the job. She also believes that if good people like her refuse to work in the industry, only those who support landmine use will work there and that will translate into little internal opposition. As evidence, Jan has convinced her employers that they need to work on a trigger mechanism that biodegrades within one year after deployment, so that no orphan mines would be active a year after military operations cease.

Is Jan behaving morally?

The answer is informed by Jan's conscience, so it can only be answered by Jan. Does she really want to change the organization from within, or is this merely a rationalization for her source of income? Few of us work in organizations that allow complete autonomy. Institutional and organizational demands are overlain onto the engineer's professional responsibilities. The problem arises when professional integrity is sacrificed for the organization (witness the Shuttle *Challenger* disaster when certain engineers put the company and agency desires above their own engineering calculations).

Another interesting aspect of this thought experience is that of mitigation and risk trade-offs. Jan's work could improve the technology, making it safer in the long run. Rather than being dangerous for decades, modifying the mine's peril would be limited if she is successful in the biodegradable design. Some might argue that the best and most ethical approach is never again to manufacture any landmines. While this is true, and is arguably the positive paradigm (i.e., the most ethical course of action), it is also highly unlikely to occur soon. Jan is proposing a pragmatic improvement to the design, albeit not the textbook paradigm. Sometimes ethical improvement is incremental, just as technological advancement. Revolution can be great, but evolution is usually the norm.

The conundrum has recently been raised in discussions about emergency technologies. It is an ethical decision that must balance lost opportunities (e.g., better drugs, devices, and biosystems) against future risks (e.g., unknown health effects).

Often, defenders of the right of conscience are swayed by the issue at hand. What if, for example, Betty and Jan were mechanical engineers working on devices that could be used to improve methods used in abortions or assisted suicides?

Thought Experiment

The late-term abortion procedure, dilation and extraction (D&X), commonly known as partial birth abortion, is a highly controversial medical procedure. Many people strongly oppose this procedure due to the aggressive and active means of terminating the life of the unborn child. By 1999, the legislatures of 29 states had banned the procedure. In 2000, the US Supreme Court in a 5-4 ruling in the case of Stenberg *v.* Carhart decided that the State of Nebraska's ban was unconstitutional. The grounds for the rejection were twofold:

1. The law did not clearly distinguish between partial birth abortion (D&X) and the most commonly used late-term abortion procedure (dilation and evacuation (D&E)); and
2. The law did not provide an exception for the health of the mother.

Now, let us consider a hypothetical case:

Although this procedure has not been banned, many doctors refuse to provide it. Bill is employed by a biomedical device company that develops nanomachinery (i.e., artificial molecular machines created through molecular manufacturing). One of these angstrom-scale devices is being adapted to deliver drugs to tumors. Bill has been modifying the design to deliver psychotropic drugs *in utero* to treat brain disorders in the third trimester. He was recently asked by the vice president (VP) for research and development if the nanomachinery could also deliver a neurotoxin (i.e., a substance that destroys or harms nerve cells) to an unborn child's brain just before extraction. The VP implied that this could be a way to prevent the legal problems of partial birth abortions, since the child would not be alive when the extraction occurs. Bill can see no structural difference in delivering either chemical and even suggests that this might be an untapped market.

Sitting in the same meeting is another engineer, Mary, who returned to work last year after two years away from the company. She had gone through a very difficult pregnancy and her daughter was born six weeks prematurely. The experience has reinforced her belief that life begins at conception and that all human life is sacred. In fact, she was one of the advocates for the orginal *in utero* research because she foresaw the importance of neural fetal nanoscale devices. She now feels betray by her company. When she raises her concerns, other meeting participants reassure her that their company is merely a device company and that it is entirely up to the attending physician to decide what is medically and morally appropriate. In fact, he quotes the canon from the engineering code of ethics:

Engineers shall perform services only in areas of their competence.

And, he reminds Mary that engineers are beholden to the client (loyal and "faithful agents").
Mary feels outnumbered, so she acquiesces.

What should Mary have done? Was the discussion at the meeting ethical? Is the response that it is not
the engineer's concern correct? What does this say about an engineer's right of conscience? It is difficult
to say whether the discussion was appropriate, but many of us would be offended by the "technology
transfer" from delivering an ethical drug to delivering a toxin. One of the themes of this book is the
need to consider the entire life cycle of an engineering decision. We must be systematic in our ethics.
Kant's categorical imperative asks us to consider the effects if the decision is universalized. Is that not
what Mary is doing and her colleagues are failing to do?

The argument that engineers must be competent is not an invitation to ignore important aspects of a design,
including possible misuses. For example, a biomedical engineer working on a team design may notice some
mechanical or structural miscalculations. Although the engineer's "area of competence" is biomedical, he
or she still has the responsibility to point out problems (or opportunities) if they are known.[15]

Actually, other canons of the codes of engineering ethics support Mary's position, including the
admonition that the safety, health, and welfare of the public are paramount, and that we avoid deceptive
acts. The arguments involve immoral means (e.g., developing technologies to be used to facilitate harmful
outcomes) to support convenient outcomes (e.g., increasing company profits). By extension, one type of
deception is lying to oneself, including deluding ourselves about possible misuse of technologies. Mary
provides a case study of the need to think about possible downsides and malicious use of beneficial
technologies. The President's Council on Bioethics put it this way:

> But, once available, powers sought for one purpose are frequently usable for others. The same technological
> capacity to influence and control bodily processes for medical ends may lead (wittingly or unwittingly)
> to non-therapeutic uses, including "enhancements" of normal life processes or even alterations in "human
> nature."[16]

It suffices to say that sometimes codes are merely guides that provide general moral direction.

GROUPTHINK AND THE RIGHT OF CONSCIENCE

Sometimes institutions and organizations militate against the right of conscience. For example, many
engineering failures occurred because of institutional mechanisms that stifled dissent and buried "bad
news." The techniques used by organizations to control their membership have been logged by psychologist Irving Janis, who coined the term "groupthink" to describe the collective set of biases held by and
perpetuated by a group.[17]

Two concepts about groupthink are particularly pertinent to engineers. Bias is simply a systematic
error. We see things wrongly, but consistently. And, affiliation is both vital and dangerous to professionals. We need to affiliate with like-minded people, but we must take care not to rely completely on
exclusive thinking.

The video *Incident at Morales* provides a very useful, hypothetical composite of how groupthink
can affect an engineer and compete with professional conscience. In the video, Fred Martinez is a

young licensed professional engineer who has recently joined the professional staff at Phaust Chemicals, a United States division of a French multinational corporation. Fred is asked to design a system to reformulate a paint-stripping product to meet new regulations. A realistic constraint, however, is that a competing firm has a similar product. In the process, Fred identifies numerous environmental and technical problems that threaten the health safety of workers and nearby residents. Although Fred tries to be true to his professional training and his own moral compass, design and construction compromises lead to failures.

A number of lessons can be found in this story. One is that the changes and shortcutting can be quite gradual. A fitting analogy is the frog who jumps in fright from boiling water when exposed instantaneously, but who willingly remains in luke warm water that is gradually increased in temperature until the frog is boiled. Another lesson is that institutions can have priorities that are different and opposed to professional responsibilities. The engineer may have a strong core set of beliefs, but these can be stretched away from the center (Figure 4.3). The elements of groupthink include:

- Illusion of invulnerability of group
- "We feeling" very strong (i.e., outsiders are the enemy)
- Rationalization (mainly to shift responsibilities)
- Illusion of morality
- Self-censorship (don't rock the boat)
- Illusion of unanimity (silence means consent)
- Direct pressure (against those who disagree)
- Mind guarding (or gatekeeping)

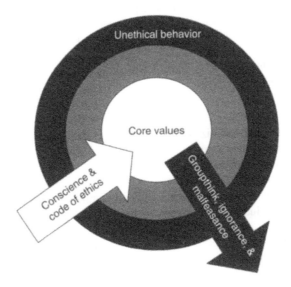

Figure 4.3 Vectors that influence ethical decisions by the engineer. The white arrow indicates that the engineer's conscience has been informed by moral development and that the engineering norms articulated in the code of ethics support ethical core values. Opposing forces (black arrow) include institutional pathologies, such as groupthink, and the individual professional's own compromises.

To help understand the power of groupthink and the role it can play in bioethics, let us reconsider the thought experiment introduced in the Prologue:

Thought Experiment

A number of public health and social sciences researchers, from various universities, government agencies, and companies, have been collaborating on an interdisciplinary project that is trying to characterize the factors that lead to terrorism, especially in hopes of finding ways to prevent it. The UN is the major funding source for the research.

Bob, the only biomedical engineer on the research team, is charged with aiding the others in finding "low-tech" devices to extract and store fluid samples and to assist behavioral scientists in data management and informatics. Sid, one of Bob's colleagues, had an opportunity to visit Pakistan. While there, Sid interviewed young people in a terrorist training camp, using a questionnaire developed by the investigators. Sid also decided to retain a local health worker to draw blood and take urine samples from the subjects to compare hormone levels with age-matched controls in the United States. Before Sid left, however, he neglected to ask the young people for their consent to participate in the research project. Sid had previously relied exclusively on anonymous databases from which he extracted data and he had not had experience in getting consent.

Sid asked Bob to take a look at the protocol for taking and handling the samples. Bob gave Sid some technical advice that greatly improved the local health worker's extraction process and would double the shelf-lives of the samples to allow for lab analyses of better resolution and sensitivity upon the team's return to the United States. Bob, however, had enough experience to know that Sid should have gotten advance consent to obtain fluid samples, but didn't think it was any of his business since this was an ancillary study separate from the larger study and Bob's role was clearly to provide engineering, not policy, advice.

After Sid returned home, he discussed his findings with the eight coprincipal investigators at his university. Most of his colleagues wanted to publish right away, with minority wanting to wait until the UN approved the results. Disagreements began to arise about the appropriate journal where the data could be published, most wanting the journal that best fit their own respective academic discipline. When word got out that there was a new emphasis on endocrine differences, one key collaborator from another university became angry about being excluded.

The grant project period has two more years remaining. Besides the authorship issues, some important bioethical issues are illuminated by this case.

Analysis:

1. What are the technical and organizational issues in this case?
2. What are ethical issues in this case?
3. What should Bob, the engineer, have done?
4. What should he do now?
5. What is different about ethical expectations of Bob compared to those of his colleagues?
6. How has groupthink influenced the actions of these otherwise moral people?

ANIMALS AND ENGINEERS

Bioethics Question: What is the moral standing of nonhuman species?

The engineering codes of ethics barely address ecological and research welfare questions, especially those related to nonhuman species, although the American Society of Civil Engineers mentions sustainability.

The codes do not specify what if any responsibility engineers owe to nonhuman animals, plants, or places. The only concern would be that the actions of the engineers not diminish the welfare of the (human) public. If an engineering project causes the demise of an animal or plant species, the principal concern is not for that plant or animal, but for future humans who may not be able to enjoy looking at this species, or obtaining some beneficial use from it. This bothers many ecologists, especially those who subscribe to so-called deep ecology. However, it also disturbs those who view the environment as so completely interconnected that a small change, such as the loss of a species, profoundly affects all aspects of the environment, including the well-being of people.

Thus, ethical consistency is sometimes extended to nonhuman species. To consider the range of bioethical thinking in this regard, let us slightly modify a thought experiment introduced in Chapter 1.

Thought Experiment

Dante is a mechanical engineer working for ICI and a vegetarian, opposed to confined animal feeding operations (CAFOs). He believes this manner of animal "husbandry" to be cruel and inhumane. Although Dante is a vegetarian, he accepts that many people will eat meat, but wishes that livestock and poultry were treated compassionately and with dignity (e.g., free ranging and mother-child bonding).

Unbeknownst to Dante, his injection system is being adapted to feed animals in CAFOs. A swine company has been ordered by the State Department of Agriculture to change its feed delivery system to segregate water and food. If the company does not do so, it will be closed down in six months. This week, Dante gets word that ICI is being contracted by the swine company to adapt the device and that for the past six months he has been working indirectly to keep the swine company in business. Dante is tempted to delay the project, so that the swine company closes the CAFO.

Is such a delay morally permissible?

This is an example of competing values. In an earlier thought experiment, the competition is between advancing the state of the science and the sanctity of life. The latter value has primacy. In the subsequent thought experiment, there is a competition between compassion for animals and professional honesty. Ironically, if Dante's system works as designed, it will be an incremental improvement to the animal's conditions in the CAFO. At least that is what the agriculture inspector recommends. So, if the company stays in operation and does not use the new injection system, the animals would be worse off than if Dante provides a workable design.

Another ethical consideration is Dante's rights of conscience. As discussed, engineers have a moral right to exercise professional judgment. This right must be balanced against obligations to employers and colleagues.[18]

Teachable Moment: Confined Animal Feeding Operations and the Moral Standing of Animals

René Descartes (1596–1650), the French philosopher and one of the founders of modern science and mathematics, saw animals as mindless machines.[19] This can mean that treating animals any way humans please should not raise any ethical concerns whatsoever. Descartes' view was challenged by contemporaries, notably by Pierre Gassendi (1592–1655), a Roman Catholic priest who in his own right helped to shape modern scientific thought. Gassendi argued that the moral focus should be on animals' ability to perceive and to suffer, rather than Descartes' conclusion that they lack human capacity to reason and to speak. Further, Gassendi argued that since humans can reason, we have a duty to be humane and to exercise compassion towards animals rather than to dominate and exploit animals.

According to the US Environmental Protection Agency (EPA), animal feeding operations (AFOs) are "agricultural operations where animals are kept and raised in confined situations," and that "congregate animals, feed, manure and urine, dead animals, and production operations on a small land area."[20] In such operations, the animals have little room (often just a few inches) in which to move their entire lives. This is a very efficient means of providing turkey, pork, beef, and other high-protein food.

Questions

1. Which position, Descartes' or Gassendi's, is closer to yours regarding the bioethics of AFOs? Why?
2. Which position is closer to yours regarding animal testing of possible carcinogens? Explain the moral justifications of your positions, and the reasons for any differences in your positions regarding AFOs and cancer research.

The moral standing of animals has changed since the Renaissance. The evolution in attitudes about animals is evident in the biographies of key nineteenth- and twentieth-century figures. Four are noted in the discussion box "Four Persons Who Changed the Way We Think about Nature."

MAKING ETHICAL DECISIONS IN ENGINEERING

Engineering solutions are predictable. A mechanical engineer designing an automobile can use the heat transfer equations in calculations, with confidence in their accuracy of applicability. If there is a difference in temperature, the heat will be transferred from the hot side to the cold side, and the rate of that transfer is governed by the properties of the material. This will always work. No exceptions. And if 100 engineers did the calculation, 100 of them would get the very same answer.

But this is not how ethics works. If 100 engineers are confronted by a problem in ethics, one would expect at least 100 suggested solutions (and possibly a lot more if some people waffle). All that can be said is that some of these solutions are better than others. In many instances, none is wrong, and none is right. Just some are better than others. So the question is, how can we design a system for arriving at answers to engineering ethical problems that will more often than not identify alternatives that most of us will consider to be acceptable?

One suggestion is to use a stepwise procedure for ethical decision making:

1. Become familiar with the relevant facts:
   Some problems disappear when the facts are all in. Getting the facts can also avoid grave embarrassment. The factual premise is the first component of a logical syllogism. It is also the first step in any ethical, including bioethical, analysis.

2. Determine and characterize the moral issues:
   What exactly is bothering you? What wrong has been done or may be done? Is this a problem in engineering ethics, or is this a question of personal morality? If it is engineering ethics, is it a breach of the engineering code of ethics, or something more complex?

3. Identify those directly and indirectly affected by the decision you have to make:
   Include your own family, your friends, and others who will be affected by your final decision.

4. Identify reasonable alternatives to address the issue:
   Here is where you want to be creative and "think outside of the box." Perhaps you can come up with some imaginative alternatives that will not harm anyone and will not compromise your own integrity.

5. Identify and describe the expected outcomes of each possible action:
   We cannot, of course, predict accurately what the future will hold and what people will do. What is important here is that you differentiate between those actions that will undoubtedly occur and those that may occur. Actions by people are not always predictable, and yet in this phase of decision making it is good to try to make such predictions.
   Sometimes it is impossible to prove that something has happened, such as proving that the company knowingly released a dangerous device years ago. You would simply not be able to prove that intent and hence going to regulators with the story might result in pitting your word against that of others – a no-win situation. However, a dispassionate disclosure of events might be possible.

6. Characterize the personal costs associated with each possible action:
   We all have an obligation to do the right thing, but this obligation is limited by the costs we might incur. For example, if a certain ethical action will more than likely result in your losing your job, this is a large cost, but it may be acceptable if the situation demands it. On the other hand, if the probable cost is the loss of your life, then most rational people will agree that this cost is too high except in highly unusual circumstances (e.g., saving other people's lives).

7. Get help:
   Chances are good that your problem is not unique. Someone has had the same concern and had considered the ethical ramifications of the various actions. Getting help from more experienced and wiser people is the best way to calibrate your own notions on what you ought to do. Another source of help is the code of ethics of your profession or discipline, and from the board of reports who oversee the code. If one of your alternatives is clearly and unequivocally listed as being unethical, then you know that this is not a reasonable alternative.

8. Considering the moral issues, practical constraints, possible costs, and expected outcomes, decide on the best action:
   This process includes quantitative, semiquantitative, and qualitative information. The decision will always includes objective elements, but often also includes subjective aspects.

Discussion Box: Four Persons Who Changed the Way We Think about Nature[21]

Ethics, especially bioethics, is concerned about what makes us do right things. One distinction is between "nature and nurture." Interestingly, the term nature took on new meaning in the nineteenth century, combining elements of both nature (genetics) and nurture (environment). The lives of four people illustrate the transition.

John Muir

John Muir came to America at eleven years of age. He and his family settled in Wisconsin. He suffered an eye injury that made him temporarily blind in 1867. When he recovered he had gained a newfound appreciation for things visual and decided to turn his eyes to fields and woods. He walked from Wisconsin to the Gulf of Mexico, then sailed the Caribbean and the West Coast of North America, landing in San Francisco in 1868.

Muir walked all over the country and started to write about his travels, eventually publishing over 300 articles and 10 books. He became an "environmental activist" (before there were such designations) and partly due to his efforts and the assistance of President Teddy Roosevelt with whom he had become friends, the Sequoia, Mount Rainier, Petrified Forest, and Grand Canyon national parks were established. His single greatest disappointment was the flooding of the Hetch Hetchy Canyon in California. However, the area would later become the Yosemite Park. In 1892 he and his friends established the Sierra Club, which remains one of the most influential environmental organizations in the United States. The club has become almost synonymous with the term "environmentalism."

Rachel Carson

After graduating from the Pennsylvania College for Women (now Chatham College) in 1929 Rachel Carson (1907–1964) worked at Woods Hole Marine Biological Laboratory and continued her education with a master"s degree in zoology from Johns Hopkins University in 1932. She was a writer, scientist, and in 1962 wrote perhaps the most influential book ever published in the environmental field, *Silent Spring*. The title comes from what she foresaw as the death and destruction of birds due to the extensive use at that time of chlorinated pesticides and she called for an end to their indiscriminate use. The reaction by the chemical industry to Carson's book was immediate and vitriolic. She was branded as anything from a flake to a communist sympathizer. The book elicited quite an emotional response, as evidenced by the following 1973 quote by Parke C. Brinkley, then President of the National Agricultural Chemicals Association[22]:

> The great fight in the world today is between Godless Communism on the one hand and Christian Democracy on the other. Two of the biggest battles in this war are the battle against starvation and the battle against disease. No two things make people more ripe for Communism. The most effective tool in the hands of the farmer and in the hands of the public health official as they fight these battles is pesticides.

But in the end, it was clear that her cause was just, and as an increasing amount of evidence grew on how the upper food chain was being affected by these nonbiodegradable pesticides, she became a hero in the environmental movement. She did not, unfortunately, live to see her life and work honored by the world.

The battle about what needs to be done still rages.

Christopher Stone

Just because an animal, or a tree, or even a place cannot hire a lawyer and argue its case in court, does this nonhuman then have an absence of standing? Can a case be argued on behalf of the animal, or tree, or other nonhuman?

This was the question that came before the Supreme Court in 1967 in a famous case, *Storm King vs. Federal Power Commission*. The Federal Power Commission wanted to lease out some federally owned land for a ski area, but the Sierra Club objected, believing that the development would harm the forests. But the problem was that no single member of the Sierra Club could prove that he or she was being directly harmed by the construction of the ski slopes. The harm to each person would be small, and collectively the harm would be great, but there was no single person who would have standing in the court. The damage to an individual is just not great enough to warrant taking up the court's time.

One of the lawyers on the case was Christopher Stone (1937–), a professor on the faculty of the University of Southern California law school. Stone, had been active in the Sierra Club and helped them on numerous occasions. In support of the Sierra Club's case, Stone wrote an article cleverly entitled "Should Trees Have Standing?" in which he argued that natural objects have every right to be represented in court if the damage is sufficiently great. The article appeared in the law review and became part of the brief submitted to the Supreme Court. Justice William O. Douglas, an avid outdoorsman, was quite sympathetic to the destruction of natural habitats. Using Stone's arguments, Douglas was able to sway the court in favor of the plaintiffs, and to stop the development of the ski area. The case fueled and encouraged the environmental movement that grew later in the 1960s, and set the stage for the rush of environmental legislation in the last quarter of the twentieth century.

Gaylord Nelson

Perhaps President Bill Clinton said it best when he presented the highest civilian award, the Presidential Medal of Freedom to Gaylord Nelson:

> As the father of Earth Day, he is the grandfather of all that grew out of that event: the Environmental Protection Agency, the Clean Air Act, the Clean Water Act, and the Safe Drinking Water Act.

Gaylord Nelson was a state legislator, a two-time governor of Wisconsin, and US senator, serving a total of eighteen years in that capacity. In 1969 he conceived of Earth Day. The national event, during the time of dissent over the Vietnam war, drew over 20 million participants and put into motion the string of legislation that forms the backbone of our environmental law today.

In 1961, while governor of Wisconsin, he created the Outdoor Recreation Acquisition Program, funded by a penny-a-pack tax on cigarettes, the acquired million acres of parkland in Wisconsin. While in the US Senate, he authored the legislation that preserved the 2100-mile-long Appalachian Trail. After leaving the Senate he served for many years as a consultant to the Wilderness Society. He died in 2005, leaving a legacy of environmental protection and care of the planet. The bioethical implications of sustainability can be summed up in one of Nelson's quotes:

> The ultimate test of man's conscience may be his willingness to sacrifice something today for future generations whose words of thanks will not be heard.

None of these four individuals was an engineer, but collectively the artist, biologist, lawyer, and politician represented the traits of good engineering: moral conviction, foresight, creativity, and sound judgment.

The engineering code of ethics is acceptable as a first, and very rough, cut at making ethical decisions in engineering. Often when engineers are confronted by ethical problems, a quick glance at an engineering code of ethics is enough to encourage a decision that the engineer can live with. But bioethical problems are seldom straightforward, and the right actions are not obvious. There is a great deal of subtlety in ethics, and any set of guidelines such as a code of ethics cannot hope to cover all cases.

NOTES AND COMMENTARY

[1] E.C. Hughes, "Professions," in *Ethical Issues in Professional Life*, ed. J.C. Callahan (New York, NY: Oxford University Press, 1988).
[2] The Uniform Code of Military Justice applies to all military personnel. The code can be accessed at http://www.au.af.mil/au/awc/awcgate/ucmj.htm.

3 This distinction between caveat and trust can be found in Everett C. Hughes' excellent article "Professions," in *Ethical Issues in Professional Life*, ed. J.C. Callahan (New York, NY: Oxford University Press, 1988). Hughes is quoted at the beginning of this chapter. In one form or another, I mention this in my engineering and professional ethics courses and lectures. The reading was originally recommended to me by my esteemed colleague (now twice retired from Duke and Bucknell), P. Aarne Vesilind.

4 Quoted by Ian Jackson, *Honor in* Science, p. 7. Sigma Xi, Research Triangle Park, NC, 1956, from J. Bronowski, *Science and Human Values* (New York: Messner, 1894), 73.

5 C.E. Harris Jr., M.S. Pritchard, and M.J. Rabins, *Engineering Ethics: Concepts and Cases* (Belmont, California: Wadsworth Publishing, 1995).

6 I ask the following question: Is DDT bad or good? By and large, the initial response of liberal arts and engineering students alike is that it is "bad." This makes for an energetic discussion, as some of the facts in this box are shared. Older adults, especially of the World War II era, are more likely to consider DDT to be "good." It is interesting how an arrangement of carbon, hydrogen, and chlorine atoms can be either good or bad (perhaps "beneficial" and "harmful" are preferable). This is a profound example of anthropomorphism.

7 R. Carson, *Silent Spring* (Boston, MA: Houghton Mifflin, 1962).

8 P.H. Müller, 1948, "Dichloro-diphenyl-trichloroethane and Newer Insecticides," *Nobel Lecture*, 11 December 1948, Stockholm, Sweden.

9 The term "organic" is sometimes unclear in various environmental and medical contents. In this usage, the term means that these pesticide compounds contain at least one carbon-to-carbon or carbon-to-hydrogen covalent bond. In contemporary usage, the term *organic* can also mean the opposite of *synthetic* or even *natural*, such as pesticides and nutrients that are derived from plant extracts, like pyrethrin from the chrysanthemum flower or herbs that provide homeopathic benefits. This is another example of how even within the scientific community, we are not clear in what we mean, making bioethical dialogue difficult.

10 J.D. Graham and J.B. Wiener, "Confronting Risk Tradeoffs," in *Risk versus Risk: Tradeoffs in Protecting Health and the Environment*, ed. J.D. Graham and J.B. Wiener (Cambridge, MA: Harvard University Press, 1995).

11 C. Vee and R.M. Skitmore, Professional Ethics in the Construction Industry, *Engineering Construction and Architectural Management* 10, no. 2 (2003): 117–27. This quote includes two embedded citations to Cardyn Whitbeck, *Ethics in Engineering Practice and Research* (Cambridge, UK: University Press, 1998).

12 R.M. Gula, *Reason Informed by Faith: Foundations of Catholic Morality* (New York, NY: Paulist Press, 1989).

13 Ibid.

14 S. Freud, "The Origin and Development of Psychoanalysis," *American Journal of Psychology* 21, no. 2 (1910): 196–218.

15 The video "Professional Ethics and Engineering" produced by Duke's Program in Science, Technology and Human Values presents a case of a foundation engineer who knew that the soil type and mechanical properties were unacceptable to support the foundation. However, the engineer said nothing because it "was not his job." As a result, a wall failed, killing and injuring a number of people. The case is complicated because the engineer was trained and practiced in a former East Bloc communist system,

where people who did not mind their own business were severely punished. This does not excuse the myopia, but it does explain it to some extent.

16 The President's Council on Bioethics, "Working Paper 1, Session 4: Human Cloning 1: Human Procreation and Biotechnology," (17 January 2002).

17 I. Janus, *Groupthink: Psychological Studies of Policy Decisions and Fiascoes*, 2nd ed. (Boston, MA: Houghton Mifflin Company, 1982).

18 M.W. Martin and R. Schinzinger, *Ethics in Engineering* (New York, NY: McGraw-Hill Higher Education, 2005), 164.

19 R. Descartes, *The Philosophical Writings of Descartes*, trans. J. Cottingham, R. Stoothoff, and D. Murdoch (Cambridge: Cambridge University Press, 1991).

20 US Environmental Protection Agency, 2006, Region 7: Confined Animal Feeding Operations: http://www.epa.gov/region7/water/cafo/index.htm (accessed 13 July 2006).

21 The source for these biographies is D.A. Vallero and P.A. Vesilind, 2006, *Socially Responsible Engineering: Justice in Risk Management*. (Hoboken, NJ: John Wiley & Sons).

22 Quote from F. Graham, "The Mississippi River Kill," in *Environmental Problems*, ed. W. Mason and G. Fokerts (New York, NY: William C. Brown, 1973).

Chapter 5

Bioethical Research and Technological Development

Research is not an organized profession in the same way as law or medicine. Researchers learn best practices in a number of ways and in different settings. The norms for responsible conduct can vary from field to field. Add to this the growing body of local, state, and federal regulations and you have a situation that can test the professional savvy of any researcher.

Nicholas H. Steneck,[1] Emeritus Professor, University of Michigan

Much of the concern about bioethics can be found in research. Any research, no matter the particular scientific discipline, demands integrity. Thus bioengineering and biomedical research in many ways is no different. It must be conducted in a way that it can be trusted. Bioresearchers cannot lie, cheat, exaggerate results, change data, fabricate findings, plagiarize, steal other people's intellectual property, or in any other manner, "cook the books."

Beyond this, since bioresearch deals in the essence of life, the ethical demands transcend the core of scientific soundness to include the larger focus of avoiding unacceptable risks to society. In other words, the engineering profession owes additional responsibilities to society beyond the collective of its membership. Both the individual engineering researcher and the engineering profession have a large stake in ensuring that any bioresearch is conducted responsibly and that such research advances the good of society.

Many Ph.D. research programs now have what are known as RCR programs for their students. Students are taught about "responsible conduct of research." The Office of Research Integrity in the US Department of Health and Human Services lists nine RCR themes:

1. Data Acquisition, Management, Sharing and Ownership
2. Conflict of Interest and Commitment
3. Human Subjects
4. Animal Welfare
5. Research Misconduct

6. Publication Practices and Responsible Authorship
7. Mentor/Trainee Responsibilities
8. Peer Review
9. Collaborative Science

These themes cover many of the areas confronted by the individual researcher. At this stage, the RCR programs are designed to prevent misconduct more so than to instill ethics: That is, they aim to reduce the negatives that have occurred in academic and institutional research. "Misconduct in science" is defined as:

> . . . fabrication, falsification, or plagiarism in proposing, performing, or reviewing research, or in reporting research results. It also sets the legal threshold for proving charges of misconduct.[2]

Misconduct includes cheating/plagiarism, harassment, patent violations, human research subject abuses and misuses, and improper animal research. It does not include honest error or honest differences in interpretations or judgments of data. Nor does misconduct take into account the societal risk of the research (Figure 5.1).

Bioethical choices can present unique challenges to the practicing engineer and researcher. No engineer will go untouched by the rapidly emerging developments in the biological sciences. Some have characterized these choices during the practice of engineering by the individual professional to be "microethics." As such, the individual engineer stands as sentry to ensure the public is protected from unwarranted risks.

Figure 5.1 Biomedical and other cutting-edge biological research is fraught with moral decisions and challenges. The researcher may follow microethical rules, such as respecting intellectual property, avoiding conflicts of interests, and properly using data. Meanwhile, the general public is at risk as the research community is in conflict about how to balance specific research objectives within an acceptable ethical framework.

Kant's categorical imperative raises the question as to whether the codes of ethics, if correct and appropriately enforced, will be enough to ensure that the engineering profession will prevent risks to society? The answer is clearly no. The profession, even if populated with an entirely ethical membership, will need to do more to prevent risks to society posed by the products and by-products of emergent technologies. Such prevention, known as "macroethics," must begin in the research phase.

The need for the macroethical perspective is not shared by all within the biomedical and biosystem engineering research community. As evidence, I had an illuminating discussion with a very bright second year biomedical engineering Ph.D. candidate. She was convinced that a training program to help Ph.D. students appreciate societal risks is a waste of time. And, worse she feared that it is wrong thinking since presenting both sides (e.g., benefits and risks of nanotechnology) makes it appear that both positions have equal moral standing. Her "faith" was in the scientific method and that the scientists are the only ones, or at least the principal ones, who should decide on right actions in their areas of expertise. This view is similar to the rationalists, expiricists, and utilitarians beginning with Descartes, Hume, and Mill. It can also represent a type of research arrogance that closes the researcher's mind about possible ethical problems. Researchers have a great amount of pressure placed on them to "stay the course" even when they might otherwise be morally opposed. For example, would a researcher recognize a possible dangerous event resulting from a misuse, and if so, would the researcher be willing to stop this line of research at great personal costs, such as adding months or years to completing the Ph.D.?

BEYOND REGULATION

Morality extends beyond legality. Doing what is "right" is more than simply avoiding what is "wrong." Much as peace is more than the absence of war, research integrity is more than avoiding immoral acts. It requires a proactive and preventive approach. Bioethics must be integrated throughout the research project. The regulated environment where researchers work can tempt us to simply "follow the rules." Certainly, the rules are important, but legality and morality are not completely inclusive (Figure 5.2).

INTEGRITY

It is generally agreed that scientists have two traditional duties: first, the duty of seeking the truth; second, the duty to communicate to all who need it the knowledge gained in their search.

Jay Orear (1969)[3]

A common theme of the thought experiment analyses in Chapter 1 is that professionalism is built upon a foundation of trust. Such trust is foremost between the patient and the physician. For engineers, the trust is with the individual client, but is also extended to the general public. Obviously, honesty is crucial to such trust. Our clients need to be reassured that we can be trusted. And, trust is something that we have to cultivate and tend carefully.

Much of the modern framework for research and scientific ethics has grown out of negative paradigms; that is, acts that are examples of unquestionably immoral behavior. The odious war crimes of the mid-twentieth century Nazi "medical" practitioners elicited a bioethical response to the blatant mistreatment

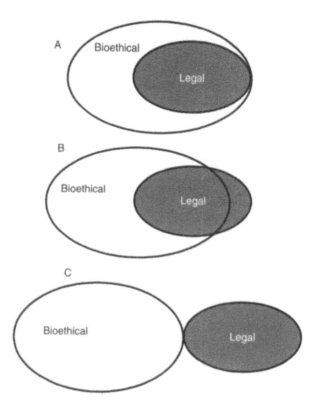

Figure 5.2 Ethical decisions differ from legal decisions. Legality can be a subset of bioethical decisions (A). An example is a bioethical researcher's responsibility to abide by copyright and intellectual property laws. In addition, bioethics also includes extralegal responsibilities, such as maintaining competence in the scientific discipline. Other situations may arise where a certain amount of legal behavior is, in fact, unethical (B). An example is the paying of bribes or gratuities or conducting research without appropriate informed consent of the subjects, which in certain cultures may be legal or even mandated by regulations. On rare occasions, the laws may be completely unethical (C), such as slavery, mandatory abortions, or "ethnic cleansing" (such as in the Darfur region of the Sudan). Many would argue, that embryonic stem cell and blastocyst research, cloning, and animal experimentation also fall in this category. Ecoterrorists in a similar but distorted way attempt to rationalize their extralegal activities, e.g., spiking trees to prevent logging activities or torching facilities that they consider ecologically immoral.

of persons in experimentation. The resulting Nuremberg Code set standards for medical experimentation on humans. The first rule of the code states that: "The voluntary consent of the human subject is absolutely essential."[4] Thus, bioethics requires that persons knowingly agree, that is, they give their consent to participate based upon credible information. From Socrates on, ethical norms must always consider antonomy. The Nuremberg Code's implied informed consent requirement was later formalized by the World Medical Association in its Declaration of Helsinki, which reiterated the need for voluntary consent, but added that participants must be fully and completely informed before deciding to participate.[5] In the United States, informed consent has been codified in the Code of Federal Regulations.[6] Integrity in bioethics was further articulated in the 1979 release of *The Belmont Report: Ethical Principles and*

Guidelines for the Protection of Human Subjects of Research,[7] which lays out three basic and inviolable principles for treating people as research subjects:

1. Respect for persons
2. Beneficence
3. Justice

The Nazi trials and other notorious unethical investigations are evidence of disrespecting people. Respect for persons necessitates intellectual honesty and truthfulness with human subjects. The *Belmont Report* states that "respect for persons demands that subjects enter into the research voluntarily and with adequate information." The widely used concept of "informed consent" is introduced in the report, when it is given as a criterion for selection of people who will be part of a research study. Actually, informed consent is a balance between respect for persons and advancing knowledge. For example, certain aspects of full and complete disclosure of *a priori* consent may be waived in a study if a particular disclosure is likely to render the research invalid. Of course, this implies an overriding medical or other scientific exigency. Timing of disclosure is also a factor. In numerous instances, the subjects can honestly be informed that they are being invited to participate in research that embodies certain facets that will not be made known until the study has been completed. The waiver is granted in accordance with three conditions:

1. Incomplete disclosure is truly necessary to accomplish the goals of the research
2. Undisclosed risks to subjects are no more than minimal
3. A plan for sufficiently debriefing subjects is prepared, when appropriate, along with a protocol for distributing the research results to them

Thus, informed consent consists of both the relevant information about the research and the actual and cognizant consent.

Some of the modifiers used in the foregoing discussion will bother those who need detailed information, including most engineers. For example, when is the amount of information "adequate"? When are undisclosed risks "minimal"? When is a debriefing "appropriate"? Such is the open-ended nature of research and the quandary between a practical profession and scientific discovery. It is also an important reason that bioethics is so important to engineering research.

If you have had a medical procedure recently, you were probably asked to sign a form that says that you understand the risks associated with that procedure. The probability of each adverse outcome (harm) is delineated with a percentage or some expression of odds. For example, your operation may have a 1–5% chance of fever and extended hospital stay, a 0.1% chance of some hearing loss, and a 0.0001% chance of death. In other words, epidemiologists have found that complications in the type of surgery you are about to receive results in death of one in ten thousand cases. This is not the same as your risk, which is a function of your own vulnerabilities and strengths. It is likely a general reflection of all cases. If you are twenty-five years of age and in good health, your individual risk of death is much lower than an 89-year-old cancer patient. So, your stratum of the population may have a one in a million chance of death and the elderly, ill stratum may have a one in five hundred chance. However, there is still a chance that the older guy will live and you won't. In statistics, you don't matter.

So, if such information is so easily misunderstood, why is it given to everyone (old, young, educated, illiterate, citizen status, etc.)?[8] The answer is "informed consent." Besides Nazi Germany's unethical treatment of prisoners of war, especially the "medical" procedures to which they were subjected, other unethical medical treatments took place in the United States under the banner of "research," such as the withholding of treatment and dishonesty of researchers in treating syphilis in African-Americans in Tuskegee, Mississippi (which actually began before World War II in 1930 and lasted forty-two years), mistreatment of mentally handicapped patients, and sterilization of poor women of childbearing age. Such inhumane and inhuman practices cried for increased scrutiny and accountability of medical and other scientific research involving human subjects. One of the mandates to become well established is that a patient, client, or subject be thoroughly informed about any risks. Risk assessment and management are quite complicated, however, even for those of us in the business of risk. An ongoing and contentions debate about "acceptable risk" is raging, no matter the venue; be it medicine, environment, education, sports, or product design.

Certainly, people need to be informed about the risks of important decisions, and medical and public health decisions are right near the top of most decisions we make. However, what is the sufficient amount of information needed to make such a decision? If we want a third grader to be included in a clinical trial, prudence and practical experience tells us that he or she will not be sufficiently prepared intellectually and morally for such a decision. We delegate such decisions to the guardian, as defined by regulation and the courts. The unsettling question is whether even the guardian is sufficiently informed on the third grader's behalf.

Another seemingly unnoticed problem is that many people may needlessly avoid certain necessary procedures because they see these percentages. The risk disclosure process may unduly remove patients and subjects from needed treatment and participation in a project that they would have willingly joined, if the health care professionals were communicating and ensuring that the patients understood the details sufficiently well. When is professional care compromised in the name of "full disclosure"? Full disclosure is only valuable if the information is properly and precisely interpreted by the person making the decision.

Many people are easily impressionable and susceptible. Is it possible that upon seeing the odds of harm some persons may simply give up or be so overcome with fear that it compromises the medical procedure? Of course, this is not an argument for the "ignorance is bliss" approach, but it is a warning that physicians (and engineers) are in a position of trust and have an enormous impact on their patients' (and clients') attitudes and outlooks. Honesty about risk means that they should not be underestimated, nor should they be exaggerated. Either way, the patient and the subject are misinformed. Aristotle's "golden mean" where in the most ethical approach lies between two extremes may well apply here.

Informed consent spills over into all of our engineering specifications and accountability. Valves that leak, clearance systems that fail prematurely, noisy pumps, and pharmaceuticals that lead to previously unexpected health effects are all fuel for the public's discontent. We often find that the engineers indeed made known what they believed to be the best data and knowledge, but subsequent events show that they missed a few things. Sometimes, these are important and the difference between success and failure. But, the public may believe that their consent was breached and that what they actually received was "misinformed consent."

The current uproar over the advice from public health officials immediately following the World Trade Center explosions and collapses is an example of the public perception of having been misinformed. The Administrator of the US Environmental Protection Agency had issued information about the air quality in lower Manhattan that led some to believe that there was no health threat. It is likely that the Administrator was sharing her conclusions based on data that were limited to a few instruments with relatively high levels of detection (i.e., lower, yet dangerous concentrations of contaminants like asbestos and dioxins would not be detected until weeks later). Higher concentrations present acute hazards, whereas lower concentrations of the same substances can lead to long-term, chronic illnesses (e.g., asthma, mesothelioma, or cancer). Some claim that these proclamations, however well-intended, led to less care in wearing proper personal protection, especially respirators.[9] The courts will in all likelihood make the final decisions in these cases about whether the pubic was satisfactorily informed.

The bioethical lesson for engineers is that there are unexpected consequences from even the most thoughtful decisions. Thus, designs must be adaptable as new information comes to light.

Teachable Moment: The Therapeutic Misconception

Paul Appelbaum, a psychiatry professor at the University of Massachusetts Medical School, and his colleagues coined the term "therapeutic misconception" to refer to situations where on the surface a patient, client, or subject appears to be giving informed consent, but is completely convinced that they are really receiving treatment and not a placebo.[10]

According to Appelbaum, despite explanation, people strongly believe that research protocols are designed to benefit them personally and directly instead of testing the effectiveness of treatment methods. This effect is not even strongly associated with understanding experiments and the scientific method, since potential subjects who understand technical concepts like placebos, randomization, and double-blind study designs are still apt to maintain that the doctor/researchers will give them the best treatment based on the subject's particular needs. In other words, the hospital or medical facility conveys such an atmosphere of trust that the subject expects the professional therapeutic relationship will transfer to the clinical trial. The physician is perceived to be a professional first and a researcher second. In other words, the patient seems to be thinking: "I know this is a study that includes a placebo, but my physician would never keep the real thing from me; I trust her."

Questions

1. Why is it difficult for potential subjects to see physicians as researchers?
2. What are some of the likely reasons leading to the therapeutic misconception in research subjects (hint: recall the second thought experiment in Chapter 1)?
3. What can be done to improve informed consent in light of the therapeutic misconception?
4. How might the therapeutic misconception affect the enginer–client relationship? Give two examples of the therapeutic misconception in engineering practice.
5. What lessons does this medical research hold for engineering research?

THE EXPERIMENT

The examination of the bodies of animals has always been my delight; and I have thought that we might thence not only obtain an insight into the ... mysteries of Nature, but there perceive a kind of image or reflex of the omnipotent Creator Himself.

William Harvey (1578–1657)[11]

Experimental science is an essential part of engineering. Engineering is the application of the physical sciences, so it is completely dependent on the reliability of the underpinning research. Engineers rely on findings in laboratory and field settings as a foundation for daily design decisions. The experiment, however, is a relatively recently accepted requirement for scientific and medical investigation. The modern experiment has its roots in the Renaissance, championed by the likes of Frances Bacon, Galileo Galilei, William Harvey, and Robert Boyle. Arguably, it was Boyle and the Royal Society in the seventeenth century that raised the need for *a posteriori* knowledge (i.e., inferring causes from observed effects) to prominence in scientific inquiry. Evidence and empirical reasoning are now requirements for what will be accepted in engineering practice.

Ironically, although he can easily be considered one of the first biomedical engineering researchers, William Harvey's experiments were not accepted by the scientific community of his day and it was not until after his death that his correct version of the circulatory system was adopted by prevailing medical science. In fact, during his lifetime, Harvey's research and conclusions were assailed as counter to the powerful paradigm of its time; that is, Galen's[12] contention that venous and arterial systems are separate. In 1649, one of Harvey's contemporary critics, Jean Riolan, was sufficiently forceful in *Opuscula anatomica* so as to induce Harvey in *Exercitatio anatomica de circulatione sanguinis* to show that the Galen system was not supported by experimental data and empirical reason.

From the Renaissance on, science has demanded that cause and effect relationships be based upon an empirical perspective. Since this is both a scientific and an ethical text, we need to define the various connotations of empiricism. The term is rooted in the concept of "experience." The Greek word, *empeiria*, and its Latin translation, *experientia*, which indicate knowledge based upon practice (of course, a defining feature of both medicine and engineering). However, in the seventeenth and eighteenth centuries, the term also grew to characterize the practice of quacks and charlatans.

Empiricism must be distinguished from rationalism, which holds that human reason apart from experience is a basis for some kinds of knowledge. A key aspect of rationalism is *a priori* knowledge. Such knowledge is gained solely using reason, even before experience. Conversely, *a posteriori* knowledge is based on experience. Thus, a statement like "a white chair is white" is an example of *a priori* knowledge. It also appears to be a needless redundancy (what logicians call a tautology). The statement, "a chair painted with titanium oxide paint is white" is an example of *a posteriori* knowledge. Experience, including experimentation, tells us the conditions where *a posteriori* knowledge applies.

Rationalists hold that knowledge is gained by extracting *a priori* truths using deduction. Empiricists usually reject such pure reason in scientific discovery, deferring to sensory experiences to derive *a posteriori* knowledge using induction.

THE HYPOTHETICO-DEDUCTIVE METHOD

It is interesting and revealing that scientists and engineers rely on a protocol posited by an English "Renaissance Man" (or *Homo universalis*, if one prefers the Latin moniker). William Whewell (1794–1866) introduced the scientific method that we apply as contemporary researchers and practitioners.[13] The hypothetico-deductive method requires that testable hypotheses be identified, followed by genuine attempts to falsify them. Although contemporary research is often presented as an attempt to confirm a study's hypothesis by supporting it with experimental evidence, the research must undergo peer review. Hopefully, diligent reviews include considerations of possible ways that the research conclusions are wrong. Any hypothesis that survives stringent and relentless testing is considered to be "confirmed." Actually, the twentieth-century philosopher, Karl Raimund Popper (1902–1994), argued that such hypotheses are simply "corroborated."

There are two reasons that scientists and engineers are reluctant to confirm most hypotheses. First, the experiment is tightly defined and restricted to the conditions under which it is conducted. Particularly in engineering practice, numerous variables are not fully considered until they are tested in the real world. That is why prototypes are so important as is follow-up to experimentation. Second, induction is a difficult process for confirming anything. Inductive reasoning or inductive logic is a way of reasoning where the premises of an argument support the conclusion but do not guarantee that the conclusion is correct. Thus, unlike deductive reasoning, induction leaves uncertainty no matter how much evidence supports a conclusion. In other words, we are drawing a general conclusion from our specific data and findings.

Professional judgment depends heavily on inductive reasoning, such as when diagnosing cancer in a patient. The oncologist combines a number of factors in characterizing the patient's status. This includes pathological reports (itself inductive, since the pathologist judges whether the cell is cancerous based on visual and other examinations of the cells, such as whether they have distinct nuclei), biomarkers (e.g., enzymes), and other biomedical data. These empirical results, which in themselves may appear unrelated, are integrated by the physician using inductive reasoning to come to general conclusions. This same process is used by engineers in most design applications, where seemingly disparate information is used to arrive at a workable design.

RESEARCH CONFLICT OF INTEREST

Bioengineering research has many of the same conflicts of interests that practitioners have, but they come in different forms. For example, researchers may be tempted to be less than honest or at least not as diligent in pursuing research that does not best serve their research purposes and agenda. These conflicts can be financial or ideological (see Teachable Moments: Truth and Turtles), and sometimes both.

It suffices to say that bioethical debates must be based on facts that are unassailable. Many divisive elements separate the factions of bioethical debates, but facts should not be one of them. Honesty should never be sacrificed for convenience. It is at times difficult for researchers immersed in science to be objective. Even when we are looking at the same facts, we can come to very different moral conclusions. Joe Herkert of North Carolina State University has observed that one division is between types of rationality employed by engineers and other technical experts and the rationality employed by social scientists and cultural experts. Some of the differences are shown in Table 5.1.

Perhaps the best advice to engineers regarding research integrity and avoiding intellectual and other conflicts of interest comes from Richard Feynman:

> The first principle is that you must not fool yourself – and you are the easiest person to fool. So you have to be very careful about that. After you've not fooled yourself, it's easy not to fool other scientists. You just have to be honest in a conventional way after that.

Table 5.1
Decision-making factors employed by engineering and other technical experts compared to factors used by social scientists and cultural experts

Factor	Technical rationality	Social or cultural rationality
Trust based on:	Scientific evidence	Political culture
Appeal to:	Expertise	Traditions and peer groups
Type of analysis:	Narrow, reductionist boundaries	Broad boundaries
Risks are:	Depersonalized	Personalized
Emphasis on:	Statistical risk	Impacts on family and community
Focus:	Appeal to consistency and universality	Particularity

Source: J. Herkert, 2004, "Engineering Ethics: What It Is and Why It Matters," Earthquake Engineering Symposium for Young Researchers, Charleston, SC, 5–8 August 2005; and S. Krimsky and A. Plough, 1988, *Environmental Hazards: Communicating Risks as a Social Process*, Auburn House Publishing Company, Oxford, UK.

I would like to add something that's not essential to the science, but something I kind of believe, which is that you should not fool the layman when you're talking as a scientist. . . . I'm talking about a specific, extra type of integrity that is not lying, but bending over backwards to show how you're maybe wrong, that you ought to have when acting as a scientist. And this is our responsibility as scientists, certainly to other scientists, and I think to laymen.[14]

I certainly agree, Dr. Feynman!

Teachable Moment: Truth and Turtles

Not all ethics is situational. Certainly, the context does help to define what a right or wrong ethical decision is. However, there are indeed absolute and objective realities. There are certain values that must be guarded fiercely. Professionals earn trust. One of the ways we do this is through honesty.

At a recent workshop of the National Academy of Engineering, John Ahearne, who served as Executive Director of Sigma Xi, the Scientific Research Society, for seven years, warned that scientists should never be arguing about the facts. Facts simply exist. We certainly must debate their meanings and give appropriate attention to their various interpretations, but we must not redefine facts simply to fit our paradigms. We may not like what the facts are telling us, but we must be objective if we are to follow the scientific method. This is particularly crucial for engineers.

We must not "cherry-pick" the science and mathematics upon which an ethical decision is based. I have seen this in recent bioethical discussions on stem cells, where the need and value is based on selected facts. An illustrative example of the conflict between good science and advocacy has to do with sea turtles. This debate may seem trivial or even comical to those outside of the endangered species community, but it is quite telling of how myopic we scientists can be. Last year, I was asked to participate in a discussion with a group of environmental science students at Duke. Their professor, Lynn Maguire, asked the students to consider the ethics of the findings of a journal article on the pros and cons of, well to be blunt, lying to the public. Some students sympathized with Stephen Schneider, a Stanford University climatologist and former government advisor, who argued some years ago that scientists ought to "offer up scary scenarios, make simplified dramatic statements, and make little mention of any doubts we may have. Each of us has to decide what the right balance is between being effective and being honest."[15] Such a distorted version of "utilitarian science" is dangerous. It violates the first canon of science, honest inquiry and reporting. Philosopher of science, C.P. Snow, articulated this principle some fifty years ago:

> The only ethical principle which has made science possible is that the truth shall be told all the time. If we do not penalise false statements made in error, we open up the way, don't you see, for false statements by intention. And of course a false statement of fact made deliberately is the most serious crime a scientist can commit.[16]

Not every scientist and environmental professional embraces Snow's principle. The debate that precipitated the Duke student's quandary was actually over the seemingly innocuous field of biological taxonomy. The dialogues exposed the acceptance by some of the justification of using morally unacceptable means to achieve the greater good.[17] The journal *Conservation Biology* published a number of scientific, philosophical, and ethical perspectives on whether to misuse science to promote the larger goal (conservation) to protect the Black Sea turtle. Even though the taxonomy is scientifically incorrect, i.e., the Black Sea turtle is not a unique species, some writers called for a "geopolitical taxonomy."[18] The analogy of war was invoked as a justification, with one writer declaring that "it is acceptable to tell lies to deceive the enemy." The debate moderators asked a telling question: "Should legitimate scientific results then be withheld, modified, or 'spun' to serve conservation goals?" Continuing with the war analogy, some scientists likened the deceptive taxonomy to propaganda needed to prevent advances by the enemy. The problem is that, as Snow would put it, once you stop telling the truth, you have lost credibility as scientists, even if the deception is for a noble cause.[19] Two writers, Kristin Shrader-Frechette and Earl D. McCoy, emphasized that credible science requires that "in virtually all cases in professional ethics, the public has the right to know the truth when human or environmental welfare is at issue."[20]

Perhaps the most troubling part of this debate is that it is occurring at all. Once we start treating the truth as a commodity, who can blame the public for losing confidence in science and its practitioners? Certainly not C.P. Snow.

Questions

1. Is it ever morally permissible to be less than honest about scientific findings?
2. Why do you believe that philosophers of science, such as Snow and Feynman, are so adamant about truth telling?
3. How might Kant and Mill differ in their views about obligations regarding intellectual honesty in bioethical decision making?

PROFESSIONALISM

Engineering disciplines require a combination of trust and competence, reflected in our codes of ethics. As mentioned, society cedes a substantial amount of trust to a relatively small group of experts, the professionals, in increasingly complex and complicated disciplines that have grown out of the technological advances that began in the middle of the twentieth century and grew exponentially in its waning decades. Key among these advances are those related to biomedicine. In fact, biomedicine is at the crossroads between the engineering and health care professions. Thus, professionalism in biomedical contexts is quite complex, especially when it comes to the development and application of technologies.

TECHNOLOGY: FRIEND AND FOE

> I love technology, but not as much as you, you see . . . But I STILL love technology. . . . Always and forever.
>
> Kip, *Napoleon Dynamite* (2004)[21]

Engineers have been called technological optimists.[22] Engineers love technology. We fashion it, we enhance it, we expand it, and we use it. Technology has two basic definitions. First, it is the application of science. Second, it the product and apparatus resulting from such an application. For example, when we use biological information to address medical problems, this is medical information technology (IT). Technology is also a term that describes the equipment that is derived from the application, such as storage devices, readers, telemetric devices, and imaging interfaces with radiography, sonography, computed tomography (CT), positron imaging tomography (PET), magnetic resonance imaging (MRI) and functional MRI (fMRI). Technology also encompasses intellectual (soft) tools such as informatics to mine for patterns and meanings, such as means for deciphering transcriptions of DNA traces in criminal investigations or epidemiological statistical tools to scrutinize possible causes of disease outbreaks.

The first definition of technology is quite similar to that of engineering. In fact, engineering is the application of science to achieve some practical end. Thus, engineers often refer to technology by its second definition. The stuff of engineering is the technology. Engineers create, adapt, refine, and use technology. So, the technology itself is often neutral from the standpoint of ethics, but engineering is not. How we use technology determines the morality.

The neutrality of technology was on display in 2005, for example, in the bioethical debate surrounding Theresa Marie "Terri" Schiavo (1963–2005). In 1990, after suffering from respiratory and cardiac arrest, Schiavo fell into a coma and was eventually diagnosed to be in a persistent vegetative state (PVS). A person in a vegetative state is characterized by a physician to be unaware and unconscious. However, a person in a vegetative state differs from a person in a coma since the former will show partial consciousness (e.g., swallowing, smiling, crying, grunting, moaning, or screaming), but these actions are not connected to an external stimulus. After about a month, a physician may classify the person in such a condition to be in a PVS.[23]

Technology took a prominent place in this debate, since in previous centuries or decades, a person in Shiavo's condition would likely have died because of the inability to provide fluids and nutrition.

However, feeding tubes, medical monitoring devices, and other technologies now allow PVS patients to survive.

> Anyone who has studied the history of technology knows that technological change is always a Faustian bargain: Technology giveth and technology taketh away, and not always in equal measure. A new technology sometimes creates more than it destroys. Sometimes, it destroys more than it creates. But it is never one-sided.
> Neil Postman (8 March 1931–5 October 2003), educator, media theorist and cultural critic[24]

Postman's caution about the dichotomous, even bifurcated, nature of technology is especially relevant to bioethics. The good stuff is readily apparent to both professionals and patients. A visit to the local general practitioner's office, let alone a health clinic or hospital, provides evidence of the rapid pace at which biomedical engineers have added to the practice of health care. Blood oxygen level readings can be taken by devices that are completely noninvasive (light penetration). At the ophthalmologist's office, glaucoma tests are on the ready and lasers can be used to reduce eye pressure. Epoxies and composites used by the dentist to repair and to preserve teeth that would have been lost a decade ago are a tribute to advances in materials sciences.

The potentially bad stuff may not be so apparent, but it can be found. Information technologies and networks that allow for timely and reliable health care have also introduced new threats to patients' privacy. Misuse of new treatments is possible, especially if the new technologies are not accompanied by adequate training and oversight. Unforeseen risks and complications are often only apparent after a device is introduced into the general patient population.

Teachable Moment: Medical Device Risk

The convexo-concave (C-C) tilting disk heart valve (Figure 5.3), designed and manufactured by Björk–Shiley, has received much attention as a case of failure of a medical device.[25] Few engineered products have undergone the scrutiny that has focused on the C-C valve, for good reason. The C-C valve was designed as an improvement to a replacement valve approved in 1969. Among the reasons for designing the new valve was to reduce the risk of thromboembolism. The new valve received premarket approval in 1979. Unfortunately, the "improvement" to the design was a factor in its calamitous failure. The design changes included repositioning the outlet-strut weld and changing the angle of orientation of the outlet strut to the ring.

The first failure of the outlet strut occurred during the premarket approval process, followed by other failures addressed during the company's quality assurance process. However, the underlying cause of the failure was not identified. This was later found to be an unexpected load on the outlet strut's tip. The increased loading was the result of the redesign aimed at reducing thromboembolism. Subsequently, in 1984, the design modifications to address these loadings essentially returned the outlet strut in the valve to the original configuration. Monitoring was also upgraded, with an examination of the potential failure site in each valve by scanning electron microscopy. The loss of life appears to have been preventable had the design process included measures to address possible risks from seemingly small changes.

Figure 5.3 Photograph of Björk–Shiley convexo-concave (C-C) tilting disk prosthetic heart valve. Tilting disk is held in place between fixed inlet strut (foreground) and welded smaller outlet strut (background). Lighter-colored outer material is sewing ring.

Arguably, a much more reliable approach on a similar device was applied to the Omniscience bileaflet heart valve, manufactured by Baxter. The company readily withdrew the valve from the market when its postmarket surveillance program showed failures. In addition, the company doggedly searched for possible causes of failure and implemented modifications to address and to eliminate factors leading to the failures. As a result, few deaths occurred and serious litigation costs were avoided. In addition, the company employed wide-ranging postmarket flow studies that showed cavitation, which was subsequently addressed in all subsequent heart valve designs.

Questions

1. What lessons can be learned from the Björk–Shiley convexo-concave valve?
2. How close are the company's actions to the negative and positive ethical paradigms?
3. What role can institutional culture play in the difference between the Björk–Shiley convexo-concave valve and Omniscience bileaflet heart valve corrective actions?

RISK HOMEOSTASIS AND THE THEORY OF OFFSETTING BEHAVIOR

Any engineer's paramount concern, ethical and otherwise, is design failure. Product liability is one of the public's responses to past failures and a means to induce researchers and designers to build in

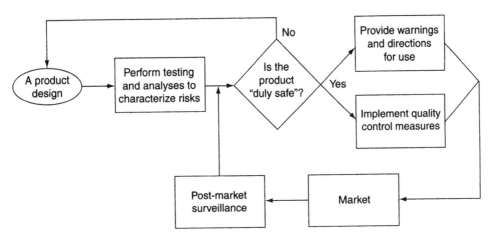

Figure 5.4 Licensing Process of the US Food and Drug Administration.
Adapted from: B. Fishhoff and J.F. Merz, 1994, "The Inconvenient Public: Behavioral Research Approaches to Reducing Product Liability Risks," in: National Academy of Engineering. *Product Liability and Innovation: Managing Risk in an Uncertain Environment*, National Academy Press, Washington, DC.

safeguards to prevent future failures. One example of product liability oversight is that of the US Food and Drug Administration (FDA). Figure 5.4 is a simplified model of FDA's licensing process.

The reasons that products fail are twofold. Either they have flaws or they are misused. Engineers clearly have an ethical responsibility to prevent the first cause and arguably much of the second cause. Good engineering practice prevents the first, but what about the second? At what point are engineers to blame for misuse of their products?

Blame is an interesting ethical and psychological concept. Sometimes, when it is clearly their own fault, people will willingly accept the blame for not using a device or drug properly. For example, they may readily see that they ignored a warning label to avoid taking certain drugs while consuming alcohol or to quit playing contact sports after recurrent concussions. Increasingly more likely, however, they are looking to blame someone else. This inclination can result both cognitively and motivationally, as noted by Baruch Fishhoff, a psychologist, and Jon Merz, an engineer:[26]

> Cognitively, injured parties see themselves as having been doing something that seemed sensible at the time, and not looking for trouble. As a result, any accident comes as a surprise. If it was to be avoided, then someone else needed to provide the missing expertise and protection. Motivationally, no one wants to feel responsible for an accident. That just adds insult to injury, as well as forfeiting the chance for emotional and financial redress.
>
> Of course, the natural targets for such blame are those who created and distributed the product or equipment involved in an accident. They could have improved the design to prevent accidents. They should have done more to ensure that the product would not fail in expected use. They could have provided better warnings and instructions in how to use the product. They could have sacrificed profits or forgone sales, rather than let users bear (what now seem to have been) unacceptable risks.

Another explanation of misuse may fall under the first definition as a design flaw. It is known as "risk homeostasis"[27]. Basically, users defeat built-in factors of safety by asserting new way to use the products. If such aggressiveness occurs, it may simply be because consumers want more from their products. Some argue that the instructions and warnings accompanying products should be given sufficient consideration, and if a user willfully ignores this information, then this constitutes the user's autonomy in a rational decision made with "informed consent." Others disagree holding that liability extends beyond labeling and reasonable use.[28]

Economists have found that policies designed to protect the public may inadvertently lead to "attenuation and even reversal of the direct policy effect on expected harm ... because of offsetting behavior (OB) by potential victims as they reduce care in response to the policy."[29] Much of the research has been related to transportation, especially road safety. For example, drivers with antilock brake systems tend to tailgate more closely and the use of helmets has not been commensurate with expected injury prevention in cycling. It is logical that such behavior would also extend to medical devices.

The offsetting behavior can even be a factor in macroethics. For example, birth control devices and drugs have clearly affected sexual and drug use behaviors. Even the public response to the acquired immunodeficiency syndrome (AIDS) epidemic was colored by offsetting behavior. For example, many did not question the morality of multiple partners so much as the failure of medical technology (e.g., a drug or a vaccine or a device) to alleviate the sexual- and drug-related transmissions of the disease. To many, it was not an ethical problem so much as a technological one. Drug users needed cleaner syringe needles and potential, sexual partners needed quicker and more reliable human immunodeficiency virus (HIV) screening technologies. The offsetting behaviors resulting from reliance of technologies (birth control pills, intrauterine devices, etc.) were clearly a factor in the changes in social mores following the 1960s, and an indirect societal challenge in trying to address the impending epidemic.

ARTIFACTS

Engineers are vulnerable to artifacts. That is, their designs may lead to impact down the road. Biomedical risk assessment should not be limited to biomedical engineers. For example, biomedical enterprises can be harmed by failure in any engineering discipline. One of the most notorious biomedical device failures had little to do with biomedical engineering and much to do with computer science and software engineering. A linear accelerator (i.e., linac) known as Therac-25 was used in the 1980s for X-ray medical radiation therapy. In the United States and Canada, six accidents associated with this device were reported between June 1985 and January 1987. These accidents led to substantial overdoses resulting in deaths and serious injuries. The accidents resulted from a design artifact:

> Between the patient and the Therac-25's radiation beam was a turntable that could position a window or an X-ray-mode target between the accelerator and patient, depending on which of two modes of operation was being used. If the window was positioned in the beam's path with the machine set to deliver radiation through the X-ray-mode target, disaster could result because software errors allowed the machine to operate in this configuration.[30]

In the Therac-25 case, the offsetting behavior was not that of the patient, but of the technical user. Users evidently became overly reliant on the software. In particular, the computer-aided system eliminated the previous Therac model's need for the operator to enable independent protective circuitry, as well as mechanical interlocking devices to prevent radiation overdosing. The "improved" model relied to a much larger extent on software to provide such protections. What made matters worse was that software was not thought to be the problem, so as the accident reports rolled in, the manufacturer could not replicate the accidents, leading to the assumption that the hardware needed a few minor enhancements, after which the company pronounced that the Therac-25 model was greatly improved. In fact, the real problem went unrecognized. Designers remained unaware that embedded systems – those where the microprocessor is not accessible to reprogramming by the user – increase the risks from inadequate quality assurance. Software risk assessment is much more difficult than hardware risk assessment.

The engineering canon that an engineer must be competent within an area of expertise is not an excuse not to consider possible flaws that result down the road. In fact, just the opposite is true. That is, part of the competence requires consideration of offsetting behavior. So, then what led to these artifacts? The FDA and others found the following:[31]

1. *Overconfidence in software* – Engineers must overcome the temptation to assume that problems are usually about hardware.
2. *Confusing reliability with safety* – The Therac software worked thousands of times without error, but even these probabilities are unacceptable when consequences include death and serious injury.
3. *Lack of defensive design* – Redundancies as well as self-checks, trouble-shooting, and error detection and correction systems are vital. A worst-case design scenario would have been appropriate.
4. *Failure to address root causes* – Causes were misidentified and the corrections were solving problems other than the ones leading to the real failures.
5. *Complacency* – The Therac-25 was a victim of the linac's own success. Much like the Santillan case, the cause that led to the artifact was part of a routine process. Continued success does not mean due diligence can be shortcut.
6. *Unrealistic risk assessment* – The failure analysis assumed independence and missed key factors. This can be likened to Kuhn's paradigm shifts or Janus' groupthink.

Seemingly mundane or benign technologies can have adverse effects. For example, video and internet games may be fun or even instructive in certain scenarios. However, they can also expose susceptible persons, especially children, to violence and antisocial examples. Even nonviolent games can be damaging by creating opportunity risks, such as the time a child (or adult) loses in actual physical activity and the indirect link to the current problem of obesity in the United States. The games may also render children less alert. My daughter, a teacher, has noted that she can almost always tell when a child has been playing video games by the child's lack of acuity and interest.

Thus, technology itself may be ethically neutral, but the risk homeostasis and offsetting behaviors that it spawns are not. This is another type of the consequence that must be considered in an engineer's design.

AUTOMATION AND MECHANIZATION OF MEDICINE

An emerging technological challenge for engineers is being manifested by the changing work environment. Medicine is increasingly dependent on automated systems (e.g., information and recordkeeping, diagnostics, and treatment). Numerous clinical and biomedical engineers are employed by firms, agencies, and institutions that also play a role in how an engineer can manage the automated and mechanized work environment. And, the engineers themselves are adapting these technologies to suit the needs of the organization.

The engineering codes of ethics recognize this by helping to remind the engineer that he or she serves numerous interests (i.e., the so-called conflict of interest and nondisclosure clauses, as well as the previously mentioned faithful agent provisions). The new conflicts of commitment and interests will become increasingly more complicated in the coming decades. The workforce has undergone significant change since the 1980s, with a greater number of "contractors" and fewer actual employees in many organizations. This means that the new engineer will need to be abundantly more self-sufficient than even a decade ago.

The Future Engineer (FE) and Professional Engineer (PE) certification processes will become even more important. Many engineers do not follow this process formally, but can benefit greatly by adhering to its principles. The professional will need a whole host of mentors, a tenet of the PE. The interdisciplinary nature of bioethics requires mentoring in each of the disciplines and perspectives. The actual amount of tutelage will vary considerably. If an engineer seeks to design medical devices, ongoing advice from the surgeons and attending physicians who use such devices must be a key part of the engineer's lifelong learning. If the engineer is more concerned about drug delivery systems, some time with a pharmacological modeler and in the laboratory a biochemist who conceive of the drugs to be administered would be worthwhile.

Mentorship in an automated and mechanized work environment must be both vertical and horizontal. Vertical integration of technical knowledge comes from the formal line-and-staff heirarchy, as well as from experts throughout the organization. Horizontal integration requires that the engineer seek out the advice, perspectives, and wisdom from those technicians who operate the equipment and manage the systems. Most good designs are the result of the designer's appreciation of user needs.

These relationships must augment the already complex continuing education from within one's specialty in the engineering profession. And, in both cases, after the initial experience, a career-long relationship with these mentors should be maintained. The mix of inputs from trusted mentors and colleagues could make for a solution very different from one where only handbooks are consulted. This mentor–learner model also helps to ensure that the knowledge and wisdom of this generation are passed on to the next, i.e., a means of providing a way to preserve "corporate" memory in the ever-changing fields of engineering.

PROFESSIONAL CONSIDERATION: DO ENGINEERS HAVE PATIENTS?

The various engineering codes of ethics clearly state that engineers have "clients." The fourth canon of the National Society of Professional Engineers[32] (NSPE) states:

> Engineers shall act for each employer or client as faithful agents or trustees.

Furthermore, the preamble to the NSPE code affirms:

> Engineering has a direct and vital impact on the quality of life for all people.

The code also cautions the engineer to know the limits of specialization. The second canon states that engineers shall "perform services only in areas of their competence." Thus, the code recognizes the high degree of specialization and that, even within the engineering profession, few fellow engineers will fully grasp the expertise of certain engineering specialties. This is particularly relevant to biological specialties within engineering.

Engineers design and test devices to be used to treat diseases and to ameliorate the quality of life of individual patients. Engineers conceive of engineering systems to be used by physicians in numerous health care scenarios. These physicians are the clients of the engineer. The engineer is the trustee of the public insofar as these devices and systems must hold paramount health, safety and welfare. But, does this mean engineers have "patients"?

Most engineers would agree that they have an indirect obligation to the patients. John Fielder, a philosophy professor at Villanova, explored this obligation one step further by considering whether engineers have a direct obligation to patients.[33] Fielder refers to the rather common medical situation of a hip replacement. Engineers have successfully designed artificial hip systems for decades. There is also a track record for physicians to use the engineered systems ethically and in the best interests of the recipient of the technology, the patient. Fielder specifically mentions that the physician:

- recommends the optimal medical treatment for the problem; in this instance a deteriorated hip;
- sufficiently informs the patient of his or her options (including not receiving an artificial hip system); and
- manages the patient's course of treatment.

Likewise, Fielder points out that the engineers outside of the health care facility who design and fabricate the artificial hips are beholden to their own engineering codes of ethics. These codes require that the designs be such that the products are safe (i.e., they have acceptable risk) and competently designed (i.e., they meet specified tolerances and performance criteria). Moreover, the codes require that the engineers completely and properly disclose any flaws, limitations, use restrictions, and other dangers of the designed products as soon as such shortcomings are known. Fielder argues that for bioengineers who work within health care and research settings, the ethical obligations are less clear, since they are immersed in the medical milieu.

The future engineer will have to design products in a way that is more "transparent" and "coherent" than those of the previous generation.[34] The process will be transparent because future patients, clients, and the affected public (and their attorneys!) can watch them as they are being developed due to the openness of the Internet, the increased technical savvy of larger numbers of nonengineers, and the growing comfort of what had heretofore been the sole domain of the scientists, engineers, and technicians. So, at a minimum, engineers are part of the team that supports the physician, who in turn treats the patient. In this regard, the clinical engineer is held responsible to both the client (physician) and patient (recipient of the engineered system).

TECHNOLOGICAL RELIABILITY

The design process will be coherent in that the public will expect consistency of cause and effect. The public needs to see concrete products (plans and programs are merely means to the end), with an understanding of the potential and actual risks and the steps that have been taken to ensure safety. They also will expect to understand how the products will improve the quality of life. A counterexample would seem to be some of the television ads that implore the viewing audience to ask their doctor to prescribe some drug, although the commercials seem to be clearer about the side effects than about the reason for taking the drug in the first place. Coherence, then, calls for engineers to design monitoring and ongoing assessment into any design. How well does it perform over its useful life?

The reliability of the device or system being designed by an engineer determines in part whether an engineer has behaved competently. Also, an engineer who uses or oversees the use of a particular technology must understand the workings, limitations, and extensions of applications to various scenarios. If this understanding is vital to the mission (e.g., a clinical engineer's recommendations to medical professionals regarding an insulin pump in an "at-risk" patient), then not knowing what reasonably should be known is unethical.[35] The decision as to whether the engineer "should have known" is complex. The perspective influences the perception. For example, consider the situation where the physician makes a professional decision based on a clinical engineer's recommendation that a certain pump designed for adults has limited applications and may pose significant risks when used in a small child, based on the engineer's knowledge of fluid mechanics, and the physician decides to use the pump anyway. Most reasonable people would agree that the attending physician is entitled to make that decision so long as the risks are reasonable and acceptable. However, if the engineer's calculations of the fluid mechanical properties and performance are incorrect, most reasonable people would consider such a mistake to be unethical.

Even more unethical would be if the engineer's miscalculations were influenced by a conflict of interest, such as the desire to use the pump in risky applications because the engineer has an undisclosed financial relationship with the pump manufacturer (e.g., spouse works for a subsidiary company). This type of behavior moves from the realm of a "mistake" to a "misdeed." Incidentally, this is an example of how disclosure could have prevented an escalation of wrongdoing. If the engineer had disclosed formally and to the attending physician that his spouse was employed by the company that manufactured the pump and that the company had asked him to extend the application to children, the misdeed, but not the mistake, would have been mitigated to a great degree (and the onus on deciding to use the advice of the engineer again rests with the physician). The onus would have shifted to the physician who must decide if the engineer's conflict of interest should proclude using the pump. Then, if this decision is wrong the gravity of the mistake would be the principal focus of ethics. For example, if the calculation errors were based on an assumption that was unreasonable (e.g., it violates the laws of motion or the Bernoulli principle), it is likely that the ethical review would be more critical of the engineer's performance than if the miscalculation were more arcane (e.g., the difference between the pharmacodynamics of a child and that of an adult is nonlinear, but the engineer assumed it to be linear). However, if the engineer does not know enough to include the most important variables, factors, and equations in the calculations, he is likely going to be found lacking in another engineering ethical principle; that is, engineers shall perform services only in areas of their competence. In both cases, the standard is really not that of a "reasonable person," but of a "reasonable engineer."

So then, how are engineers held accountable for technological performance? Part of the answer is the reliability of the system and devices being used. Thus, reliability is a direct measure of design performance and an indirect measure of ethical performance. Like risk, reliability is an expression of likelihood, but rather than conveying something bad, it tells us the probability of a good outcome. Reliability is the extent to which something can be trusted. A system, process, or item is reliable to the extent that it performs the designed function under the specified conditions during a certain time period.

Thus, reliability means that something will not fail prematurely. Or, stated more positively, reliability is expressed mathematically as the probability of success. Thus reliability is the probability that something that is in operation at time 0 (t_0) will still be operating until the designed life [time = (t_t)]. People who will receive an implanted device want to be assured that it will work and will not fail. If $t_t = 100$ years, most patients would be happy. If $t_t = 30$ days for absorbable sutures, this may be acceptable, but very likely not to be acceptable for nonabsorbable sutures (depending on the actual use by the physician). Thus, "failure" depends on the expected life of the engineered system.

The probability of a failure per unit time is the "hazard" rate, a term familiar to risk assessment, but many engineers may recognize it as a "failure density," or $f(t)$. This is a function of the likelihood that an adverse outcome will occur, but note that it is not a function of the severity of the outcome. The $f(t)$ is not affected by whether the outcome is very severe (e.g., complete failure to provide oxygen) or relatively benign (e.g., muscle soreness or very small amount of leakage of a nontoxic lubricant). It is up to the designer to decide which factors are to be tested for reliability.

The likelihood that something will fail at a given time interval can be found by integrating the hazard rate over a defined time interval:

$$P\{t_1 \leq T_f \leq t_2\} = \int_{t_1}^{t_2} f(t)dt \tag{5.1}$$

where T_f = time of failure.

Thus, the reliability function $R(t)$ of a system at time t is the cumulative probability that the system has not failed in the time interval from t_0 to t_t:

$$R(t) = P\{T_f \geq t\} = 1 - \int_0^t f(x)dx \tag{5.2}$$

One major point worth noting from the reliability equations is that everything we design will fail. Engineers can improve reliability by extending the time (increasing t_t). And, this is done by making the system more resistant to failure. For example, proper engineering design of a barrier between tissues can decrease the migration of a fluid. However, the barrier does not completely eliminate failure, i.e., $R(t) = 0$; it simply protracts the time before the failure occurs (increases T_f). Reliability can be affected by societal factors. For example, health studies in much of the twentieth century focused on adult white males to a great extent. It is possible to expect a decrease in T_f because certain attributes of children, women, and minorities were not properly considered in the original design of a device (e.g., the presence of different chemicals in the endocrine, neural, and immune systems, which changes the performance of a device).

A discipline within engineering, i.e., reliability engineering, looks at the expected or actual reliability of a process, system, or piece of equipment to identify the actions needed to reduce failures, and once a failure occurs, how to manage the expected effects from that failure. Thus, reliability is the mirror image of failure. Since risk is really the probability of failure (i.e., the probability that our system, process, or equipment will fail), risk and reliability are two sides of the same coin. A device leaking chemicals into the bloodsteam is an engineering failure, as is exposure of people to medical monitoring and sensing that has not been properly tested. A system that protects one group of people at the expense of another is also a type of failure (i.e., injustice), such as a device that works well in the average male but has an inordinate failure rate in the average female. This failure only occurs if the design limitation is not properly disclosed. For example, the system would be a failure if physicians used the device expecting the same performance in women as in men. The designer must clearly denote the limitations and shortcomings of any design. One size seldom fits all. So, if we are to have reliable engineering we need to make sure that whatever we design, build, and operate is done with an eye toward possible misapplications. Otherwise, these systems are, by definition, unreliable.

The most common graphical representation of engineering reliability is the bathtub curve (Figure 5.5). The curve is U-shaped, meaning that failure is more likely to occur at the beginning (infant mortality) and near the end of the life of a system, process, or equipment. Actually, failure can occur even before infancy. In fact, many problems in injustice occur during the planning and idea stage. A great idea may be shot down before it is born. Or, certain groups may be excluded out of hand even before the design process begins. Such exclusion can be explicit, such as stating that the company will target the richest 5% (e.g., for cosmetic surgical devices) or implicit, such as when trials are based on old selection criteria that continue to exclude certain members of society from focus groups.

Error can gestate even before the engineer becomes involved in the project. This "miscarriage of justice" follows the physiological metaphor closely. Certain groups of people have been historically excluded from preliminary discussions, so that if and when they do become involved the design process has advanced well beyond the "power curve" and they have to play catch-up. The momentum of a project, often being pushed by the project engineers, makes participation very difficult from some groups. So, we can modify the bathtub distribution accordingly (Figure 5.6).

Note that in engineering and other empirical sciences there is another connotation of "reliability," which is an indication of quality, especially for data derived from measurements, including health data. In this use, reliability is defined as the degree to which measured results are dependable and consistent with respect to the study objectives (e.g., percentage of pumps performing as designed after five years). This

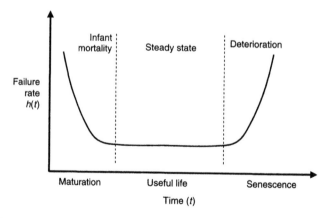

Figure 5.5 Prototypical reliability curve, i.e., the bathtub distribution. The highest rates of failure, *h(t)*, occur during the early stages of adoption (analogous to infant mortality) and when the systems, processes, or equipment become obsolete or begin to deteriorate. For well-designed systems, the steady-state period can be protracted, e.g., for decades. Note that time (*t*) varies by engineered system it may be days for topical or temporary devices to years for an engineered joint to even longer for implanted systems (e.g., artificial heart valves).

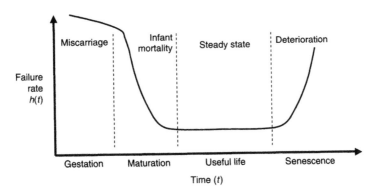

Figure 5.6 Prototypical reliability curve with a gestation (e.g., idea) stage. The highest rate of failure, $h(t)$, occurs even before the system, process, or equipment has been made a reality. Exclusion of people from decision making or failure to get input about key scientific or social variables can create a high hazard.

specific connotation is sometimes called "test reliability" in that it indicates how consistent measured values are over time, how these values compare to other measured values, and how they differ when other tests are applied. Test reliability, like engineering reliability, is a matter of trust. As such, it is often paired with test validity; that is, just how near to the true value (as indicated by some type of known standard) the measured value is. The less reliable and valid the results, the less confidence scientists and engineers have in interpreting and using them. This is very important in engineering communications generally, and risk communications specifically.

To solve problems, scientists, engineers, and decision makers need to know how reliable and valid the data are. And, this information must be properly communicated to those potentially or actually being affected. This includes candid and understandable ways to describe all uncertainties. Uncertainties are ubiquitous in risk assessment. The engineer should take care not to be overly optimistic, nor overly pessimistic about what is known and what needs to be done. Full disclosure is simply an honest rendering of what is known and what is lacking for those listening to make informed decisions. But, remember, a word or phrase can be taken many ways. Engineers should liken themselves to physicians writing prescriptions. Be completely clear, otherwise confusion may result and lead to unintended, negative consequences.

LOW TECH ENGINEERING

One of the differences between engineers and numerous other professions is the role of technology. First, engineers are behind the conception and design of many emerging technologies. Second, engineers (like medical doctors) are increasingly dependent upon technologies to do their jobs. But what exactly is technology?

Biomedical and other engineering disciplines will continue to improve technology, giving rise to better global health and improved health care. Sharing of technologies among nations is an ethical obligation, especially "low-tech" solutions that can be applied in remote and rural regions of developed countries. Such solutions need to be developed and shared with developing countries, where possible, since these are more likely to be successful when resources are scarce and access to highly trained practitioners is limited.

INFORMATION TECHNOLOGY

The sheer amount and complexity of data and information are enormous at present and will continue to grow. Bioengineers must be comfortable with ambiguity, since every professional decision is made under uncertainty, often a great deal of it. A lot of what scientists and engineers do does not always seem logical to most observers. And explaining the meaning of data can be very challenging. This is in part due to the incompleteness of the understanding of the methods used to gather data. Even well-established techniques like chromatography have built-in uncertainties. I like these scientific and methodological uncertainties to Plato's Allegory of the Cave in *The Republic*. Recall that Plato argued when we are ignorant of how things work (actually Plato was referring to people untutored in the Theory of Forms, but let us not worry about that). We are like prisoners chained inside of a cave, who are not allowed to turn their heads. So, what we see is exclusively is the wall of the cave. Behind us burns a fire, and between the fire and us is a parapet, on which puppeteers are walking. The puppeteers behind us periodically hold up puppets that cast shadows on the wall of the cave. We cannot see these puppets (i.e., "reality"). What we do see and hear are the shadows cast and echoes from the objects (i.e., measurements).[36] Recall from statistics that the definition of accuracy is how close we are to the "true value" or reality. Thus our instruments and other sensors are to some extent what we see on the cave wall and not the puppets themselves. In chromatography, for example, we are fairly certain that the peaks we are seeing represent the molecule in question, but actually depending on the detector, all we are seeing is the number of carbon atoms (e.g., flame ionization detection) or the mass-to-charge ratios of molecular fragments (e.g., mass spectrometry), but not the molecule itself. Add to this instrument and operator uncertainties and one can see that even the more accepted scientific approaches are biased and inaccurate, let alone an approach like mathematical modeling, where assumptions about initial and boundary conditions, values given to parameters, and the propagation of error render our results even more uncertain.

The first ethical principle of the engineering profession is to hold paramount the health, safety, and welfare of the general public. However, the public itself is changing, as exemplified by two seemingly contradictory trends. The first is that there seems to be a great divergence between the technologically literate and those not conversant in technical matters. Those who are trained in the technical fields appear to be becoming more highly specialized and steeped in "techno-jargon," leaving the majority of people dependent upon whatever the technologists say. Simultaneously, there seems to be a second trend of greater fluency with what were formerly specialized engineering terms in the larger public arena.

The first trend, the technological literacy gap, as it applies to engineering is not the same as the so-called digital divide, i.e., the difference in the access and use of information technology (IT) by groups of different socioeconomic status (SES), race, and gender. The literacy gap is more fundamental than any single issue. Some fear that the future citizenry is ill-prepared and undereducated to be able to participate fully in an increasingly complex, technology-rich future. In the United States, this concern is in part manifested by so many students' inadequate preparation in math and science upon their entry into the present and future workforce. This is the mirror image of the problem of engineers' inadequate training in and appreciation for the humanities and social sciences. Engineers will definitely have to enhance their reach to include a greater number of perspectives in their projects and, simultaneously, we need to help the members of the general public increase their appreciation for things technological. This confluence will not be easy. For example, the National Center for Education Statistics[37] reports that in 1999 the United States lagged behind much of the developed world, and even behind a number of developing nations, in its middle school and high school students' achievement in mathematics and science (Table 5.2).

Interestingly, the tests used in these comparisons place heavy emphasis on the math and science underpinning engineering. The Trends in International Mathematics and Science Study (TIMSS) measures aptitude in fractions and number sense, algebra, geometry, data representation, analysis, and probability, measurement, earth sciences, life sciences, physics, chemistry, environmental science, scientific inquiry, and the nature of science.

The other trend is that previously complicated and highly technical concepts and jargon are becoming increasingly mainstreamed. So, while many students do not seem motivated to participate fully in the increasing technological demands of society, they somehow are gaining a large repertoire of scientific expertise in their everyday lives.[38]

Few things have changed science and engineering more than the computer. In fact, engineering has traditionally been among the "first adopters" of many new technologies, and this has certainly been the case for IT. But, like any tool, it can either be well used or misused. Two quotes seem to capture the range of thinking on whether the computer is a blessing or a curse:

In a way not seen since Gutenberg's printing press that ended the Dark Ages and ignited the Renaissance, the microchip is an epochal technology with unimaginably far-reaching economic, social, and political consequence.

Michael Rothchild[39]

While all this razzle-dazzle connects us electronically, it disconnects us from each other, having us "interfacing" more with computers and TV screens than looking in the face of our fellow human beings. Is this progress?

Jim Hightower[40]

At one extreme we have IT being a panacea and at the other end the bane of modern civilization. Arguably, most engineers and technicians come closer to Rothchild's perspective, as strong advocates for the application of IT in every aspect of engineering. After all, engineers need to apply the sciences, so we must avail ourselves to the best tools and methods to accomplish this. And, the adoption of these new technologies by society has been phenomenally rapid and the rate of adoption has increased with each new technology. For example, it only took sixteen years for the personal computer to be adopted by 25% of US households and merely seven years for them to accept Internet access (Table 5.3).

There are also downsides, such as the temptation for engineering students to observe the world through their computer screens, preferring virtual solutions to virtual problems over actual solutions to real-world problems. But there is no arguing that computer modeling has become an essential part of almost any assessment, design, or engineering activity.

Is this an oxymoronic situation? The future profession will have to reconcile any technical deficiencies indicated by the math and science gaps with the creeping technological savvy of the general public.

Table 5.2
Mathematics and science achievement of eighth graders in the 1999 Trends in International Mathematics and Science Study (TIMSS), formerly known as the Third International Mathematics and Science Study

Mathematics		Science	
Nation	Average	Nation	Average
Singapore	604	Chinese Taipei	569
Korea, Republic of	587	Singapore	568
Chinese Taipei	585	Hungary	552
Hong Kong SAR	582	Japan	550
Japan	579	Korea, Republic of	549
Belgium-Flemish	558	Netherlands	545
Netherlands	540	Australia	540
Slovak Republic	534	Czech Republic	539
Hungary	532	England	538
Canada	531	Finland	535
Slovenia	530	Slovak Republic	535
Russian Federation	526	Belgium-Flemish	535
Australia	525	Slovenia	533
Finland	520	Canada	533
Czech Republic	520	Hong Kong SAR	530
Malaysia	519	Russian Federation	529
Bulgaria	511	Bulgaria	518
Latvia-LSS	505	**United States**	**515**
United States	**502**	New Zealand	510
England	496	Latvia-LSS	503
New Zealand	491	Italy	493
Lithuania	482	Malaysia	492
Italy	479	Lithuania	488
Cyprus	476	Thailand	482
Romania	472	Romania	472
Moldova	469	Israel	468
Thailand	467	Cyprus	460
Israel	466	Moldova	459
Tunisia	448	Macedonia, Republic of	458
Macedonia, Republic of	447	Jordan	450
Turkey	429	Iran, Islamic Republic of	448
Jordan	428	Indonesia	435
Iran, Islamic Republic of	422	Turkey	433
Indonesia	403	Tunisia	430
Chile	392	Chile	420
Philippines	345	Philippines	345
Morocco	337	Morocco	323

TIMSS was developed by the International Association for the Evaluation of Educational Achievement to measure trends in students' mathematics and science achievement. The regular four-year cycle of TIMSS allows for comparisons of students' progress in mathematics and science achievement.

Table 5.3
Years needed for 25% of US households to adopt new technologies

Technology	Years to reach 25% of US households
Automobiles	55
Electricity	45
Telephone	35
Radio	27
Television	25
Personal computers	16
Cellular phones	13
Internet (World Wide Web)	7

Source: S. Baase, 2003, *A Gift of Fire: Social Legal, and Ethical Issues for Computers and the Internet*, Second Edition, Prentice-Hall, Upper Saddle River, NJ.

Perhaps, the best strategy is to be ready to explain complicated engineering concepts in a straightforward manner, but at the same time, be prepared for a public that expects high-tech solutions to their problems.

THE ETHICS OF NANOTECHNOLOGY

Approaching emergent technological developments in an ethical way is a daunting challenge. It is a balance between irresponsible, societal risk-taking and irrational anxieties that lead to loss or delays of opportunities. For example, research at the nanoscale is likely to change biomedicine profoundly. The state of the science of biomedicine has entered a new era (Figure 5.7). Atomic scale instruments are now available to study ways to affect materials and processes; however, it may soon be such that manipulations by nanorobots, nanomachinery, and other nanoscale systems will take place beyond what is discernible. Table 5.4 describes the similarities and differences in ethical considerations between conventional technologies and nanotechnologies.

According to the National Science Foundation,[41] nanoscale science and engineering will lead to seemingly countless advances, including:

- Manufacturing: The nanometer scale is expected to become a highly efficient length scale for manufacturing. At this scale, engineering will lead to materials with high performance, unique properties, and functions that traditional chemistry could not create.
- Electronics: Nanotechnology is projected to yield an annual production of about $300 billion for the semiconductor industry and a few times more for global integrated circuit sales within 10 to15 years.
- Improved health care: Nanotechnology will help to extend the life span, improve its quality, and extend human physical capabilities.

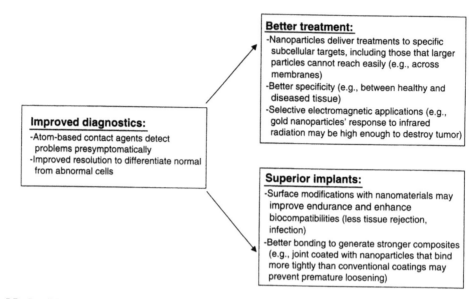

Improved diagnostics:
-Atom-based contact agents detect problems presymptomatically
-Improved resolution to differentiate normal from abnormal cells

Better treatment:
-Nanoparticles deliver treatments to specific subcellular targets, including those that larger particles cannot reach easily (e.g., across membranes)
-Better specificity (e.g., between healthy and diseased tissue)
-Selective electromagnetic applications (e.g., gold nanoparticles' response to infrared radiation may be high enough to destroy tumor)

Superior implants:
-Surface modifications with nanomaterials may improve endurance and enhance biocompatibilities (less tissue rejection, infection)
-Better bonding to generate stronger composites (e.g., joint coated with nanoparticles that bind more tightly than conventional coatings may prevent premature loosening)

Figure 5.7 Possible improvements to medicine made possible by nanoscale research.
Adapted from: A.P. Alivisatos, 2001, Less Is More in Medicine, *Scientific American*, September, pp. 59–65.

Table 5.4
Ethical considerations of conventional technological and nanotechnological development

Conventional approach to ethics	Ethical aspects of nanotechnology
Identify and respond to stakeholders who may have an interest or who are affected	Unable to identify clear moral leadership directing the course of research
Identify issues of justice that may result	Certain components will be unseen and inaccessible
Establish guidelines governing moral rules and principles	No sensual human-to-technology interface (e.g., earphones, laptop computer, mouse)
Assess rights (human and nonhuman species), consequences (e.g., harm), and risks	Targeted for multiple and varied applications, including biomedical, food, cosmetics, devices, and monitoring systems (ambiguous places to assess ethical aspects)
Determine associated values and beliefs that could be threatened or protected	Swift and dangerous rate of technological development (little time to consider possible consequences)
Outline duties and responsibilities possibly associated with developing the technologies	Powerful role of personal belief, cultural myth, and human imagination to conceptualize and justify nanotechnological developments

Source: R.W. Berne, 2006, *Nanotalk: Conversations with Scientists and Engineers about Ethics, Meaning, and Belief in the Development of Nanotechnology*, Lawrence Erlbaum Associates, Mahway, NJ.

- Pharmaceuticals: About half of all production will be dependent on nanotechnology, affecting over $180 billion per year in 10 to15 years.
- Chemical plants: Nanostructured catalysts have applications in the petroleum and chemical processing industries, with an estimated annual impact of $100 billion in 10 to 15 years.
- Sustainability: Nanotechnology will improve agricultural yields for an increased population, provide more economical water filtration and desalination (such as the flowthrough capacitor with aligned carbon nanotube electrodes), and enable renewable energy sources (such as highly efficient solar energy conversion); it will reduce the need for scarce material resources and diminish pollution leading to a cleaner environment. For example, in 10 to 15 years, projections indicate that such nanotechnology-based lighting advances have the potential to reduce worldwide consumption of energy by more than 10%, reflecting a savings of $100 billion dollars per year and a corresponding reduction of 200 million tons of carbon emissions (as well as significantly reduced releases of toxics like mercury, dioxins, and oxides of nitrogen and sulfur).

Emerging technologies will continue to challenge the engineering profession. Staying abreast of these revolutions and enhancing the engineer's understanding of the risks and benefits to society, at least at some basic level, will serve the profession well.

NOTES AND COMMENTARY

[1] N.H. Steneck, *Introduction to Responsible Conduct of Research*, Office of Research Integrity (Washington, DC: Department of Health and Human Services, 2004).

[2] Office of Science and Technology Policy, Executive Office of the President quoted by N.H. Steneck, *Introduction to Responsible Conduct of Research*, Office of Research Integrity (Washington, DC: Department of Health and Human Services, 2004).

[3] J. Orear, 1968, Letter to the journal *Physics Today*, pertaining to a proposed amendment to the American Physical Union's constitution. Cited in M.W. Friedlander, *The Conduct of Science* (Englewood Cliffs, NJ: Prentice-Hall, 1972).

[4] Trials of War Criminals before the Nuremberg Military Tribunals under Control Council Law No. 10, vol. 2, 181 Washington, DC: US GPO, 1949.

[5] World Medical Association, Declaration of Helsinki (1989).

[6] 21 C.F.R. §50.02 (2003).

[7] US Department of Health, Education and Welfare, National Commission for the Protection of Human Subjects of Biomedical and Behavioral Research, 1979, *The Belmont Report: Ethical Principles and Guidelines for the Protection of Human Subjects of Research*, 18 April 1979. The notice states:

> The Belmont Report attempts to summarize the basic ethical principles identified by the Commission in the course of its deliberations. It is the outgrowth of an intensive four-day period of discussions that were held in February 1976 at the Smithsonian Institution's Belmont Conference Center supplemented by the monthly deliberations of the Commission that were held over a period of nearly four years. It is a statement of basic ethical principles and guidelines that should assist in resolving the ethical problems that surround the conduct of research with human subjects. By publishing the Report in the Federal Register, and providing reprints upon request, the Secretary intends that it may be made readily available to scientists, members of Institutional Review Boards, and Federal employees.

8   This goes well beyond mere translation. For example, it is very common in emergency departments of hospitals to hear the physician or nurse conversing in Spanish themselves or via a translator with a patient or family member. Spanish has many dialects and idioms that, if incorrectly translated, can lead to a miscommunication of symptoms and risks. Other tools are used, such as graphical devices (e.g., amount of pain ranging from 0 or no pain is shown as a smiling face to an obviously painful grimace, or the need for preventing the spread of germs with cartoons of people wearing masks and lathering soap). One cannot help concluding, however, that much of the more subtle and complex medical information is lost on the subjects and their attending medical staffs. For example, does nodding by a person from Japan mean the same as that of a person from Ecuador or Mozambique (or East LA *versus* Beverly Hills, for that matter)?

9   I participated in air quality monitoring near Ground Zero in the fall of 2001 and in fact noticed some workers not wearing their respirators (some police officers, for example, "wore" them, but not on their faces, but around their necks). However, this was in spite of the presence of stations operated by the Occupational Safety and Health Administration and local occupational hygiene offices. Our team, for example, was fitted with respirators in Research Triangle Park and certified to use them well before arriving in New York. Although my sample is unscientific, the majority of workers that I observed near Ground Zero, at least at times nearest the collapses, were wearing protective gear. However, it is another matter whether they used the equipment properly and whether it was correctly fitted.

10 P.S. Appelbaum, L.H. Roth, C.W. Lidz, P. Benson, and W. Winslade, "False Hopes and Best Data: Consent to Research and the Therapeutic Misconception," *Hasting Center Report* 17, no. 2 (1987): 20–4.

11 William Harvey was one of the first modern medical researchers and is credited with being the first to explain the movement of blood in the circulatory system. Harvey reported this breakthrough in 1616. Twelve years later the treatise, *Excercitatio Anatomica de Motu Cordis et Sanguinis in Animalibus* (*Anatomical Exercise on the Motion of the Heart and Blood in Animals*), provided the scientific basis and description of how blood is pumped throughout the body by the heart within a closed system wherein the blood is recirculated and reoxygenated. For more information, see R. French, *William Harvey's Natural Philosophy* (New York, NY: Cambridge University Press, 1994).

12 Claudius Galenus of Pergamum (AD 131–202) was the Greek physician, Galen, whose concept of medicine was dominant for a millennium. In fact, Galen's biomedical findings in twenty-two volumes were based upon his practice, but also extrapolated largely from animal experiments that would be considered cruel by contemporary ethical standards. Galen expanded his knowledge partly by experimenting with live animals. For example, Galen publicly dissected a live pig, cutting its nerve bundles one at a time to demonstrate the function of the laryngeal nerve (now also known as *Galen's Nerve*). When this nerve was cut, the pig stopped squealing. In other experiments, he cut spinal cords in animals to illustrate paralysis and bound the ureters of living animals to demonstrate that their source is the kidneys. For additional information, see A. Guerrini, *Experimenting with Humans and Animals: From Galen to Animal Rights* (Baltimore, MD: Johns Hopkins University Press, 2003).

13 Whewell's major works that articulate the scientific method are *History of the Inductive Sciences* (1837) and *The Philosophy of the Inductive Sciences, Founded upon Their History* (1840).

14 R. Feynman, 1974, "Cargo Cult Science," Commencement Address at California Institute of Technology.

15 Quoted in J. Schell, "Our Fragile Earth," *Discover* (October 1987): 47.

16 C.P. Snow, *The Search* (New York, NY: Charles Scribner's Sons, 1959).

17 The principal source for this discussion is B. Cooper, J. Hayes, and S. LeRoy, "Science Fiction or Science Fact? The Grizzly Biology behind Parks Canada Management Models," *Frasier Institute Critical Issues Bulletin* (Vancouver, BC: 2002).

18 Articles included S.A. Karl and B.W. Bowen, "Evolutionary Significant Units *versus* Geopolitical Taxonomy: Molecular Systematics of an Endangered Sea Turtle (genus *Chelonia*)," *Conservation Biology* 13 (1999): 990–9; P.C.H. Pritchard, "Comments on Evolutionary Significant Units *versus* Geopolitical Taxonomy," *Conservation Biology* 13 (1999): 1000–3; J.M. Grady and J.M. Quattro, "Using Character Concordance to Define Taxonomic and Conservation Units," *Conservation Biology* 13 (1999): 1004–7; K. Shrader-Frechette and E.D. McCoy, "Molecular Systematics, Ethics, and Biological Decision Making under Uncertainty," *Conservation Biology* 13, no. 5 (1999): 1008–10; and B.W. Bowen and S.A. Karl, "In War, Truth Is the First Casualty," *Conservation Biology* 13, no. 5 (1999): 1013–16.

19 Bowen and Karl, 1015.

20 Shrader-Frechette and McCoy, 1012.

21 At the end of the movie, these words are sung by Napoleon's older brother, Kip (played by Aaron Ruell), to his new bride, Lafawnduh (Shondrella Avery), after they are pronounced husband and wife. Another memorable conversation occurs between Kip and his technologically illiterate Uncle Rico (Jon Gries):

Rico: Kip, I reckon . . . you know a lot about . . . cyberspace? You ever come across anything . . . like time travel?
Kip: Easy, I've already looked into it for myself.
Rico: Right on . . . right on.

22 This moniker is in sharp contrast to that of economists, who have been called the dismal scientists. Some of this is due to the difference in outlooks for the future. For example, many economists (and ecologists, for that matter) hold to the convictions of Malthusian economics, best characterized as the law of diminishing returns. While engineers know quite well the laws of thermodynamics, especially that matter and energy exist in balances and that neither is created nor destroyed, we also know that many limits that now exist can be mollified or eliminated with proper design. As such, many technological breakthroughs, like agricultural technologies, improved strains of crops and better farming implements, were unforeseen by the doomsayers and dismal scientists who saw "overpopulation" to begin when the world hit 1 billion people (we are now well over 6 billion and many of the problems have nothing to do with total amount of food, but in the fairness of its delivery to people in need). On the other hand, our optimism can be overdrawn, so we must not assume that new technologies will completely "solve" such problems as global warming or unsustainable energy demands. Technology will be part of the solution, but must be coupled with wise stewardship of the resources made available to humankind and balanced with ecological well-being, as expressed by biodiversity, productivity, and sustainable development.

23 See B. Jennett, *The Vegetative State: Medical Facts, Ethical and Legal Dilemmas* (New York: CUP, 2002); B. Jennett and F. Plum, Persistent vegetative state after brain damage. A syndrome in search of a name. *Lancet* no. 7753 (1972): 734–7; and Multi-Society Task Force on PVS, Medical aspects of the persistent vegetative state. *New England Journal of Medicine* no. 330 (1994): 1499–508.

24 N. Postman, *Crazy Talk, Stupid Talk: How We Defeat Ourselves by the Way We Talk and What to Do About It* (New York, NY: Delacorte Press, 1976).

25 The principal source for information about the C-C valve is W.J. Blot, M.A. Ibrahim, T.D. Ivey, D.E. Acheson, R. Brookmeyer, A. Weyman, J. Defauw, J.K Smith, and D. Harrison, "Twenty-Five-Year Experience With the Björk-Shiley Convexoconcave Heart Valve: A Continuing Clinical Concern," *Circulation* 111 (2005): 2850–7.

26 B. Fishhoff and J.F. Merz "The Inconvenient Public: Behavioral Research Approaches to Reducing Product Liability Risks" in: National Academy of Engineering. *Product Liability and Innovation: Managing Risk in an Uncertain Environment* (Washington, DC: National Academy Press, 1994).

27 G.J.S. Wilde, "The Theory of Risk Homeostasis: Implications for safety and health," *Risk Analysis* 2 (1982), 209–225.

28 See, for example: P. Slovic, and B. Fischhoff, "Targeting Risks: Comments on Wilde's "Theory of Risk Homeostasis"," *Risk Analysis* 2 (1983), 227–34.

29 J. C. Hause, "Offsetting Behavior and the Benefits of Safety Regulations," *Economic Inquiry* 44, no. 4 (2006), 689–98.

30 G. Johnson, "Reliable software for protection systems," *Science and Technology Review* (March 1998), 21–3.

31 N. Leveson. "Medical Devices: The Therac-25," in: *Safeware: System Safety and Computers.* (New York, NY: Addison-Wesley Professional Publications, 1995).

32 National Society of Professional Engineers, 2006, *NSPE Code of Ethics for Engineers,* http://www.nspe.org/ethics/eh1-code.asp (accessed 17 May 2006).

33 J. Fielder, "The Bioengineer's Obligation to Patients," *Journal of Investigative Surgery,* 5 (1992): 201–8.

34 Sometimes the best resources come from the strangest places. The discussion on transparency and coherence must be credited to an article by columnist George Will that I read in the 8 September 2005 edition of the Durham (NC) *Herald Sun* newspaper, p. A-11. Will was particularly curious as to the success of the Entertainment and Sports Programming Network (ESPN). One reason, according to Will, is that unlike most endeavors in contemporary society, sports provide closure almost immediately. One knows who won and who lost. This is coherence. The other is that the spectator is included in the process. What is going on is completely transparent. There are 4 minutes and 15 seconds left in the game and the Chiefs are ahead by 11 points. The Cardinals are leading the Cubs 4 to 2, with two outs in the bottom of the ninth inning, with Albert Pujols on deck. Engineers will increasingly be working in such environments as a result of the Internet, sunshine laws, and, I believe, an increased technical literacy of a greater number of people. Not everyone wants to know the score of the Cardinals game, but they can find it immediately if they wish. They can even find the box scores of every game played this year, or the statistics of every minor league player on the Cardinal's triple-A farm club. Likewise, not everyone wants to know the details of the pump you may be designing for a biomedical company, but they can find them almost immediately if they so choose. So, unlike only a few years ago, the well-advised engineer should be ready for queries from both within and outside of the organization.

Unfortunately, the technical literacy is not evenly distributed throughout the population. For example, the so-called digital divide does exist, but it is rapidly changing. For example, according to Sara Base, *A Gift of Fire: Social Legal, and Ethical Issues for Computers and the Internet,* 2nd ed. (Upper Saddle River, NJ: Prentice-Hall, 2003), at the beginning of the 1990s, about 10% of Internet users were women, but by the end of the decade about half the users were women. However, in 1999, lower SES groups were half as likely as the general population to have Internet access in their residences. And, there is also a racial divide, as African-American and Hispanic households were half as likely as the general population to have a personal computer. Sometimes, even against such odds, people will have access. For instance, Alaska has the highest percentage of homes with Internet access of all the states of the United States. In this case, with its greater isolation and large expanses, the need for remote access overrides even some very large cultural and SES barriers.

[35] This is another example of the complexity and uncertainty of the "reasonable person" legal fiction.

[36] B. Jowett (translator), *The Republic by Plato*, Oxford University Press, Oxford, UK. Another way of looking at the incompleteness of our understanding of how things work is St Paul's likening of the way that we see the world to our "... seeing through a glass, darkly...." (1 Corinthians 13:12). As humans, we will always be limited in our understanding of even the most fundamental aspects of the tools we use.

[37] US Department of Education, National Center for Education Statistics, 2003, J.D. Sherman, S.D. Honegger, and J.L. McGivern, Comparative Indicators of Education in the United States and Other G8 Countries: 2002, Report No. NCES 2003–026, Washington, DC.

[38] In a discussion, Aarne Vesilind argued that there is a problem in terminology, especially in what is labeled "technology." For example, just because one plays a video game or watches television does not make that person technologically advanced. On the other hand, if the person writes codes for video games or understands the circuitry of the TV, this is an indication of technological literacy.

[39] M. Rothchild, "Beyond Repair: The Politics of the Machine Age Are Hopelessly Obsolete," *The New Democrat* (July/August, 1995: 8–11).

[40] J. Hightower, quoted in R. Fox, "Newstrack," *Communications of the ACM* 38 no. 8 (1995): 8–11.

[41] N.C. Roco, 2000, "Government Nanotechnology Funding: An International Outlook," National Science Foundation; presentation made at the Cornell Nanofabrication Center; accessed 30 May 2006 at http://www.nsf.gov/crssprgm/nano/reports/roco_vision.jsp.

Chapter 6

Bioethical Success and Failure

Engineers are perpetually seeking to define how well we are doing. We compare our designs to the standards and specifications of certifying authorities, such as building codes and regulations, design guidebooks, and standards promulgated by international agencies (e.g., the International Standards Organization (ISO)) and national standard-setting bodies (e.g., the American Society for Testing and Materials (ASTM)).

Teachable Moment: Engineering Measurement

Ethics consists of two parts. The Ancient Greeks considered human excellence to consist of *ethike aretai*, sometimes translated as "skill of character." Thus, excellence requires both technical skill and adherence to moral principles. Engineers must produce scientifically sound products and systems in ways that are fair and just. Fairness and justice require inclusion of diverse perspectives, especially of those most directly or indirectly affected by our decisions. Thus, engineers must be able to communicate effectively in order to arrive at adequate designs, to ensure that these technically sound designs are accepted by clients and stakeholders, and to convey sufficient information to users so that the designs are operated and maintained satisfactorily.

Technical communication can be seen as a critical path, where the engineer sends a message and the audience receives it (Figure 6.1). The means of communication can be either perceptual or interpretive.[1] Perceptual communications are directed toward the senses. Human perceptual communications are similar to that of other animals; that is, we react to sensory information (e.g., reading body language or assigning meaning to gestures, such as a hand held up with palms out, meaning "stop," or a smile, conveying approval).

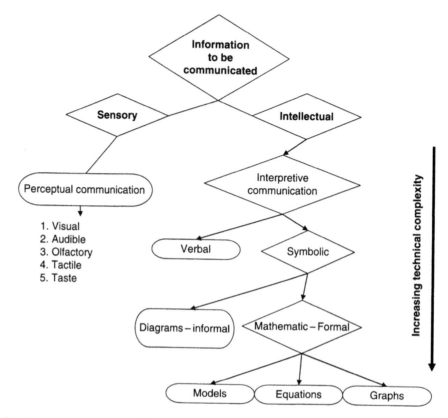

Figure 6.1 Types of communication. All humans use perceptual communication, such as observing the body language of an engineer or smelling an animal-feeding operation. The right side of the figure is the domain of technical communication. The public may be overwhelmed by perceptive cues or may not understand the symbolic, interpretive language being used by an engineer. Thus, the type of communication in a scientific seminar is quite different from that of a public meeting or a briefing for a patient or client.

Adapted from: M. Myers and A. Kaposi, The First Systems Book: Technology and Management, 2nd ed. (London, UK: Imperial College Press, 2004); and T.R.G. Green, "Cognitive Dimensions of Notations," in People and Computers V, ed. A. Sutcliffe and L. Macaulay (Cambridge, UK: Cambridge University Press, 1989).

Interpretive communications encode messages that require intellectual effort by the receiver to understand the sender's meanings. This type of communication can be either verbal or symbolic. Scientists and engineers draw heavily on symbolic information when communicating amongst themselves. If you have ever mistakenly walked into the seminar where experts are discussing an area of science not familiar to you, using unrecognizable symbols and vernacular, this is an example of symbolic miscommunication. In fact, the experts may be using words and symbols that are used in your area of expertise, but with very different meanings. For example, psychologists speak of "conditioning" with a meaning very different than that understood by an engineer (e.g., conditioning a matrix). Thus, the engineer must be aware of the venue of communication to apply designs properly.

Question

Discuss a current bioethical issue with a friend. Record the key points of the conversation and log the types shown in Figure 6.1.

1. Were your and your friend's communications of similar types?
2. Did one use comparatively greater amounts of sensory or intellectual communication?
3. How far down the intellectual communication flow did your conversation migrate?
4. Did the passion you hold about the issue influence the types of communications you used? Explain.

MEASUREMENTS OF SUCCESS AND FAILURE

For engineers, society may use the words success and failure to evaluate the performance of a design. Does it "work" (effectiveness)? Is it the best way to reach the end for which we strive (efficiency)? Then we consider whether or not it will likely continue to "work" (reliability) and we consider what adverse implications we might face (risk). Thus the "risk" associated with a design is used to refer to the possibility and likelihood of undesirable and possibly harmful effects. Errors in engineering can range from those that are merely annoying, like when a concrete building develops cracks that mar it as it settles, to those that are seemingly unforgivable like the collapse of a bridge, which causes human death.[2] For example, consider the Ford Pinto, a subcompact car produced by the Ford Motor Company between 1971 and 1980. The car's fuel tank was poorly placed in such a way that it increased the probability of a fire from fuel spillage by a rear collision. The ensuing adverse implications manifested themselves in the series of injuries that resulted from the defect. The C–C valve discussed in Chapter 5 has similarities to such faulty decision making.

Likewise, while many of the engineering errors associated with today's technology might seem new, the idea failure of products of engineering is far from new. If one were to consider this concept of "risk" approximately 4000 years ago in the kingdom of Babylon, the concept takes on a different meaning. Here, the "risk" associated with engineering transcended the failures of the product itself, extending to the physical well-being of the engineer himself and his family.

The risk objective in Babylon takes on a new meaning for engineering, as those who were designing and building assumed personal risk to themselves and to their families. Should they face failure, their lives were at stake, which is clearly different from today's protocols. Consequently, it is important to consider the evolution of the meaning of the term "risk" for engineers over time.

TECHNOLOGICAL SUCCESS AND FAILURE

Safety is a key expectation of all biomedical endeavors. It is part of both the physician's and the engineer's practice, although the focus is somewhat different. Safety and risk are also critical components of whether a professional or research approach is considered sound. When patients, subjects, or the public are exposed to unacceptable risk, the practitioner or researcher is deemed to have performed in an unethical manner.

The fifth principle of medical ethics of AMA states:

> A physician shall continue to study, apply, and advance scientific knowledge, maintain a commitment to medical education, make relevant information available to patients, colleagues, and the public, obtain consultation, and use the talents of other health professionals when indicated.[3]

The seventh AMA principle states:

> A physician shall recognize a responsibility to participate in activities contributing to the improvement of the community and the betterment of public health.[4]

Likewise, the first fundamental canon of the engineering profession, as articulated by the National Society of Professional Engineers, requires that "engineers, in the fulfillment of their professional duties, shall hold paramount the safety, health and welfare of the public."[5] To emphasize this professional responsibility, the engineering code includes this same statement as the engineer's first rule of practice.

These principles and canons indicate that safety and risk are not only technical concepts, but are also ethical concepts. A professional cannot behave ethically and simultaneously ignore the direct and indirect risks stemming from one's practice. Thus, competence and character demand an appreciation of risks, both real and perceived. For example, perceived risks may be much greater than actual risks, or they may be much less. So then, how is it possible to square technical facts with pubic fears? Like so many engineering concepts, timing and scenarios are crucial. What may be the right manner of saying or writing something in one situation may be very inappropriate in another. Communication approaches will differ according to whether we need to motivate people to take action, alleviate undue fears, or simply share our findings clearly, no matter whether they are good news or bad. For example, some have accused certain businesses of using pubic relations and advertising tools to lower the perceived risks of their products. The companies may argue that they are simply presenting a counterbalance against unrealistic perceptions.

Engineering success or failure is largely what we do measured against what our profession "expects" us to do. Safety is a fundamental facet of our professional duties. Thus, we need a set of criteria that tells us when designs and projects are sufficiently safe. Four safety criteria are applied to test engineering safety:[6]

1. The design must comply with applicable laws.
2. The design must adhere to "acceptable engineering practice."
3. Alternative designs must be sought to see if there are safer practices.
4. Possible misuse of the product or process must be foreseen.

The first two criteria are easier to follow than the third and forth. The well-trained designer can look up the physical, chemical, and biological factors to calculate tolerances and factors of safety for specific designs. Laws have authorized the thousands of pages of regulations and guidance that demark when acceptable risk and safety thresholds are crossed, meaning that the design has failed to provide adequate protection. Engineering standards of practice go a step further. Failure here is difficult to recognize. Only other engineers with specific expertise can judge whether the ample margin of safety as dictated by sound engineering principles and practice has been provided in the design. Identifying alternatives and predicting misuse requires quite a bit of creativity and imagination. For example, as discussed in Chapter 5, engineering can be a victim of its own success, as consumers take more risks because of safety features (known as offsetting behavior).

But, can risks really be quantified? Risk assessors and actuary experts would answer with a resounding "yes." Medical practitioners routinely share risks with patients in preoperative preparations, usually in the form of a probability. However, the general public's perception is often that one person's risk is different from another's and that risk is in the "eye of the beholder." Some of the rationale appears to be rooted in the controversial risks associated with tobacco use and other daily decisions, such as choice of modes of transportation. What most people perceive as risks and how they prioritize those risks is only partly driven by the actual objective assessment of risk, i.e., the severity of the hazard combined with the magnitude, duration, and frequency of the exposure to the hazard. For example, the smoker may be aware that cigarette smoke is hazardous, but she not directly aware of what these are (e.g., carbon monoxide, polycyclic aromatic hydrocarbons, and carcinogenic metal compounds). She has probably read the conspicuous warning labels many times as she held the pack in their hands, but these really have not "rung true" to her. They may have never met anyone with emphysema or lung cancer, or they may not be concerned (yet) with the effects on the unborn (i.e., *in utero* exposure).[7] Psychologists also tell us that many in this age group have a feeling of invulnerability. Those who think about it may also believe that they will have plenty of time to end the habit before it does any long-term damage. Thus, we should be aware that what we are saying to people, no matter how technically sound and convincing it is to us as engineers and scientists, may be simply a din to our targeted audience.

The converse is also true. We may be completely persuaded based upon data, facts, and models that something clearly does not cause significant damage, but those we are trying to convince of this finding may not buy it. They may think we have some vested interest, or that they find us guilty by association with a group they do not trust, or that we are simply "guns for hire" for those who are sponsoring our research or financially backing the product development. The target group may not understand us because we are using jargon and are not clear in how we communicate the risks. So, the perception of risk will not match the risk being quantified.

This chapter deals with concepts important to all engineering fields. The principal value added by engineers is in the improvement in the quality of human health. Engineers add value when we decrease risk, a crucial concern of bioethics. By extension, reliability tells us and everyone else just how well our designs are performing by reducing overall risk. What we design must continue to serve its purpose throughout its useful life.

RISK AS A BIOETHICAL CONCEPT

As is generally understood, risk is the chance that something will go wrong or that some undesirable event will occur. Every time we get on a bicycle, for example, we are taking a risk that we might be in an accident and may damage the bike, get hurt, injure others, or even die in a mishap. The understanding of the factors that lead to a risk is called risk analysis and the reduction of this risk (e.g., by wearing a helmet and staying on bike paths) is risk management. Risk management is often differentiated from risk assessment, which comprises the scientific considerations of a risk.[8] Risk management includes the policies, laws, and other societal aspects of risk.

Engineers constantly engage in risk analysis, assessment, and management. Engineers must consider the interrelationships among factors that put people at risk, suggesting that we are risk analysts. Engineers provide decision makers with thoughtful studies based upon the sound application of the physical sciences and, therefore, are risk assessors by nature. Engineers control things and, as such, are risk managers.

We are held responsible for designing safe products and processes, and the public holds us accountable for its health, safety, and welfare. The public expects engineers to "give results, not excuses,"[9] and risk and reliability are accountability measures of engineers' success. Engineers design systems to reduce risk and look for ways to enhance the reliability of these systems. Thus, every engineer deals directly or indirectly with risk and reliability.

Bioethical analysis (e.g., formulating factual premises and reaching moral conclusions) embodies the concept of risk and how it can be quantified and analyzed. Bioethical analysis also considers the morality of the ways of reducing risk by conscious and intended risk management and how to communicate both the assessment and management options to those affected.[10]

SAFETY, RISK, AND RELIABILITY IN DESIGN

Probable impossibilities are to be preferred to improbable possibilities.

Aristotle

Aristotle was not only a moral and a natural philosopher (the forerunner to "scientist"); he was also a risk assessor. In the business of human health and bioethics, we are presented with "probable impossibilities" and "improbable possibilities."

To understand these two outcomes, we must first understand the different connotations of risk. Aristotle's observation is an expression of probability. People, at least intuitively, assess risks and determine the reliability of their decisions every day. We want to live in a "safe" world. But, safety is a relative term. The "safe" label requires a value judgment and is always accompanied by uncertainties, but engineers frequently characterize the safety of a product or processes in objective and quantitative terms. Factors of safety are a part of every design. Biological safety is usually expressed by its opposite term, risk.

PROBABILITY: THE MATHEMATICS OF RISK AND RELIABILITY

The syllogism is a vital tool for determining whether an argument, including an ethical one, is valid. All syllogisms consist of premises. In a valid argument, if the premises are true, then the conclusion must also be true. And, the premises of a valid argument cannot be true while its conclusion is false. Any argument not meeting these conditions is invalid; meaning that it is possible for their conclusions to be false even when their premises are true. The first premise of an ethical syllogism is a factual statement. The second premise connects the factual premise to a statement of morality. Finally, a moral conclusion can be derived from an argument where reliable facts are linked to a correct evaluative (moral) premise.

This logical construct needs facts that are not vague and not ambiguous. Vagueness means that the factual statement or words in the statement lack precision. Even one term that is vague makes it almost impossible to establish whether a premise is true or false. Ambiguity exists when there are more than one possible connotations of a term. Both vagueness and ambiguity are problematic for bioethical decision making, since ethics is fraught with phrases like "People should not suffer." The ambiguity comes from the definition of the terms, "people" and "suffer." Are "people" synonymous with "humans?" Or, "persons?" Does to suffer mean "to hurt or to feel physical pain?" Most physiologists would consider this situation to be pathological. We need to feel pain in order to properly respond to certain dangerous situations or to sense feedback from physiological messages. Does to suffer mean "to endure?"

Even if the terms are used unambiguously, the premise can be vague. Consider the evaluative premise, "People should not *needlessly* suffer." This implies that we know the difference between needed and needless suffering. Such terms are vague.

Even defining right action is uncertain. Ethicists use what is known as deontic logic to determine duty or viture. So-called deontic operators give certain properties of moral significance in an argument. The possible operators are the following:[11]

- Obligatory, required, or called for means that the situation of the proposition ought to be the case.
- Permissible or allowed means that the situation of the proposition may be the case.
- Forbidden means that the situation of the proposition ought not to be the case.
- Omissible means that the situation in which the proposition is not the case, i.e., it is not true that the proposition ought to be the case.
- Optional or indifferent means that the situation of the proposition and the situation in which the proposition may not be the case (proposition is permissible and omissible; that is, neither obligatory nor forbidden).

These operators require certain logical and ethical outcomes, shown in Table 6.1.

The possible deontic operators assume that we know what is morally permissible and obligatory. However, the vagueness and ambiguity of bioethical information is highly varied. This uncertainty can be categorized into three types of knowledge, known as "epistemic states:"[12]

- *Certainty, sureness (Latin: certitudo).* I am sure that p is true without fear that it is not the case.
- *Belief, opinion (Latin: opinari).* I believe that p is true, but with some fear that p is false.
- *Wavering (Latin: dubitare).* I vacillate in regard to the truth of p in two cases: (a) when I neither believe that p is true nor that p is false, or I merely surmise that p is true.

Such variation calls for expressions of certainty and likelihood. One means of addressing vagueness is to state terms probabilistically. Probability is the likelihood of an outcome. The outcome can be bad or good, desired or undesired. The history of probability theory, like much of modern mathematics

Table 6.1
Possible deontic operators that indicate certain properties of morally significant situations, especially a possible action or omission (p stands for any proposition)

If	Then
p is obligatory	not-p is not permissible; and not-p is forbidden
not-p is obligatory	not-p is obligatory; and p is forbidden
p is not obligatory	not-p is permissible and p is omissible
not-p is not obligatory	p is permissible; and not-p is omissible
p is optional	p is permissible and omissible, and p and not-p are permissible

Source: W. Redmond, 1998, "Conscience as Moral Judgment: The Probabilist Blending of the Logics of Knowledge and Responsibility." *Journal of Religious Ethics* 26 (2), 389–405.

and science, is rooted in the Renaissance. Italian mathematicians considered some of the contemporary aspects of probability as early as the fifteenth century, but saw no need to or were unable to devise a generalized theory. Blaise Pascal and Pierre de Fermat, the famous French mathematicians, developed the theory after a series of letters in 1654 considering some questions posed by the nobleman, Antoine Gombaud, Chevalier de Méré, regarding betting and gaming. Other significant Renaissance and post-Renaissance mathematicians and scientists soon weighed in, with Christian Huygens publishing the first treatise on probability, *De Ratiociniis in Ludo Aleae*, which was specifically devoted to gambling odds. Jakob Bernoulli (1654–1705) and Abraham de Moivre (1667–1754) also added to the theory. However, it was not until 1812 with Pierre LaPlace's publication of *Théorie Analytique des Probabilités* that probability theory was extended beyond gaming to scientific applications.[13]

Probability is now accepted as the mathematical expression that relates a particular outcome of an event to the total number of possible outcomes. This is demonstrated when we flip a coin. Since the coin has only two sides, we would expect a 50–50 chance of either heads or tails. However, scientists must also consider rare outcomes, so there is a very rare chance (i.e., highly unlikely, but still possible) that the coin could land on its edge, i.e., the outcome is neither heads nor tails. A confluence of unlikely events is something that engineers must always consider, such as the combination of factors that led to the suffering and death as a result of a seemingly inconsequential design flaw in a medical device, or unforeseen side effects from an ethical drug. Rare combination of events has led to major disasters like the "perfect storm" of a Hurricane Katrina and release of a toxic plume in Bhopal, India, or the introduction of a seemingly innocuous opportunistic species that devastates an entire ecosystem, or the interaction of one particular congener of a compound in the right cell in the right person that leads to cancer.

Randomized and double-blind trials have become the benchmark for biomedical research. Any new therapy must first undergo tests of safety and efficacy based on clinical trials. The data from these trials form the factual premises that are used by the biomedical community to determine "no or no go" decisions on devices and drugs. Bioethical decisions, therefore, are based on an assessment of risk. This assessment is rooted in probabilities.

As engineers, we also know that the act of flipping or the characteristics of the coin may tend to change the odds. For example, if for some reason the heads side is heavier than the tails side or the aerodynamics is different, then the probability could change.

The total probability of all outcomes must be unity, i.e., the sum of the probabilities must be 1. In the case of the coin standing on end rather than being heads or tails, we can apply a quantifiable probability to that rare event. Let us say that laboratory research has shown that one in a million times $(1/1\,000\,000 = 0.000001 = 10^{-6})$ the coin lands on edge. By difference, since the total probabilities must equal 1, the other two possible outcomes (heads and tails) must be $1 - 0.000001 = 0.999999$. Again, we are assuming that the aerodynamics and other physical attributes of the coin give it an equal chance of being either heads or tails, then the probability of heads $= 0.4999995$ and the probability of tails $= 0.4999995$.

Stated mathematically, an event (e) is one of the possible outcomes of a trial (drawn from a population). All events, in our coin toss case, heads, tails, and edge, together form a finite "sample space," designated as $E = [e_1, e_2, \ldots, e_n]$. The lay public is generally not equipped to deal with such rare events, so by convention, they usually ignore them. For example, at the beginning of overtime in a football game, a tossed coin determines who will receive the ball, and thus has the first opportunity to score and win. When the referee tosses the coin, there is little concern about anything other than "heads or tails."

However, the National Football League undoubtedly has a protocol for the rare event of the coin not being a discernible heads or tails. In medical and epidemiological studies, *e* could represent a case of cancer. Thus, if a population of 1 million people is exposed to a pesticide over a specific time period, and one additional cancer is diagnosed that can be attributed to that pesticide exposure, we would say that probability of *e*, i.e., $p\{e\}$, is 10^{-6}. Note that this was the same probability that we assigned to the coin landing on its edge.

Returning to our football example, the probability of the third outcome (a coin on edge) is higher than "usual" since the coin lands in grass or on artificial turf, compared to landing on a hard flat surface. Thus, the physical conditions increase the relative probability of the third event. This is analogous to a person who may have the same exposure to a carcinogen as the general population, but who may be genetically predisposed to develop cancer. The exposure is the same, but the probability of the outcome is higher for this "susceptible" individual. Thus, risk varies by both environmental and individual circumstances.

Events can be characterized in a number of ways. Events may be discrete or continuous. If the event is forced to be one of a finite set of values (e.g., six sides of a die), the event is discrete. However, if the event can be any value, e.g., the size of tumor (within reasonable limits), the event is continuous. Events can also be independent or dependent. An event is independent if the results are not influenced by previous outcomes. Conversely, if an event is affected by any previous outcome, then the event is a dependent event.

Joint probabilities must be considered and calculated since, in most biomedical scenarios, events occur in combinations. So, if we have *n* mutually exclusive events as possible outcomes from *E* that have probabilities equal to $p\{e_i\}$, then the probability of these events in a trial equals the sum of the individual probabilities:

$$p\{e_i \text{ or } e_2 \ldots \text{ or } e_k\} = p\{e_1\} + p\{e_2\} + \cdots + p\{e_k\} \tag{6.1}$$

Further this helps us to find the probabilities of events e_i and g_1 for two independent sets of events, *E* and *G*, respectively:

$$p\{e_i \text{ or } g_i\} = p\{e_i\}p\{g_i\} \tag{6.2}$$

Another important concept for interpreting data in a factual premise is that of conditional probability. If we have two dependent sets of events, *E* and *G*, the probability that event e_k will occur if the dependent event *g* has previously occurred is shown as $p\{e_k|g\}$, which is found using Baye's theorem:

$$p\{e_k|g\} = \frac{p\{e_k \text{ and } g\}}{p\{g\}} = \frac{p\{g|e_k\}p\{e_k\}}{\sum\limits_{i=1}^{n} p\{g|e_i\}p\{e_i\}} \tag{6.3}$$

A review of this equation shows that conditional probabilities are affected by a cascade of previous events. Thus, the probability of what happens next can be highly dependent upon what has previously occurred. For example, the cumulative risk of cancer depends on the serial (dependent) outcomes. Similarly, reliability can also be affected by dependencies and prior events. Thus, characterizing any risk and determining how reliable our systems are expressions, at least in part, of probability.

Risk itself is a probability (i.e., the chance of an adverse outcome). So, any calculation of harm will likely be based on some use of probabilities. Also, event relationships are vital to ethical decision making.

A decision leads to consequences that lead to decision points. Every consequence has a probability, so a decision maker should have a good estimate of the probability of an outcome.

But risk is more than a probability. It is an ethical concept. (see Discussion Box: Choose Your Risk).

Discussion Box: Choose Your Risk

Old man, look at my life, twenty four and there's so much more . . .
Give me things that don't get lost.
Like a coin that won't get tossed . . .

Old Man, Neil Young (b. 1945)

Perhaps, since Young was about twenty-five years old when he wrote *Old Man*, he was displaying cognitive dissidence; on the one hand, he wanted to take risks but, on the other, he recognized risk avoidance as a necessity in some matters (i.e., a coin that won't get tossed). An interesting phenomenon that supports this view seems to be taking place on today's college campuses. From some anecdotal observations, it would appear that students are more concerned about some exposure pathways and routes than about others. It is not uncommon at Duke University or the University of North Carolina, for example, to observe a student park her bicycle to smoke a cigarette and take a few drinks of branded bottled water.

At first blush, one may conclude that the student has conducted some type of risk assessment, albeit intuitive, and has concluded that she needs to be concerned about physical fitness (biking) and the oral route of exposure to contaminants, as evidenced by the bottled water. For some reason, the student is not as concerned about the potential carcinogens in tobacco smoke as the contaminants found in tap water. Or is it simply taste . . . or mass marketing?

With a bit more analysis, however, the apparent lack of concern for the inhalation route (tobacco smoke) may not be the case. The behavior may be demonstrating the concept of risk perception. The biking, smoking, and drinking activities seem to illustrate at least two principles regarding increased concern about risk: whether the student maintains some control over risk decisions and whether the exposures and risks are voluntary or involuntary. The observation may also demonstrate the lack of homogeneity in risk perception. For example, risk perception appears to be age dependent. Teenagers and young adults perceive themselves to be invincible, invulnerable, and even immortal more frequently than does the general population. Like most decisions, risk decisions consist of five components:

1. An inventory of relevant choices
2. An identification of potential consequences of each choice
3. An assessment of the likelihood of each consequence actually occurring
4. A determination of the importance of these consequences
5. The synthesis of this information to decide which choice is the best[15]

It has been said that philosophy, which includes ethics, is reasonably successful if it allows us to ask the right questions. In fact, the questions are preferable to answers. Answers are found on warning labels, bike assembly instructions, and bottled water contents. But, what are the questions? What are the choises? What are we worried about or, ever more importantly, what should we be worried about?

These perceptions change with age, as a result of experiences and physiological changes in the brain. However, like in the case of the risks associated with a lack of experience in driving an automobile, the young person may do permanent damage while traversing these developmental phases. In fact, this mix of physiological, social, and environmental factors in decision making is an important variable in characterizing hazards. In addition, the hazard itself influences the risk perception. For example, whether the hazard is intense or diffuse or whether it is natural or human induced (Figure 6.2) is a determination of public acceptance of the risks associated with

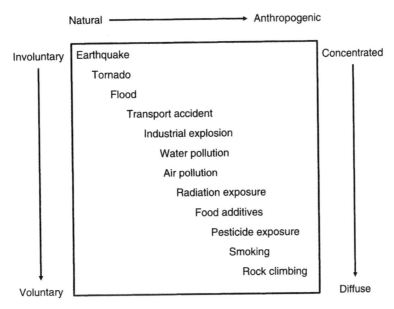

Figure 6.2 Spectrum of hazards.
Adapted from: D.A. Vallero, 2005, *Paradigms Lost: Learning from Environmental Mistakes, Mishaps, and Misdeeds*, Butterworth-Heinemann, Burlington, MA, adapted from K. Smith, 1992, *Environmental Hazards: Assessing Risk and Reducing Disaster*, Routledge, London, UK.

the hazard. People tend to be more accepting of hazards that are natural in origin, voluntary, and concentrated in time and space.[16]

Other possible explanations are risk mitigation and sorting of competing values. The biker may well know that smoking is a risky endeavor and is attempting to mitigate that risk by other positive actions, such as exercise and clean water. Or, he or she may simply be making a choice that the freedom to smoke outweighs other values like a healthy lifestyle (students have reported that biking may well be simply a means of transportation and not a question of values at all).

It is likely that all of these factors affect different people in myriad ways, illustrating the complexities involved in risk management decisions.

Demographics is a determinant of risk perception, with certain groups more prone to "risk taking" and averse to authority. Teenagers, for example, are often shifting dependencies, e.g., from parents to peers. Later, the dependencies may be transitioning to greater independence, such as that found on college campuses. Eventually, these can lead to interdependent, healthy relationships. Engineers have to deal with these dynamics as a snapshot. Although the individual is changing, the population is often more static. There are exceptions, for example, if the mean age of target group is undergoing significant change (e.g., getting younger), then there may a concomitant change in risk acceptance and acceptance of controls (e.g., changes in acceptance of invasive technologies).

What people perceive as risks and how they prioritize these risks is only partly driven by the actual objective assessment of risk, i.e., the severity of the hazard combined with the magnitude, duration, and frequency of the exposure to the hazard. For example, the young person may be aware that cigarette smoke contains some nasty compounds, but is not directly aware of what these are (e.g., benzo(a) pyrene, cadmium, and carbon monoxide). Those diseases mentioned on the side of the pack of cigarettes are risks to others, or at least will not be a risk until the teenager is "old." So, no matter how good the risk numbers are, they must be understood correctly by those assuming the risk.

Engineers tend to think of mathematics as universally accepted. The laws of mathematics generally are, but people seem to have their own "mathematics" when it comes to risk. If you visit a local hospital, you are likely to see a number of patients gathered near the hospital entrance in designated smoking areas. Here they are hooked up to IVs, pumps and other miracles of medical technology and are simultaneously engaged in one the most potent health hazards: smoking. Of course, this is ironic. On the one hand we are assigning the most talented (and expensive) professionals to treat what is ailing them, yet those patients have made a personal decision to engage in very unhealthful habits. It seems analogous to training and equipping a person at great social costs to drive a very expensive vehicle, all the time knowing that the person has a nasty habit of tailgating and weaving in traffic. Even if the person is well trained and has the best car, this habit will increase the risk. However, there is another way to look at the smoking situation; that is, the person has decided (mathematically) that the "sunken costs" are dictating the decision. Intuitively, the smoker has differentiated long-term risks from short-term risks. The reason the person is in the hospital (e.g., heart disease, cancer, emphysema) is the result of risk decisions the person made years, maybe decades, ago. The exposure is long in duration and the effect is chronic. So the person may reason that the effects of today's smoking will only be manifested twenty years hence and thus has little incentive to stop engaging in the hazardous activity. Others see the same risk and use a different type of math. They reason that there are X odds that they will live a somewhat normal life after treatment, so they need to eliminate bad habits that will put them in the same situation twenty years from now. They both have the same data, but reach very different risk decisions.

Another interesting aspect of risk perception is how it varies in scale and scope. I recall attending a meeting in the late 1970s among a group of highly trained engineers and scientists from the US Environmental Protection Agency (EPA) and the Kansas Department of Health and Environment. Mind you, these are the two principal federal and state agencies, respectively, charged with protecting the environment. Ironically, the meeting was called to determine the appropriate ways to reduce the ambient concentrations of pollutants, especially particulate matter and carbon monoxide (CO). The meeting was held in a small room and almost every person, except me, was smoking. The room was literally "smoke-filled." The surreal (at least to me) discussion was exploring options to reduce ambient pollutant levels to the parts per million range (high by today's standards, but at detection limits in the 1970s) in a room where one could almost report it as a percent (i.e., 10^{-6} *versus* 10^{-2}). The irony was completely lost on the participants, probably for good reason. They were not making personal decisions; they were making policy decisions. This is akin to saying "Do as I say, not as I do." Such compartmentalization is not foreign to engineers. Many of our own homes would not meet many of the standards that we require for our corporate and institutional clients' structures (e.g., accessibility, water and air quality, egress/exit, and signage).

The compartmentalization concept was brought home some years back in a story shared by a former decision maker in the Office of Management and Budget (OMB). He had been in a budget meeting earlier in the morning, where he had been discussing a few multimillion dollar projects and recommending some million dollar increases in a number of them. Later, his wife called to update him on the new house they were building. One item was that the window contractor needed an additional $200 above the original estimate. The OMB director was outraged and starting ranting about how important it was to hold the line on such expenses. At this point, he was struck by the irony. Again, the policy decision has a different scale and scope than the individual decision (in the OMB director's case, 10^8 *versus* 10^2). But is this right? Is vigilance a function of size (dollars, risk, number of people) or must it be independent of scale? Future users or patients do not care that you have a bigger project or one with much more societal import. This is their only body or that of their only child or other loved one, and the risks are the ones that they will have to abide.

Another consideration is whether the scope flavors what we consider to be of value. For example, how do we value the life of an animal? Biomedical engineering and health studies are presently highly dependent on information gathered from comparative biology. We use models to extrapolate potential hazards of new drugs, foods, pesticides, cosmetics, and other household products. We study the effects of devices as prototypes in animals to determine their risks and efficacy prior to introducing them to humans. When we do this, we are placing a value on the utility of the results of the animal research; that is, we are applying a utilitarian ethical model. If the "greater good" is served, we conclude that the animal studies are worthwhile. Critics of this approach are likely to complain that the animals' welfare is not given the appropriate weight in such a model. For example, they argue that the suffering and pain experienced by animals with nervous systems like ours and clear capacities for higher-level social behavior are unjustifiable means to an end, albeit one that serves science well.

We do not have to go to the laboratory, however, to consider valuation as it pertains to nonhuman animals. Many communities are overrun with deer populations because suburban developments have infringed on the habitats of deer and their predators, changing the ecological dynamics. Certainly, the deer population increases present major safety problems, especially during rutting season when deer enter roadways and the likelihood of collisions increase. The deer also present nuisances, such as their invasions of gardens. They are even part of the life cycle of disease vectors when they are hosts to ticks that transmit Rocky Mountain spotted fever and Lyme's disease. In this sense, we may see deer as a "problem" that must be eradicated. However, when we come face to face with the deer, such as when we see a doe and her fawns, we can appreciate the majesty and value of these individual deer.

Recently, I had an epiphany of sorts. I am constantly trying desperately to prevent the unwelcome visits of deer to my Chapel Hill garden. I have had little success (they particularly like eating tomatoes before they ripen and enjoy nibbling young zucchini plants). In the process, I have uttered some rather unpleasant things about these creatures.

On a recent occasion, however, I was traveling on a country highway and noticed an emergency vehicle assisting a driver who had obviously crashed into something. The driver and responder were in the process of leaving the scene. Driving 20 meters further, I noticed a large doe trying to come to her feet to run into the woods adjacent to the highway, but she could not lift her back legs. I realized that the people leaving the scene must have concluded that the "emergency" was over, without regard to the deer that had been struck by the car. Seeing the wounded creature was a reminder that the individual, suffering animal had an intrinsic value. Such a value is lost when we focus exclusively on instrumental value or when we see only the large-scale problem, without consideration of the individual. Incidentally, by the time I had turned around, another person had already called the animal control authorities. Coming into personal contact with that deer makes it our deer, just as knowing that there are too many dogs and cats in the nation does not diminish our devotion to our own dog or cat.[17] When we do this, we move from a utilitarian model to one of empathy. This certainly applies to most matters of bioethics. Our ability to empathize about the risks and benefits of a given drug, device, or system is a tool for predicting the necessary moral action.

Another lesson here is our need to be aware about what we are saying to people. No matter how technically sound and convincing our communications seem to us technical types, they may be wholly unconvincing to our targeted audience, or even an affront to the values they cherish. Their value systems and the ways that they perceive risk are the result of their own unique blend of experiences. So, do not be surprised if the perception of risk does not match the risk you have quantified.

RELIABILITY: AN ETHICS METRIC

A system, process, or item is reliable so long as it performs the designed function under the specified conditions during a certain time period. As mentioned in Chapter 5, in most engineering applications, reliability means that what we design will not fail prematurely. Reliability is the probability that something that is in operation at time 0 (t_0) will still be operating until the designed life (time $t = (t_t)$). As such, it is also a measure of the engineer's social accountability. People receiving a medical device want to know if it will work and not fail.

The probability of a failure per unit time is the failure density, or $f(t)$ which is a function of the likelihood that an adverse outcome will occur, but note that it is not a function of the severity of the outcome. Recall that the likelihood that something will fail at a given time interval can be found by integrating the hazard rate over a defined time interval:

$$P\{t_1 \leq T_f \leq t_2\} = \int_{t_1}^{t_2} f(t)dt \tag{6.4}$$

where $T_f =$ time of failure.

Again, the reliability function, $R(t)$, of a system at time t is the cumulative probability that the system has not failed in the time interval from t_0 to t_t:

$$R(t) = P\{T_f \geq t\} = 1 - \int_0^t f(x)dx \tag{6.5}$$

Engineers must be humble, since everything we design will fail. We can improve reliability by extending the time (increasing t_t), thereby making the system more resistant to failure. For example, proper engineering design of a stent allows for improved bloodflow. The stent may perform well for a certain period of time, but loses efficiency when it becomes blocked. A design flaw occurs due to poor fabrication or if the wrong material is used, such as one that allows premature sorption of cholesterol or other substances to the interior lining of the stent. However, even when the proper materials are used, the failure is not completely eliminated. In our case, the sorption still occurs, but at a slower rate, so 100% blockage (i.e., $R(t) = 0$) would still be reached eventually. Selecting the right materials simply protracts the time before the failure occurs (increases T_f).[18] Ideally, if possible the reliability of a device or system reflects an adequate margin of safety. So, if the stent fails after 100 years, few reasonable people would complain. If it fails after ten years, and was advertised to prevent blockage for five years, there may still be complaints, but the failure rate and margin of safety were fully disclosed. If it fails after ten years, but its specifications require twenty years, this may very well constitute an unacceptable risk. Obviously, other factors must be considered, such as sensitive subpopulations. For example, it may be found that the stent performs well in the general population, but in certain subgroups (e.g., blood types) the reliability is much less. The engineer must fully disclose such limitations.

Equation 6.2 illustrates built-in vulnerabilities, such as the inclusion of inappropriate design criteria, like cultural bias, which may translate into a shortened time before failure. If we do not recognize these inefficiencies upfront, we will pay by premature failures (e.g., law suits, unhappy clients, and a public that has not been well served in terms of our holding paramount their health, safety, and welfare). So, if we are to have reliable engineering we need to make sure that whatever we design, build, and operate is done with fairness and openness. Otherwise, these systems are, by definition, unreliable.

The U-shaped curves (Figures 5.5 and 5.6) are a good way to picture a reliability life cycles. Another good way to visualize reliability as it pertains to bioethics is to link potential causes to effects. Cause and effect diagrams (also known as Ishikawa diagrams) identify and characterize the totality of causes or events that contribute to a specified outcome event. The "fishbone" diagram (Figure 6.3) arranges the categories of all causative factors according to their importance (i.e., their share of the cause). The construction of this diagram begins with the failure event to the far right (i.e., the "head" of the fish),

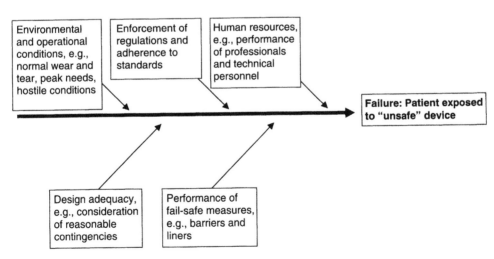

Figure 6.3 Fishbone reliability diagram showing contributing causes to an adverse outcome (exposure to an unsafe medical device).

followed by the spine (flow of events leading to the failure). The "bones" are each of the contributing categories. This can be a very effective tool in explaining failures to clients and the public. Even better, the engineer may construct the diagrams in "real time" at a design meeting. This will help to open up the design for peer review and will help get insights, good and bad, early in the design process.

The premise behind cause-and-effect diagrams like the fishbones and fault trees is that all the causes have to connect through a logic gate. This is not always the case, so another more qualitative tool may need to be used, such as the Bayesian belief network (BBN). Like the fishbone, the BBN starts with the failure (Figure 6.4). Next, the most immediate contributing causes are linked to the failure event. The next group of factors that led to the immediate causes is then identified, followed by the remaining contributing groups. This diagram helps to catalog the contributing factors and also compares how one group of factors impacts the others.

The engineering codes require us to optimize our competence. A big part of this is knowing just how reliable our designs are expected to be. And, the engineer must properly communicate this to clients and the public. This means, however discomfiting, the engineer must "come clean" about all uncertainties. Uncertainties are ubiquitous in risk assessment. The engineer should take care to not be overly optimistic or overly pessimistic about what is known and what needs to be done. Full disclosure is simply an honest rendering of what is known and what is lacking for those listening to make informed decisions. Part of the uncertainty involves conveying the meaning; we must clearly communicate the potential risks. A word or phrase can be taken many ways. As mentioned, engineers should liken themselves to physicians writing prescriptions. Be completely clear, otherwise confusion may result and may lead to unintended, negative consequences.

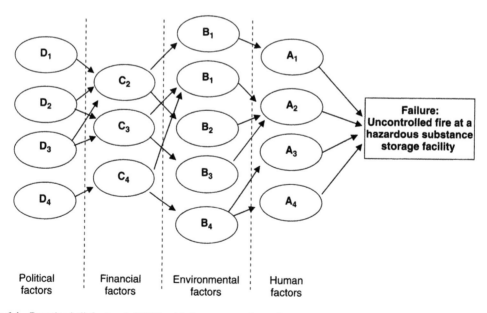

Figure 6.4 Bayesian belief network (BBN), with three groups of contributing causes leading to a failure.

The concept of safety is laden with value judgments. Thus, ethical actions and decisions must rely on both sound science and quantifiable risk assessment, balanced with an eye toward social, macroethical ramifications.

REDUCING RISKS

What constitutes an acceptable risk? A convenient standard is that a risk from a product or system should be "as low as reasonably practical" (ALARP), a concept coined by the United Kingdom Health and Safety Commission.[19] This can be envisioned as a diagram (Figure 6.5). The upper area (highest actual risk) is clearly where the risk is unacceptable. Below this intolerable level is the ALARP. Risks in this region require measures to reduce risk to the point where costs outweigh benefits disproportionately. This approach to determining an ethically acceptable level based upon risks and benefits is one form of utilitarianism. The utility of a medical device, for example, is based upon the greatest good it will engender, but this must be compared to the potential harm it may cause.

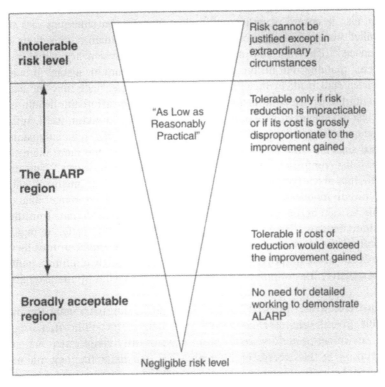

Figure 6.5 Three regions of risk tolerance.
Adapted from: United Kingdom Health and Safety Commission, 1998, http://www.hse.gov.uk/nuclear/computers.pdf; accessed on 26 May.

Another aspect of ALARP that is especially important to device and system designs is that a margin of safety should be sought. This margin is both protective and reasonable.[20] Reaching ALARP necessitates qualitative and/or quantitative measures of the amount of risk reduced and costs incurred with the design decisions:

> The ALARP principle is based on the assumption that it is possible to compare marginal improvements in safety (marginal risk decreases) with the marginal costs of the increases in reliability.[21]

One of the challenges in applying a utilitarian model to engineered systems is that financial costs and benefits are often easier, or at least more straightforward, to calculate than other (i.e., nonmonetizable) costs and benefits. Another problem is the issue of the costs of not designing a solution to a problem. For example, as mentioned in Chapters 1 and 5, the current controversies associated with conducting research at the nanoscale (near the diameter of the hydrogen atom) are sometimes rooted in fears of the potential Pandora's boxes of unknown technologies. If decisions are only made to avoid these problems, society exposes itself to opportunity risks. In other words, if we inordinately avoid designs for fear of potential harm, we may forfeit lifesaving and enriching technologies (e.g., drug delivery systems that differentiate tumor tissue from healthy tissue based on electromagnetic differences at the nanoscale).

Engineers must also be aware of vested and hidden interests. An emerging risk management technique is the so-called outrage management, coined by Peter Sandman, a consultant to businesses and governmental agencies.[22] The first step of this approach is to present a positive public image, such as a "romantic hero," pointing out all the good things the company or agency provides, such as jobs, modern conveniences, and medical breakthroughs. Although these facts may be accurate, they often have little to do with the decisions at hand, such as the need for affordable health care. Another way that public image can be enhanced is to argue that the company is a victim itself, suffering the brunt of unfair media coverage or targeted by politicians. If these do not work, some companies have confessed to being "reformed sinners," who are changing their ways. One of the more interesting strategies put forth by Sandman is that companies can portray themselves as "caged beasts." This approach is used to convince the public that even though in the past they have engaged in unethical and unfair practices, the industry is so heavily regulated and litigated against that they are no longer able to engage in these acts. So, the public is encouraged to trust that this new product is different from the ones that came previously from the company. There is obviously some truth to this tactic, as regulations and court precedents have curtailed a lot of corporate irresponsibility. But the engineer must be careful to discern the difference between actual improvement and mere spin tactics to eliminate public outrage. And, from an ethical perspective, the culture has not necessarily changed, only the extended controls on the company.

Companies must look at the financial bottom line, but holding paramount the health, safety, and welfare of the public gives the engineer no room for spin. Indeed, the public often exaggerates risks, and abating risks that are in fact quite low could mean unnecessarily complicated and costly measures. It may also mean choosing the less acceptable alternative, i.e., one that in the long run may be more costly and deleterious to public health. The risk assessment and risk perception processes differ markedly, as shown in Table 6.2. Engineering design makes use of problem identification, data analysis, and risk characterization, including cost–benefit ratios. Perception relies on thought processes, including intuition, personal experiences, and personal preferences. Engineers tend to be more comfortable operating in the

Table 6.2
Differences between risk assessment and risk perception processes

Analytical phase	Risk assessment processes	Risk perception processes
Identifying risk	Physical, chemical, and biological monitoring and measuring the event	Personal awareness
	Deductive reasoning Statistical inference	Intuition
Estimating risk	Magnitude, frequency and duration calculations	Personal experience
	Cost estimation and damage assessment	Intangible losses and nonmonetized valuation
	Economic costs	
Evaluating risk	Cost–benefit analysis	Personality factors
	Community policy analysis	Individual action

Source: Adapted from K. Smith, 1992, *Environmental Hazards: Assessing Risk and Reducing Disaster*, Routledge, London, UK.

middle column (using risk assessment processes), while the general public often uses the processes in the far right column. One can liken this to the "left-brained" engineer trying to communicate with a "right-brained" audience. It can be done so long as preconceived and conventional approaches do not get in the way.

Both "lay" groups and engineers have difficulty in parsing perceived and real risks. How we think shapes what we do and how we communicate. John Gray's bestselling book, *Men Are from Mars, Women Are from Venus: A Practical Guide to Improving Communications and Getting What You Want from Your Relationships*, First Edition, HarperCollins, 1992, New York, NY, explores the general differences between the way men and women process information to arrive at conclusions, and how very different can be what is meant *versus* what is heard. Engineers also share some general personality patterns that differ from the general public, as indicated by popular personality tests, such as the Myers Briggs® typologies. Often, engineers and scientists direct their intellectual energies toward the inner world, at least while they are on the job. They attempt to be clear about data and information in order to understand and explain what they have been studying. They trust experience (i.e., they adhere to experimental findings). Conversely, many of their clients direct their energies outwardly, speaking before they have completely and thoroughly formulated an idea. This is not necessarily "sloppiness," although scientists frequently perceive it to be. It is often an attempt to explore possible alternatives to address a problem. In other words, when it comes to science, the client is often more comfortable with ambiguity than is the engineer. Interestingly, some of the great scientists, like Einstein, Bohr, and their modern incarnations like Gould and Hawking, have evidenced a great deal of comfort with ambiguous and yet-to-be-explained paradigms. Thus, even within the scientific community there is a great amount of diversity on how we see the world.

RISK AS AN ETHICAL CONCEPT

Risk is tied to almost every technical professional decision. To demonstrate this, let us consider an illustrative hypothetical case.[23]

Cornwallis, a township in a county seat in the southeastern region of the United States, has a population of about 5000, mainly living in subdivisions within a large rural buffer surrounding the town's central business district. As such, many of the township's residents draw their drinking water from private wells. The aquifer from which the water is taken consists of unconsolidated basalt gravel that is intermixed with various minerals. Also, like much of the Southeast, the soils are quite deep, sometimes deeper than 10 meters. They have weathered in place for many millions of years (known as ultisols).[24]

For the past few years, a number of residents in the Beagle Creek Road neighborhood have complained that the water they have been drinking is linked to what they believe to be an inordinately large number of cancers in Cornwallis. The Beagle Creek Road neighborhood has about 150 dwelling units. Following the latest meeting of the County Board of Commissioners in which a number of Beagle Creek residents testified, the Board approved $21 500 to test the well water and authorized the County Health Department to conduct brief phone interviews with the residents, mainly inquiring about cancer histories (number of cases and the specific types of cancer). The county epidemiologist is hoping to have about 115 households participate in the interviews. She says that "the more people who participate in the interviews, the better we will know what is going on." The study is expected to be completed in three months.

The well testing, which is entirely voluntary, must measure pathogenic bacteria, radon, and "chemicals like pesticides," according to the Board. Residents will also receive, free of charge, kits to be used to detect radon in the indoor air of their homes. The county health department has some experience in biological and chemical testing, but has not previously conducted health surveys and interviews of this type coupled with the physical testing.

The residents report that nine residents of Beagle Creek have died of cancer in the past ten years. In addition to the cancer complaints other Beagle Creek residents complain of kidney stones. A sixty-year-old retired police officer blames his stones on water that looks clear from the tap but leaves a precipitate and stains when left to settle. They also have complaints about the aesthetics of the water, especially the discoloration of clothing that is washed, usually leaving a reddish stain. Those who have installed filters complain that they must be changed daily. Some say that the problem is relatively recent and that the laundry problems only began after a number of new residents installed wells in the last fifteen years. Others say that the problems fluctuate, where there are periods of time when the water seems fine, but then exhibits unpleasant smell, bad taste, and other problems.

The county epidemiologist is trying not to overpromise, stating that "cancer is a tough disease to study since it differs from acute diseases like those usually studied by the health department, such as gastrointestinal tract infections." The study will have to be augmented by advice from State cancer experts, particularly on how to determine the cancer incidence.[25] Also, previous environmental studies in the county have not been able to link diseases to groundwater contamination. Thus, the

County Board has required that the county epidemiologist provide more information about cancer rates not only in the county but across the state in order to put whatever is found into context.

Similar cases to this one are taking place throughout the continent, and engineers are often intimately involved. Among the risk-related issues are the need to determine the hazards, to estimate the likelihood of these hazards, and to differentiate between actual and perceived hazards and risks. The case also illustrates the different expectations of those involved in an issue. The residents have a different kind of stake in the problem than do the various professionals and government officials. Finally, the case points out that each group has its own measures of success, ranging from understanding health outcomes, linking them to possible causes, and eliminating the sources, or at least exposure to the sources, of these health problems. In fact, one possibility is that there is absolutely no linkage between the health problems and the presupposed cause, i.e., something in the water.

So, let us parse the details of the case. There are a number of elements about which everyone involved may well agree. For example, the water is more than simple H_2O. Even this, however, is not "normal," whatever that means. No drinking water is completely free of substances. The chemical composition reflects the source of the water. So, since this water supply is from an igneous rock formation, many of the rock's chemical elements will have leached into the water. The amount of contamination is directly related to each constituent's aqueous solubility, but is also affected by the quality of the water that originally infiltrated the aquifer, as well as by the characteristics of the soil through which the water has percolated.

Something to keep in mind is that geology and soil type are highly variable, even within the same formation. So, one neighbor's water quality may differ considerably from that of his or her neighbor only a few hundred meters away. Just because the neighborhood is defined by the county and state as a unit, it does not mean it is homogeneous. The neighborhood has been defined for other reasons (e.g., property purchases, zoning, subdivision regulations, and accessibility to roads and other facilities), so expecting uniform environmental quality is unrealistic. The lesson here is that the engineer should be reticent to aggregate data mindlessly.

Another important aspect of this case is that what appears to be bad or unpleasant may have no direct bearing on a particular outcome; that is, risk perception is very different from risk assessment. For example, the discoloration of the water is likely due to the presence of iron or manganese. Indeed, it will ruin a white shirt, but does this have any connection at all to cancer clusters? Also, neither iron nor manganese is linked to kidney stones. On the other hand, this does not mean the neighbors are wrong. In fact, many metals co-occur in the environment. Depending on the formation, if you see iron and manganese, you may also find a series of other metals, such as calcium, or other toxic metals such as copper, zinc, lead, mercury, and arsenic (actually a metalloid). In addition, the naturally occurring metallic deposits can contain other harmful substances, notably asbestos and radon. So, pardon the pun, but to the neighbors the iron oxide discoloration is a "red flag"![26]

It is not uncommon in medical science and engineering to use indicators, but it is quite tricky. An indicator must be both sensitive (i.e., we can see it and it should tell us that something is wrong) and specific (i.e., when we see it we know that it means one or only a few things). But iron in water is neither a sensitive nor a specific indicator of cancer-causing substances in water. It is not even a good indicator of water hardness, since the two ions that cause hardness, calcium and magnesium, may or may not substantially co-occur with iron in natural waters.

Whether advising about a failure or problem or designing a device or system, engineers must reduce uncertainty in the science that they apply. Without a readily understandable "benchmark" of measurements, people can be left with possible misunderstandings, ranging from a failure to grasp a real problem that exists to perceiving a problem even when things are better than or do not differ significantly from those of the general population. Two errors can occur when information is interpreted in the absence of sound science. The first error mentioned above is the false negative, or reporting that there is no problem when one in fact exists. In this case, if the country engineer advises the water is fine, but it is in fact one of the sources of the problems, this is a false negative. The need to address this problem is often at the core of the positions taken by public health agencies and advocacy groups. They ask questions like:

- What if the epidemiological study designed by the county shows no harm, but in fact toxic substances there are in our drinking water?
- What if the substances that have been identified really do cause cancer but the tests are unreliable?
- What if people are being exposed to a contaminant, but via a pathway other than the ones being studied (e.g., exposed to volatile organic compounds in water when showering or transformed toxic compounds when the water is used for cooking)?
- What if there is a relationship that is different from the laboratory, that is, when this substance is released into the "real world," such as the difference between how a chemical behaves in the human body by itself as opposed to when other chemicals are present (i.e., the problem of "complex mixtures")?

The other concern is, conversely, the false positive. This can be a major challenge for health agencies with the mandate to protect the public. For example, what if previous evidence shows that an agency had listed a compound as a carcinogen only to find that a wealth of new information is now showing that it has no such effect? This can happen if the conclusions were based upon faulty models, or models that only work well for lower organisms, but subsequently developed models have taken into consideration the physical, chemical, and biological complexities of higher-level organisms, including humans. Biomedical researchers see this when trying to apply comparative biology to the safety of a new drug. False positives may force public health officials to devote inordinate amounts of time and resources to deal with so-called nonproblems. False positives also erroneously scare people about potentially useful products.[27] False positives, especially when they occur frequently, create credibility gaps between engineers and scientists and the decision makers. In turn, the public, those whom we have been charged to protect, lose confidence in us as professionals. Risk assessments need to be based on high-quality, scientifically based information. In engineering language, the risk assessment process is a "critical path" in which any unacceptable error or uncertainty along the way will decrease the quality of the risk assessment and will increase the likelihood of bad decisions.

While we almost never reach complete scientific consensus when it comes to ascribing cause, technical professionals rely on "weight of evidence," much as juries do in legal matters. The difference is that in science a single data point can undo mountains of contravening evidence. Statisticians tell us that we must reduce both type I and type II errors:

1. A type I error is rejecting a hypothesis when it is true.
2. A type II error is failing to reject the hypothesis when it is false.

So, this is simply another way to say that we wish to avoid false negatives and false positives. The challenge of our Beagle Creek case is that we are unlikely to have a sufficient amount of data to allay criticisms that we have large type I and type II errors. The residents are likely to complain about false negatives should the epidemiologist hint at a "clean bill of health." Others, such as commissioners from other districts who may have to pay for corrective actions from which they receive no direct benefit may complain that the tests were flawed or misinterpreted, giving false positives (e.g., just because you find 100 ppb of a pesticide in water does not mean any of the cancers were "caused" by exposures to this pesticide).

Perhaps the best we can hope for here is to say that all life is indeed sacred, but society's role in protecting an individual is limited and finite. Unfortunately, from a fairness perspective, the protection is uneven, such as the public funds devoted to preventing cancer from a potential release of a toxic substance compared to the prevention of genocide in Rawanda or Darfur. And, the Ancients do not give us safe quarters here as they require that equals be treated equally and unequals be treated unequally (i.e., greater need calls for greater effort).

RISK-BASED ETHICS: THE SYLLOGISM REVISITED

Upon examination, the syllogism described in Chapter 2 is not necessarily as straightforward as it may first appear. In fact, the exact meanings of premises and moral conclusions have led to very vigorous debates (and lawsuits). For example, all parties may agree with the evaluative premise, for example, that releases of a toxic substance from factories or concentrations in household products should not cause cancer. However, various groups frequently disagree strongly about the "facts" underlying the factual premise, such as whether the data really show that these dosages "cause" cancer or whether they are just coincidental associations. Or, they may agree that they cause cancer, but not at the rate estimated by scientists. Or, they may disagree with the measurements and models that project the concentrations of chemical X to which people would be exposed (e.g., a conservative model may show high exposures and another model, with less protective algorithms, such as faster deposition rates, may show very low exposures). Or, they may argue that measurements are not "representative" of real exposures. There are even arguments about the level of protection. For example, should public health be protected so that only one additional cancer is expected in a population of a million or one in ten thousand? If the former (10^{-6} cancer risk) were required, the plant would have to lower emissions of chemical X far below the levels that would be required for the latter (10^{-4} cancer risk). This is actually an argument about the value of life. Believe it or not, there are "price tags" placed quite frequently on a prototypical human life, or even expected remaining lifetimes. These are commonly addressed in actuarial and legal circles. For example, Paul Schlosser in his discussion paper, "Risk Assessment: The Two-Edged Sword," states:

> The processes of risk assessment, risk management, and the setting of environmental policy have tended to carefully avoid any direct consideration of the value of human life. A criticism is that if we allow some level of risk to persist in return for economic benefits, this is putting a value on human life (or at least health) and that this is inappropriate because a human life is invaluable – its value is infinite. The criticism is indeed valid; these processes sometimes do implicitly put a finite, if unstated, value on human life.
>
> A bit of reflection, however, reveals that in fact we put a finite value on human life in many aspects of our society. One example is the automobile. Each year, hundreds or thousands of US citizens are killed in car accidents. This is a significant risk. Yet we allow the risk to continue, although it could be substantially reduced or eliminated by banning cars or through strict, nation-wide speed limits of 15 or 20 mph. But we do

not ban cars and allow speeds of 65 mph on major highways because we derive benefits, largely economic, from doing so. Hence, our car "policy" sets a finite value on human life.

You can take issue with my car analogy because, when it comes to cars, it is the driver who is taking the risk for his or her own benefit, while in the case of chemical exposure, risk is imposed on some people for the benefit of others. This position, however, is different from saying that a human life has infinite value. This position says that a finite value is acceptable if the individual in question derives a direct benefit from that valuation. In other words, the question is then one of equity in the risk–benefit trade-off, and the fact that we do place a finite value on life is not of issue.

Bioethics Question: Is all human life sacred?

One manner of determining whether a monetary value is placed on human life is to ask, "How much are we willing to spend to save a human life?" Table 6.3 provides one group's estimates of the costs to save one human life. Although the group sharing this information has a political agenda (they are opposed to much of the "environmentalist agenda,") and their bias colors these data, their method of calculating the amount of money is fairly straightforward. If nothing else, the amounts engender discussions about possible risk trade-offs since the money itself is fungible and may otherwise be put to more productive use.

Schlosser is asking, "How much is realistic?" He argues that a line must be drawn between realistic and absurd expenditures. He states:

In some cases, risk assessment is not used for a risk–benefit analysis, but for comparative risk analysis. For example, in the case of water treatment one can ask: is the risk of cancer from chlorination by-products greater than the risk of death by cholera if we do not chlorinate? Similarly, if a government agency has only enough funds to clean up one of two toxic waste sites in the near future, it would be prudent to clean up the

Table 6.3
Regulation cost of saving one life

Activity	Cost (in US dollars)
Auto passive restraint/seat belt standards	100 000.00
Aircraft seat cushion flammability standard	400 000.00
Alcohol and drug control standards	400 000.00
Auto side door support standards	800 000.00
Trenching and excavation standards	1 500 000.00
Asbestos occupational exposure limit	8 300 000.00
Hazardous waste listing for petroleum refining sludge	27 600 000.00
Cover/remove uranium mill tailings (inactive sites.)	317 000 000.00
Asbestos ban	110 700 000.00
Diethylstilbestrol (DES) cattle feed ban	124 800 000.00
Municipal solid waste landfill standards (proposed)	19 107 000 000.00
Atrazine/Alachlor drinking water standard	92 069 700 000.00
Hazardous waste listing for wood preserving chemicals	5 700 000 000 000.00

Source: P.M. Schlosser, 1997, Risk Assessment: The Two-Edged Sword, http://pw2.netcom.com/~drpauls/just.html; accessed on 12 April 2005.

site which poses the greatest risk. In both of these cases, one is seeking the course of action which will save the greatest number of lives, so this does not implicitly place a finite value on human life. (In the second example, the allocation of finite funds to the government agency does represent a finite valuation, but the use of risk assessment on how to use those funds does not.)[28]

Schlosser is using a utilitarian vernacular. Scarce resources are dictating the alleged "right action". Human beings are fallible, thus they are not always the best assessors or predictors of value. So, how do moral arguments about where to place value and the arguments made by Schlosser and Feldman fit with moral theories, such as duty-based ethics (i.e., deontology), consequence-based ethics (teleology), or social contract theory (contractarianism)? Where do concepts like John Stuart Mill's harm principle, John Rawls' veil of ignorance, and Immanuel Kant's categorical imperative come into play? How do such concepts fit with the code in one's chosen profession? How do teleological, deontological, contractarian, and rational models hold up this scrutiny?

In the process of allowing latitude in corporate and public decision making we have, by default, limited the value and sanctity of a human life. Is this different from the opposition to the state-sanctioned death penalty? Many oppose it on the grounds that the "state is us".

CAUSATION

No matter where we stand personally, practical considerations do affect morality. So engineers have a moral obligation to "get the science right." Zero risk can only occur when either the hazard does not exist or the exposure to that hazard is zero. Association of two factors, such as the level of exposure to a compound and the occurrence of a disease, does not necessarily mean that one necessarily "causes" the other. Often, after study, a third variable explains the relationship. However, it is important for science to do what it can to link causes with effects. Otherwise, corrective and preventive actions cannot be identified. So, strength of association is a beginning step toward cause and effect. A major consideration in strength of association is the application of sound technical judgment of the weight of evidence. For example, characterizing the weight of evidence for carcinogenicity in humans consists of three major steps:[29]

1. Characterization of the evidence from human studies and from animal studies individually.
2. Combination of the characterizations of these two types of data to show the overall weight of evidence for human carcinogenicity.
3. Evaluation of all supporting information to determine if the overall weight of evidence should be changed.

Note that none of these steps is absolutely certain.

People have a keen sense of observation, especially when it has to do with the health and safety of their families and neighborhoods. They can "put 2 and 2 together." Sometimes, it seems that as engineers we are asked to tell them that $2 + 2$ does not equal 4. That cluster of cancers in town may truly have nothing to do with the green gunk that is flowing out of the abandoned building's outfall. But in their minds, the linkage is obvious.

The challenge is to present information in a meaningful way without violating or overextending the interpretation of the data. If we assign causality when none really exists, we may suggest erroneous solutions. But, if all we can say is that the variables are associated, the public is going to want to know more about what may be contributing to an adverse affect (e.g., learning disabilities and blood lead levels). This was particularly problematic in early cancer research. Possible causes of cancer were being explored and major research efforts were being directed at myriad physical, chemical, and biological agents. So, there was a need for some manner of sorting through findings to see what might be causal and what is more likely to be spurious results. Sir Austin Bradford Hill (see Biographical Box) is credited with articulating key criteria that need to be satisfied to attribute cause and effect in medical research (see Chapter 3).[30] Recall that the factors to be considered in determining whether a cause elicits an effect include:

- Criterion 1: Strength of Association
- Criterion 2: Consistency
- Criterion 3: Specificity
- Criterion 4: Temporality
- Criterion 5: Biologic Gradient
- Criterion 6: Plausibility
- Criterion 7: Coherence
- Criterion 8: Experimentation
- Criterion 9: Analogy

Biographical Box: Sir Bradford Hill

In 1965 Hill made a major contribution to the science of epidemiology by publishing his famous paper, "The Environment and Disease: Association or Causation?," which included the nine guidelines for establishing the relationship between exposure and effect (see Table 3.1). Hill meant for the guidelines to be just that – guidelines – and not an absolute test for causality. A situation does not have to meet Hill's nine criteria to be shown to be causally related. In the introduction to his paper, Hill acknowledges this by suggesting that there will be circumstances where not all of the nine criteria need to be met before action is taken. He recommended that action may need to be taken when the circumstances warrant it. In his opinion, in some cases "the whole chain may have to be unraveled" or in other situations "a few links may suffice." The case of the 1853 cholera epidemic in London, concluded by John Snow to be waterborne, and controlled by the removal of the pump handle, is a classic example where only a few links were understood.

In assessing risks, some of Hill's criteria are more important than others. Risk assessments rely heavily on strength of association, e.g., to establish dose–response relationships. Coherence is also very important. Animal and human data should support one another and should not disagree. Biological gradient is crucial, since this is the basis for the dose–response (the higher the dose, the greater the biological response).

Temporality is crucial to all scientific research as it is in ethics, i.e., the cause must precede the effect. However, this is sometimes difficult to see in some instances, such as when the exposures to suspected agents have been continuous for decades and the health data are only recently available.

The key is that sound engineering and scientific judgment, based on the best available and most reliable data, be used to link cause and effect. Sometimes, the best we can do is to be upfront and clear about the uncertainties and the approaches we use.

Medical risk by its very nature addresses probable impossibilities. From a statistical perspective, it is extremely likely that cancer will not be eliminated during our lifetimes. But, the efforts to date have shown great progress toward reducing risks from several forms of cancer. This risk reduction can be attributed to a number of factors, including changes in behavior (smoking cessation, dietary changes, and improved lifestyles), source controls (less environmental releases of cancer-causing agents), and the reformulation of products (substitution of chemicals in manufacturing processes).

Risk characterization is the stage where the engineer summarizes the necessary assumptions, describes the scientific uncertainties, and determines the strengths and limitations of the analyses. The risks are articulated by integrating the analytical results, interpreting adverse outcomes, and describing the uncertainties and weights of evidence.

Risk assessment decisions are distinct from risk management decisions, where actions must be chosen to address and reduce the risks. But the two are deeply interrelated and require continuous feedback with each other. Engineers are key players in both efforts. In addition, risk communication between the engineer and the client further complicates the implementation of the risk assessment and management processes. What really sets risk assessment apart from the actual management and policy decisions is that the risk assessment must follow the prototypical rigors of scientific investigation and interpretation that are outlined in this chapter.

What we do about risks reflects what we value. One method for testing our ethics is to try to look back from a 100 years hence, such as we can do now with slavery and women's rights. What would you expect the future societies to think of our present bioethical decision making and about what we truly value?

Whenever we have drawn a moral conclusion; that is, when the behavior of certain groups was improper, unacceptable, or downright immoral, we have intuitively drawn a syllogism. Intuitive syllogisms are present every time we give credit or place blame. The best we can hope for is that we have thoroughly addressed the most important variables and with wisdom may prevent similar problems in the future.

Syllogisms can easily be inverted to fit the perception and needs of those applying them; that is, people already have a conclusion in mind and go searching for facts to support it. This is improper ethical analysis. The general public expects that its professionals understand the science and that any arguments being made are based in first principles. We must be careful that this "advocacy science" in its worst form, "junk science", does not find its way into engineering. There is a canon that is common in most engineering codes that tells us that we need to be "faithful agents." This, coupled with an expectation of competency, requires us to be faithful to the first principles of science. Because of pressures from clients and political or ideological correctness, the next generation of engineers likely will be tempted to "repeal Newton's laws" in the interest of certain influential groups! This is not to say that engineers will have the luxury to ignore the wishes of such groups, but since we are the ones with our careers riding on these decisions, we must clearly state when an approach is scientifically unjustifiable. We must be good listeners, but honest arbiters.

Unfortunately, many scientific bases for decisions are not nearly as clear as Newton's laws. They are far removed from first principles. The engineer operating at the mesoscale (e.g., a test chamber) can be fairly confident about the application of first principles of motion and thermodynamics, but the biomechanical engineer looking at the same contaminant at the nanoscale is not so confident about how to apply them. That is where junk science is sometimes able to raise its ugly head. In the void of certainty, e.g., at the molecular scale, some crazy arguments are made about what does or does not happen. This is the stuff of infomercials. The new engineer had better be prepared for some off-the-wall ideas of how the world works. New hypotheses for causes of cancer, or even etiologies of cancer cells, will be put forward. Most of these will be completely unjustifiable by physical and biological principles, but they will sound sufficiently plausible to the unscientific. The challenge of the new engineer will be to sort through this morass without becoming closed minded. After all, many scientific breakthroughs have been considered crazy when first proposed (recalling Copernicus, Einstein, Bohr, and Hawking, to name a few). But, even more were really wrong and unsupportable upon scientific scrutiny.

In addition to the differences in scientific training and understanding, the disconnections between the public and the professional can also result from differing perspective. The neighbors are living with the fear every day. So, when the scientist and the engineer correctly see the discoloration and other physical attributes of the water as suspended solids and ionic strength, the neighbors see it as an ominous warning that something is terribly wrong with their personal life support system. Success comes with sensitivity to client needs without sacrificing sound engineering principles.

NOTES AND COMMENTARY

[1] The principal sources of this discussion are M. Myers and A. Kaposi, *The First Systems Book: Technology and Management*, 2nd ed. (London, UK: Imperial College Press, 2004); and T.R.G. Green, "Cognitive Dimensions of Notations," in *People and Computers V*, ed. A. Sutcliffe and L. Macaulay (Cambridge, UK: Cambridge University Press, 1989).

[2] H. Petroski, *To Engineer Is Human: The Role of Failure in Successful Design* (New York, NY: St. Martin's Press, 1985).

[3] American Medical Association, *Code of Medical Ethics Current Opinions with Annotations, 2004–2005* (Chicago, Illinois: AMA, 2004).

[4] Ibid.

[5] National Society of Professional Engineers, *NSPE Code of Ethics for Engineers* (Virginia: Alexandria, 2003), http://www.nspe.org/ethics/eh1-code.asp (accessed 21 August 2005).

[6] C.B. Fleddermann, "Safety and Risk," in *Engineering Ethics* (Upper Saddle River, NJ: Prentice-Hall, 1999).

[7] An episode of the popular television series, *The Sopranos*, included a scene where the teenage girlfriend of the main character's son is eating fruit and smoking a cigarette at a wedding reception. She is offered an *hors d'oeuvre* that contains seafood. She rejects the offer after puffing on the cigarette, saying that she does not eat fish because it "contains toxins."

[8] The segregation of risk assessment (science) from risk management (feasibility) is a recent approach.

[9] C. Mitcham and R.S. Duval, "Responsibility in Engineering," in *Engineering Ethics* (Upper Saddle River, NJ: Prentice-Hall, 2000).

[10] Although presented within the context of how risk is a key aspect of environmental justice, the information in this chapter is based on two principal sources: D.A. Vallero, *Environmental Contaminants: Assessment and Control* (Burlington, MA: Elsevier Academic Press, 2004); and D.A. Vallero,

Paradigms Lost: Learning from Environmental Mistakes, Mishaps, and Misdeeds (Burlington, MA: Butterworth-Heinemann, 2005).

[11] W. Redmond, "Conscience as Moral Judgment: The Probabilist Blending of the Logics of Knowledge and Responsibility," *Journal of Religious Ethics* 26, no. 2 (1998), 389–405.

[12] Ibid.

[13] T.M. Apostol, *Calculus, Volume II*, 2nd ed. (New York, NY: John Wiley & Sons, 1969).

[14] C.B. Fleddermann, "Safety and Risk," in *Engineering Ethics* (Upper Saddle River, NJ: Prentice-Hall, 1999).

[15] R. Beyth-Marom, B. Fischhoff, M. Jacobs-Quadrel, and L. Furby, "Teaching Decision Making in Adolescents: A Critical Review," in *Teaching Decision Making to Adolescents*, ed. J. Baron and R.V. Brown, 19–60 (Hillsdale, NJ: Lawrence Erlbaum Associates, 1991).

[16] K. Smith, "Hazards in the Environment," in *Environmental Hazards: Assessing Risk and Reducing Disaster* (London, UK: Routledge, 1992).

[17] This calls to mind Jesus' parable of the lost sheep (Gospel of Matthew, Chapter 18). In the story, the shepherd leaves, abandons really, ninety-nine sheep to find a single lost sheep. Some might say that if we as professionals behaved like that shepherd, we would be acting irresponsibly. However, it is actually how most of us act. We must give our full attention to the patient or client individually. There are a number of ways to interpret the parable, but one is that there is the individual, inherent value to each member of society and the value of society's members is not mathematically divisible. In other words, a person in a population of one million is not one-millionth of the population's value. The individual value is not predicated on the group's value. This can be a difficult concept for those of us who are analytical by nature, but it is important to bioethics.

[18] Hydraulics and hydrology provide very interesting case studies in the failure domains and ranges, particularly how absolute and universal measures of success and failure are almost impossible. For example, a levee or dam breach, such as the recent catastrophic failures in New Orleans during and in the wake of Hurricane Katrina, experienced failure when flow rates reached cubic meters per second. Conversely, a hazardous waste landfill failure may be reached when flow across a barrier exceeds a few cubic centimeters per decade, and a few percent change in blood flow to the brain can be devastating.

[19] United Kingdom Health and Safety Commission, 1998, http://www.hse.gov.uk/nuclear/computers.pdf (accessed 26 May 2006). The Commission is responsible for health and safety regulation in Great Britain. The Health and Safety Executive and local government are the enforcing authorities that work in support of the Commission.

[20] The concept of "reasonableness" is a constant theme of this book. Unfortunately, the means of determining reasonable designs are open to interpretation. The legal community uses the reasonable person standard. Professional ethics review boards may also indirectly apply such a standard. The reasonable person is akin to the Kantian categorical imperative: That is, if the act (e.g., the approach to the design) were universalized, would this be perceived to be a "good" or a "bad" approach? This is the type of query where reviewers might ask whether the designer "knew or should have known" the adverse aspects of the design in advance.

[21] United Kingdom Health and Safety Commission.

[22] P. Sandman's advice is found in S. Rampton and J. Stauber, *Trust Us, We're Experts: How Industry Manipulates Science and Gambles with Your Future* (New York, NY: Jeffrey B. Tarcher/Putnam, 2001).

[23] This case is actually occurring in a town in the southeastern United States. The names have been changed since the actions are still pending, as well as to incorporate risk concepts from other actual cases.

[24] According to the US Geological Survey, ultisols are characterized by sandy or loamy surface horizons and loamy or clayey subsurface horizons. They are deeply weathered soils derived from underlying acid crystalline and metamorphic rock (see http://ga.water.usgs.gov/nawqa/basin4.html). Since the soils have had so much weathering, metals (especially iron) have been oxidized. Thus, ultisols are often reddish (iron oxides) and mahogany colored (magnesium oxides).

[25] Incidence is the number of new cases per reporting period, usually per year. Prevalence is the number of cases present in the reporting period.

[26] For my colleagues who may have forgotten their inorganic chemistry, some forms of iron oxide are red (rust).

[27] This is a type of "opportunity risk," such as when people worry about pesticide risks to the point where they eat less fresh fruit and vegetables, thereby increasing their risks to many diseases, including cancer. Another example of opportunity risk trade-off is that of native populations, such as the Inuits who, because of long-range transport of persistent organic pollutants (so-called POPs), have elevated concentrations of polychlorinated biphenyls and other POPs in mother's milk. However, the ill effects of not breastfeeding (e.g., immunity, neonatal health, and colostrums intake) may far outweigh the long-term risks from exposures to POPs.

[28] For a different, even contrary view, see http://www.brown.edu/Administration/George_Street_Journal/value.html. Richard Morin gives a thoughtful outline of the Allen Feldman's model and critique of the "willingness to pay" argument (very commonly used in valuation).

[29] US Environmental Protection Agency, 1986, Guidelines for Carcinogen Risk Assessment, Report No. EPA/630/R-00/004, *Federal Register* 51 (185): 33992–34003, Washington, DC.

[30] A. Bradford Hill, The environment and disease: association or causation? *Proceedings of the Royal Society of Medicine, Occupational Medicine*, 58 (1965), 295; and A. Bradford-Hill, The environment and disease: association or causation? President's Address. *Proceedings of the Royal Society of Medicine* 9 (1965), 295–300.

Chapter 7

Analyzing Bioethical Success and Failure

Men wiser and more learned than I have discerned in history a plot, a rhythm, a predetermined pattern. These harmonies are concealed from me. I can only see one emergency following another, as wave follows upon wave, only one great fact with respect to which, since it is unique, there can be no generalizations, only one safe rule for the historian: that he should recognize in the development of human destinies the play of the contingent and the unforeseen. This is not a doctrine of cynicism and despair. The fact of progress is not a law of nature. The ground gained in one generation may be lost by the next.

H.A.L. Fisher (1936)[1]

A daunting bioethical challenge for engineers is that change is so rapid. We live in a time when misreading the slightest nuance of a design or operation can be the difference between success and failure. Often, we are not immediately even sure whether we are succeeding or failing. In fact, we may not know whether the fruition of our ideas in many emerging technologies will be for good or ill until well after the research. The explosion in emergent technologies is like having countless chests before us, some are treasure troves but others are Pandora's boxes.[2]

Anything that changes rapidly is difficult to measure. Those who are engaged in the research and the practice of new technologies may not recognize the hazards and pitfalls. In fact, those directly involved in the advancement may be the worst at appraising its worth and estimating its risks. The innovators have a built-in conflict of interest, which works against the ability to serve as society's honest brokers of emerging technologies.

Fisher's advice and admonition to the historian holds for the contemporary engineer. It envisages the famous counsel of George Santayana:[3]

Progress, far from consisting in change, depends on retentiveness.... Those who cannot remember the past are condemned to repeat it.

Santayana's advice is too frequently ignored. What we remember can save us from much despair and needless waste and failure in the long run. We forget important events at our own peril. It is one thing to fail, but quite another not to refuse to learn from our failures. As engineers, we must consider the reasons and events that led to the failure in the hope that corrective actions and preventive measures will be put in place to avoid their reoccurrence. This is not easy and is almost always complicated, especially for systems as elegant and complex as the human species.

The difference between success and failure often hinges on very subtle hints in the design process. Every failure results from a unique series of events. Human factors must always be considered in any design implementation. Often, seemingly identical situations lead to very different conclusions. In fact, the mathematics and statistics of failure analysis are some of the most complicated, relying on nonlinear approaches and chaos theory, making use of nontraditional statistical methods, such as Bayesian theory.[4] This is not an excuse to acquiesce. On the contrary, the complexities require that engineering creativity and imagination can and must be used to envisage possible failures and to take steps to ensure success.

MEDICAL DEVICE FAILURE: HUMAN FACTORS ENGINEERING

Human factors engineering (HFE) addresses the interface between humans and things, especially how we use these things. Human factor studies address all of the elements at this interface:

- systems performance;
- problems encountered in information presentation, detection, and recognition;
- related action controls;
- workspace arrangement; and
- skills required.[5]

Any of these factors can contribute to failure. Sometimes the inherent use of a device is problematic, such as the repetitive actions of people using manufacturing or office equipment. From a bioethical perspective, HFE is needed to determine how devices fail and to suggest ways that such failures can be prevented. A design that does not properly account for problems in usage by medical practitioners or by the patients themselves is morally unacceptable. For example, a device is unacceptable if it does not account for variability in dexterity and finger length of practitioners in the operating room. Likewise, a pump delivering medication to a patient is also unacceptable if it requires an inordinate degree of user sophistication, especially if operator error can lead to major medical complications, and if there is an available alternate design that is more user-friendly.

Concern with emerging technological developments is not the exclusive domain of neo-Luddites. Thoughtful people are simply asking for protections and commitments to precaution before going into an unbridled mode of research and applications. Some would also argue that research in many areas, including genetic engineering and human enhancement, has already passed the point of precaution, so that any attempts to control or even moderate the direction of the societal risk is akin to changing a tire on a bus as it moves down the highway. It is physically possible, but certainly not very probable. You probably will have to stop the bus for a while. Knowing when to stop the bus requires some objective measure of success and failure.

Teachable Moment: How to Analyze a Medical Device

The US Food and Drug Administration (FDA)[6] has emphasized the importance of human factors in device risk:

> Designers need a more complete and accurate understanding of device use and approaches to include consideration of unique limitations and failure modes of device users as critical components of the device-user system. Relatively few user actions that can cause the device to fail other than the most apparent (e.g., fire or explosion), or well-known instances of use problems are considered by designers. This limitation during device design increases the likelihood of unexpected use scenarios and use-related hazards for users and patients.

Hazards typically can be chemical (e.g., toxic chemicals), mechanical (e.g., kinetic or potential energy from a moving object), thermal (e.g., high temperature components), electrical (e.g., electrical shock, electromagnetic interference (EMI)), radiation (e.g., ionizing and nonionizing), and biological (e.g., allergic reactions, bioincompatibility, and infection).[7] Human factors play at least six key roles in risks associated with these hazards, including:

- Devices are used in ways that were not anticipated.
- Devices are used in ways that were anticipated, but inadequately controlled.
- Device use requires physical, perceptual, or cognitive abilities that exceed those of the user.
- Device use is inconsistent with user's expectations or intuition about device operation.
- The use environment effects device operation and this effect is not understood by the user.
- The user's physical, perceptual, or cognitive capacities exceeds when using the device in a particular environment.

It is incumbent on the designer to consider the possible risk and work to reduce possible human errors in the use of devices. The engineer can take at least nine steps to enhance safety[8]:

1. Identify and describe use-related hazards through analysis of existing information.
2. Apply empirical approaches using representative device users to identify and to describe hazards that do not lend themselves to identification or understanding through analytic approaches.
3. Estimate the risk of each use-related hazard scenario.
4. Develop strategies and controls to reduce the likelihood or mitigate the consequences of use-related hazard scenarios.
5. Select and implement control strategies.
6. Ensure that controls are appropriate and effective in reducing risk.
7. Determine if new hazards have been introduced as a result of implementing control strategies.
8. Verify that functional and operational requirements are met.
9. Validate safe and effective device use.

Furthermore, the FDA has recommended that, at a minimum, the design engineer should ask a number of questions to address and to prevent possible hazards associated with human factors[9]:

1. Why have problems occurred with the use of other similar products?
2. What are the critical steps in setting-up and operating the device? Can they be performed adequately by the expected users? How might the user set up the device incorrectly and what effects would this have?
3. Is the user likely to operate the device differently than the instructions indicated?
4. Is the user or use environment likely to be different than that originally intended?
5. How might the physical and mental capabilities of users affect their use of the device?
6. Are users likely to be affected by clinical or age-related conditions that impact their physical or mental abilities and could affect their ability to use the device?
7. How might safety-critical tasks be performed incorrectly and what effects would this have?
8. How important is user training, and will users be able to operate the device safely and effectively if they do not have it?
9. How important are storage and maintenance recommendations for proper device function, and what might happen if they are not followed?
10. Do any aspects of device use seem complex, and how can the operator become "confused" when using the device?
11. Are the auditory and visual warnings effective for all users and use environments?
12. To what extent will the user depend on device output or displayed instructions for adjusting medication or taking other health-related actions?
13. What will happen if necessary device accessories are expired, damaged, missing, or otherwise different than recommended?
14. Is device operation reasonably resistant to everyday handling?
15. Can touching or handling the device harm the user or the patient?
16. If the device fails, does it "fail safe" or give the user sufficient indication of the failure?
17. Could device use be affected if power is lost or disconnected (inadvertently or purposefully), or if its battery is damaged, missing, or discharged?

Questions

1. Consider a medical device (e.g., Therac-25, C–C value, etc.) that has failed. Which of the FDA questions seem to have been improperly addressed?
2. What lessons can this failure give us about the importance of a systematic and life cycle viewpoint when designing a medical device?
3. Are there special ethical considerations needed when designing a medical device compared to most other engineering designs? Explain.

UTILITY AS A MEASURE OF SUCCESS

Quantitative types, like most engineers, have a strong affinity for objective measures of success. Thus, we gravitate to usefulness as a measure of success. Such utility is indeed part of any successful engineering enterprise. After all, engineers are expected to provide reasonable and useful products. In a word, we look for utility. It turns out that one of the major ethical constructs is utilitarianism. The first two dictionary[10] definitions of utilitarianism (Latin *utilis*, useful) are relevant to bioethics:

1. The belief that the value of a thing or an action is determined by its utility.
2. The ethical theory that all action should be directed toward achieving the greatest happiness for the greatest number of people.

Since utilitarianism is based on outcome, it is a form of consequentialism. Engineers are expected to provide products so such a theory is attractive to us. Furthermore, some of the tools used to determine the morality of a decision are objective and quantifiable. Chief among these is the benefit-to-cost (B/C) ratio. It is an attractive metric due to its simplicity and seeming transparency. To determine whether a project is worthwhile, one need only add up all of the benefits and put them in the numerator and all of the costs (or risks) and put them in the denominator. If the ratio is greater than 1, its benefits exceed its costs. By this simple metric, the project is "good."

One obvious problem is that some costs and benefits are much easier to quantify than are others. Some, like those associated with quality of life, are nearly impossible to quantify accurately. How does one quantify aesthetics or biological diversity or "happiness?" Further, the comparison of doing anything with doing nothing cannot always be captured with a B/C ratio. Opportunity costs and risks are associated with taking no action (e.g., loss of an opportunity to apply an emerging technology may mean delay or nonexistent treatment of diseases). Simply comparing the *status quo* to costs and risks associated with a new technology may be biased toward no action. But, costs (time and money) are not the only reasons for avoiding action. The availability of a new monitoring device may invite more tests or drugs that, if not managed properly, could interfere with quality of life of patients, could carry their own risks, and could add costs to the public and the community (e.g., "defensive medicine" to avoid liability), with little actual benefit to anyone. So, it is not simply a matter of benefits *versus* cost, it is often one risk being traded for another.

Often, addressing contravening risk is a matter of optimization, which is a proven analytical tool in engineering. However, the greater the number of contravening risks that are possible, the more complicated such optimization routines become. Risk trade-off is a very common phenomenon in everyday life. For example, local governments enforce building codes to protect health and safety. Often times, these added protections are associated with indirect, countervailing risks. For example, the costs of construction may increase safety risks via "income" and "stock" effects. The income effect results from pulling money away from family income to pay the higher mortgages, making it more difficult for the family to buy other items or services that would have protected them. The stock effect results when the cost of the home is increased, families have to wait to purchase a new residence, so they are left in substandard housing longer.[11] Such countervailing risks are common in health care decisions. Patients demanding immediate and state-of-the-art technologies could be also arguing for increased risks from income and stock effects by imposing needless costs and restraints in the health care system. Once a technology exists, it will probably be used (supply creates demand). Thus, the engineer is frequently asked to optimize two or more conflicting variables in many situations.

Bioethics Question: How do engineers fail?

The reasons for failure vary widely, but can be broadly categorized as mistakes, mishaps, and misdeeds. The terms all include the prefix "mis-" that is derived from Old English, "to miss." This type of failure applies to numerous ethical failures. However, the prefix "mis-" can connote something that is done "poorly," i.e., a mistake. It may also mean that an act leads to an accident because the original expectations were overtaken by events, i.e., a mishap. Medical and engineering codes of ethics include tenets and principles related to competence, such as only working in one's area of competence or specialty. Finally, "mis-" can suggest that an act is immoral or ethically impermissible, i.e., a misdeed. Interestingly, the theological derivation for the word "sin" (Greek: *hamartano*) means that a person has missed the mark, i.e., the goal of moral goodness and ethical uprightness. That person has sinned or has behaved immorally by failing to abide by an ethical principle, such as honesty and justice.

Bioethical failures have come about by all three means. The lesson from Santayana is that we must learn from all of these past failures. Learning must be followed by new thinking and action, including the need to forsake what has not worked and shift toward what needs to be done.

Engineering failure can be categorized into five types. Whether the failure is deemed unethical is determined by the type of failure and the circumstances contributing to the failure.

FAILURE TYPE 1: MISTAKES AND MISCALCULATIONS

Engineers have been known from time to time to make mistakes. Their works fail due to their own miscalculations, such as when parentheses are not closed in computer code, leading to errors in predicting pharmacokinetic behavior of a drug. Some failures occur when engineers do not correctly estimate the corrosivity that occurs during sterilization of medical devices (e.g., not properly accounting for fatigue of materials resulting from high temperature and pressure of autoclave). Such mistakes are completely avoidable if the physical and biological sciences and mathematics are properly applied. Avoiding such errors is reflective of engineering competence. Interestingly, such failures are sometimes referred to as technicalities. After all, how unethical can an omitted parenthesis be? Actually, according to the reasonable engineer standard it is quite unethical. To paraphrase Norman Augustine, bad technical decisions *are* unethical.[12]

FAILURE TYPE 2: EXTRAORDINARY NATURAL CIRCUMSTANCES

Failure can also occur when factors of safety are exceeded due to extraordinary natural occurrences. Engineers can, with fair accuracy, predict the probability of failure due to biophysical and biochemical processes, such as metabolism and cellular receptor activity, but "normal" thresholds can be exceeded. Engineers design for an acceptably low probability of failure – not for 100% safety and zero risk. However, tolerances and design specifications must be defined as explicitly as possible.

The tolerances and factors of safety have to match the consequences. For example, while a failure rate of 1% may be acceptable for the general use light bulb, it is grossly inadequate for the light system (laser) used in medical systems. And, the failure rate of devices may spike up dramatically during an extreme natural event (e.g., power surges during storms). Equipment failure is but one of the factors that lead to medical failure.

A recent study,[13] for example, identified the reasons for relatively high failure rates associated with 180° selective laser trabeculoplasty (SLT) to treat intraocular pressure (IOP) in glaucoma patients. In this procedure, a very intensely focused beam of light is used to treat the eye's drainage angle (the point in the eye where the iris and sclera meet). The researchers suggest that the poor results could be attributed to:

1. The patients studied had advanced glaucoma and were on an average of two to three medications.
2. The baseline IOP was lower than that reported in other studies, and success is less likely with a lower baseline IOP.
3. Seventy-five percent of eyes were affected by a prostaglandin, which increases uveoscleral outflow and may result in hypoperfusion of the trabecular meshwork, thus decreasing the possible action of SLT.
4. SLT is thought to upregulate metalloproteinases similar to prostaglandins, and the prior use of prostaglandins may decrease this action of SLT and decrease its efficacy.

Note that none of the postulated reasons for failure were associated with the equipment itself, but mainly with the physiological and anatomical features of the patient. In fact, devices should be one of the smallest contributors to risk. At the very least, the failure rate should be characterized as best as possible by the design and clinical engineers. Conditional probabilities of failure should be known. That way, backup systems can be established in the event of extreme conditions (e.g., blood pressure, hormonal activity, digestion). If appropriate contingency planning and design considerations are factored into operations, the engineer's device may still fail, but the failure would be considered reasonable under the extreme circumstances.

FAILURE TYPE 3: CRITICAL PATH

No engineer can predict all of the possible failure modes of every structure or other engineered device, and unforeseen situations can occur. A classical case is the Holy Cross College football team hepatitis outbreak in 1969.[14] Circumstances, which could not have been foreseen, occurred when water contaminated with the hepatitis virus entered a drinking water system. A water pipe connected the college football field with the town, passing through a golf course. Children had opened a water spigot on the golf course, splashed around in the pool they created, and apparently discharged hepatitis virus into the water. A low water pressure in the pipe resulted when a house caught on fire and water was pumped out of the water pipes. This low pressure back-syphoned the hepatitis-contaminated water into the water pipe. The next morning the Holy Cross football team drank water from the contaminated water line and many players contracted hepatitis. The case is memorable because it was so highly unlikely – a combination of circumstances that were impossible to predict. Nevertheless, engineers must do just that: predict the seemingly unpredictable and thereby protect the health, safety, and welfare of the public.

This is an example of how engineers can fail, yet may not be blamed for the failure. If the public or their peers agree that the synergies, antagonisms, and conditional probabilities of the outcome could not reasonably be predicted, the engineer is likely to be forgiven. However, if a reasonable person deems that a competent engineer should have predicted the outcome, the engineer is to that extent accountable. The tragic industrial accident at Bhopal, India, is one such engineering failure. In 1984 a toxic cloud drifted over the city of Bhopal, eventually killing an estimated 20 000 people and permanently injuring 120 000. Blame was placed on the company, Union Carbide, for not paying attention to details. We often talk about a failure that results from not applying the science correctly (e.g., a mathematical error and an incorrect extrapolation of a physical principle). Another type of failure results from misjudgments of human systems. Bhopal had both.

Although the Union Carbide Company was headquartered in the United States, as of 1984, it operated in thirty-eight different countries. It was quite large (thirty-fifth largest US company), and was involved in numerous types of manufacturing, most of which involved proprietary chemical processes. The pesticide manufacturing plant in Bhopal, had produced the insecticides Sevin and Carbaryl since 1969, using the intermediate product, methyl isocyanate (MIC), in its gas phase. The MIC was produced by the reaction shown in Figure 7.1.[15] This process was highly cost effective, involving only a single reaction step. The schematic of the MIC process in shown in Figure 7.2. MIC is highly water reactive (Table 7.1), i.e., it reacts violently with water, generating a very strong exothermic reaction that produces carbon dioxide. When MIC vaporizes it becomes a highly toxic gas that, when concentrated, is highly caustic and burns tissues. This can lead to scalding nasal and throat passages, blinding, tissue atrophy and necrosis, as well as death (see Figure 7.3).

On 3 December 1984, the Bhopal plant operators became concerned that a storage tank was showing signs of overheating and began to leak. The tank contained MIC. The leak rapidly increased in size, and within 1 hour of the first leakage exploded and released approximately 80 000 lbs (4×10^4 kg) of MIC into the atmosphere. The release led to, arguably, the worst chemical industrial disaster on record.[16] The human exposure to MIC was widespread, with a half million people exposed. Nearly 3000 people died within the first few days after the exposure, 10 000 were permanently disabled. Ten years after the incident, 12 000 death claims had been filed, along with 870 000 personal injury claims. However, only $90 million of the Union Carbide settlement agreement had been paid out.

Introduction of water to the MIC storage tank resulted in a highly exothermic reaction generating CO_2, leading to a rapid increase in pressure, which appears to have caused the release of 40 metric tons of MIC into the atmosphere. As of 2001, many victims had received compensation, averaging about $600 each, although some claims are still outstanding.

Political factors were important in the chain of events. The Indian government required that the plant be operated exclusively by Indian workers, so Union Carbide agreed to train them, including flying them to a sister plant in West Virginia for hands-on sessions. In addition, the company required that US engineering teams make periodic on-site inspections for safety and quality control, but these ended in 1982, when the plant decided that these costs were too high. So, instead, the US contingency was only responsible for budgetary and technical controls, but not safety. The last US inspection in 1982 warned of many hazards, including a number that have since been implicated as contributing to the leak and release.

Lack of due diligence and failure to meet standards of best engineering practice were major contributors to the disaster. From 1982 to 1984, safety measures declined, attributed to high employee turnover, improper and inadequate training of new employees, and low technical savvy in the local workforce.

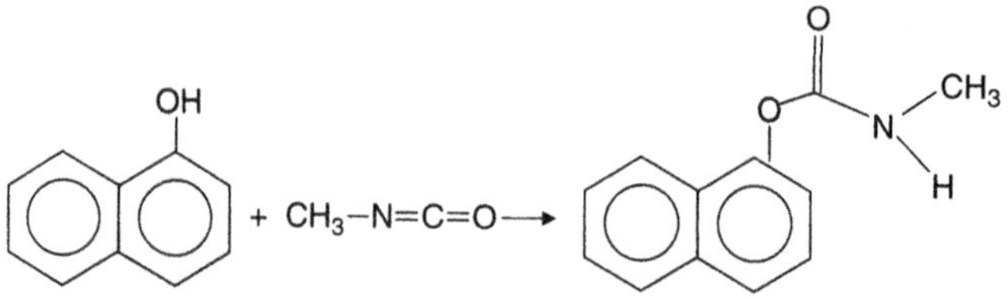

Figure 7.1 Chemical reaction producing methyl isocyanate (MIC) at the Bhopal, India, Union Carbide plant.

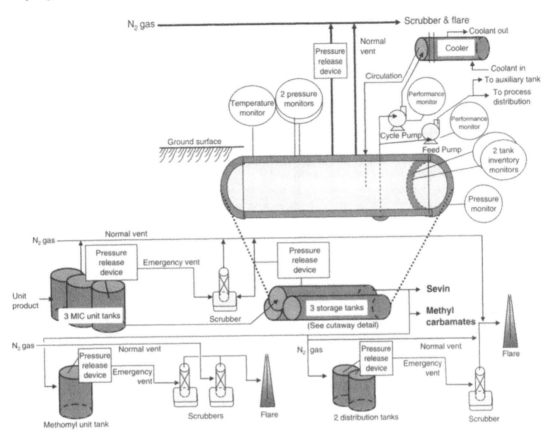

Figure 7.2 Schematic of methyl isocyanate (MIC) processes at the Bhopal, India, plant (c. 1984).
Adapted from: W. Worthy, 1985, Methyl isocyanate: The chemistry of a hazard, *Chemical Engineering News*, 63 (66), p. 29.

On-the-job experiences were often substituted for reading and understanding safety manuals (remember, this was a pesticide plant). In fact, workers would complain of typical acute symptoms pesticide exposure, such as shortness of breath, chest pains, headaches, and vomiting, yet they would typically refuse to wear protective clothing and equipment. Human factors led to this behavior. The refusal in part stemmed from the lack of air conditioning in this subtropical climate, where masks and gloves can be uncomfortable.

Indian, rather than the more stringent US, safety standards were generally applied at the plant after 1982. This likely contributed to overloaded MIC storage tanks (company manuals cite a maximum of 60% fill).

Social factors were also important. The release lasted about 2 hours, after which the entire quantity of MIC was released. The highly reactive MIC arguably could have reacted and become diluted within a certain safe distance. However, over the years, tens of thousands of squatters had taken up residence just outside of the plant property, hoping to find work or at least take advantage of the plant's water and electricity. The squatters were not notified of the hazards and risks associated with the pesticide

Table 7.1
Properties of MIC

Common name	Isocyanic acid, methylester, and methyl carbylamine
Molecular mass	57.1
Properties	Melting point: $-45°$ C; boiling point: $43-45°$ C
	Volatile liquid
	Pungent odor
	Reacts violently with water and is highly flammable
	MIC vapor is denser than air and will collect and stay in low areas. The vapor mixes well with air and explosive mixtures are formed
	May polymerize due to heating or under the influence of water and catalysts
	Decomposes on heating and produces toxic gases like hydrogen cyanide, nitrogen oxides, and carbon monoxide
Uses	Used in the production of synthetic rubber, adhesives, pesticides, and herbicide intermediates. It is also used for the conversion of aldoximes to nitriles
Side effects	MIC is extremely toxic by inhalation, ingestion, and skin absorption. Inhalation of MIC causes cough, dizziness, shortness of breath, sore throat, and unconsciousness. It is corrosive to the skin and eyes. Short-term exposures can also lead to death or adverse effects like pulmonary edema (respiratory inflammation), bronchitis, bronchial pneumonia, and reproductive effects. The Occupational Safety and Health Administration's permissible exposure limit to MIC over a normal 8-hour workday or a 40-hour workweek is $0.05 \, \text{mg m}^{-3}$

Sources: US Chemical Safety and Hazards Board, http://www.chemsafety.gov/lib/bhopal.0.1.htr); Chapman and Hall, *Dictionary of Organic Chemistry*, Volume 4, Fifth Edition, Mack Printing Company, United States of America, 1982; and T.W Graham, *Organic Chemistry*, Sixth Edition, John Wiley and Sons, Canada, 1996.

manufacturing operations, except by a local journalist who posted signs saying: "Poison Gas. Thousands of Workers and Millions of Citizens Are in Danger." Thus, a factor usually associated with land use planning, accessibility, was also a contributor to the disaster. In fact, many bioethical issues share this factor. For example, people may have few options in where they live, so the poorer strata may be over represented in higher risk neighborhoods and occupations. These social factors are considered in Chapter 8.

The Bhopal disaster is a classic instance of a "confluence of events" that led to a disaster. More than a few mistakes were made. The failure analysis found the following:

- The tank that initiated the disaster was 75% full of MIC at the outset.
- A standby overflow tank for the storage tank contained a large amount of MIC at the time of the incident (defeating its intended purpose).
- A required refrigeration unit for the tank was shut down five months prior to the incident, leading to a three- to fourfold increase in tank temperatures over expected temperatures.
- One report stated that a disgruntled employee unscrewed a pressure gauge and inserted a hose into the opening (knowing that it would do damage, but probably not nearly at the scale of what occurred).

- A new employee was told by a supervisor to clean out connectors to the storage tanks, so the worker closed the valves properly, but did not insert safety disks to prevent the valves from leaking. In fact, the worker knew the valves were leaking, but they were the responsibility of the maintenance staff. Also, the second-shift supervisor position had been eliminated.
- When the gauges started to show unsafe pressures, and even when the leaking gases started to sting mucous membranes of the workers, they found that evacuation exits were not available. There had been no emergency drills or evacuation plans.
- The primary fail-safe mechanism against leaks was a vent-gas scrubber, i.e., normally, this release of MIC would have been sorbed and neutralized by sodium hydroxide (NaOH) in the exhaust lines, but on the day of the disaster, the scrubbers were not working. (The scrubbers were deemed unnecessary, since they had never been needed before.)
- A flare tower to burn off any escaping gas that would bypass the scrubber was not operating because a section of conduit connecting the tower to the MIC storage tank was under repair.
- Workers attempted to mediate the release by spraying water 100 feet high, but the release occurred at 120 feet.

Thus, according to a post-failure audit, many checks and balances were in place but the cultural considerations were ignored or given low priority, such as when the plant was sited the need to recognize the differences in land use planning and buffer zones in India compared to those in Western nations or the difference in training and the oversight of personnel in safety programs. Every engineer needs to recognize that much of what we do is affected by geopolitical realities and that we work in a global economy. This means that we must understand how cultures differ in their expectations. For example, offsetting behaviors may lead to misuses of systems and devices in developing nations if the designs are based exclusively on Western medical venues. One cannot assume that a model that works in one setting will necessarily work in another without adjusting for differing expectations. Bhopal demonstrated the consequences of ignoring these realities.

Even in the West, versions of the Bhopal incident occur, but usually with more limited impacts. For example, two freight trains collided in Graniteville, SC, just before 3:00 a.m. on 6 January 2005, resulting in the derailment of three tanker cars carrying chlorine (Cl_2) gas and one tanker car carrying sodium hydroxide (NaOH) liquids. The highly toxic Cl_2 gas was released into the atmosphere. The wreck and gas release resulted in hundreds of injuries and eight deaths. This raises an issue important to biomedical and biosystem engineers: The value of the worst case scenario. In this case, such a scenario could assume that all of the Cl_2 will leak and all would be in the gas phase. However, this is not entirely assured. For example, some of chlorine may be in the liquid phase. Some will react with H_2O vapor and become hydrochloric acid (HCl), which is also highly toxic. But, safety may dictate assuming the worst case.

Characterizing many contingencies and possible outcomes in the critical path is an essential part of many biohazards. The Bhopal incident provides this lesson. For example, engineers working with bioreactors and genetically modified materials must consider all possible avenues of release. They must ensure that fail-safe mechanisms are in place and are operational. Quality assurance officers note that testing for an unlikely but potentially devastating event is difficult. Everyone in the decision chain must be "on board." The fact that no incidents have yet to occur (thankfully) means that no one really knows what will happen in such an event. That is why health and safety training is a critical part of the engineering process and is an ethical responsibility of the engineer.

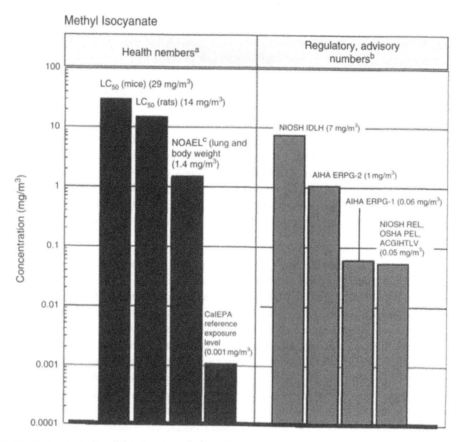

Figure 7.3   Health data for inhalation of methyl isocyanate (MIC).

Adapted from: US Environmental Protection Agency, 1986, *Health and Environmental Effects Profile for Methyl Isocyanate*, Environmental Criteria and Assessment Office, Office of Health and Environmental Assessment, Office of Research and Development, Cincinnati, OH; US Department of Health and Human Services, 1993, Hazardous Substances Data Bank (HSDB, online database), National Toxicology Information Program, National Library of Medicine, Bethesda, MD; US Department of Health and Human Services, 1993, Registry of Toxic Effects of Chemical Substances (RTECS, online database), National Toxicology Information Program, National Library of Medicine, Bethesda, MD; California Environmental Protection Agency (CalEPA), 1997, Technical Support Document for the Determination of Noncancer Chronic Reference Exposure Levels, Draft for Public Comment, Office of Environmental Health Hazard Assessment, Berkeley, CA; US Environmental Protection Agency, 1999, *Integrated Risk Information System (IRIS) on Methyl Isocyanate*, National Center for Environmental Assessment, Office of Research and Development, Washington, DC; J.E. Amoore and E. Hautala, 1983, Odor as an Aid to Chemical Safety: Odor Thresholds Compared with Threshold Limit Values and Volatilities for 214 Industrial Chemicals in Air and Water Dilution, *Journal of Applied Toxicology*, 3 (6), pp. 272–90; National Institute for Occupational Safety and Health (NIOSH), 1997, *Pocket Guide to Chemical Hazards*, US Department of Health and Human Services, Public Health Service, Centers for Disease Control and Prevention, Cincinnati, OH; Occupational Safety and Health Administration (OSHA), 1998, Occupational Safety and Health Standards, Toxic and Hazardous Substances, *Code of Federal Regulations* 29 CFR 1910.1000; American Conference of Governmental Industrial Hygienists (ACGIH), 1999, *1999 TLVs and BEIs. Threshold Limit Values for Chemical Substances and Physical Agents. Biological Exposure Indices*, Cincinnati, OH; and American Industrial Hygiene Association (AIHA), 1998, *The AIHA 1998 Emergency Response Planning Guidelines and Workplace Environmental Exposure Level Guides Handbook*.

FAILURE TYPE 4: NEGLIGENCE

Engineers also have to protect the public from its members' own carelessness. We have to assume such design defeats. The case of the woman trying to open a 2-liter soda bottle by turning the aluminum cap the wrong way, with a pipe wrench no less, and having the cap fly off and into her eye, is a famous example of unpredictable ignorance. She sued for damages and won, with the jury agreeing that the design engineers should have foreseen such an occurrence. (The new plastic caps have interrupted threads that cannot be stripped by turning in the wrong direction.)

Design engineers recognize that if something can be done incorrectly, sooner or later it will, and that it is their job to minimize such possibilities.

Another such example occurred in the early 1970s, when jet-powered airliners were replacing propeller aircraft, the fueling system at airports was not altered and the same trucks fueled both types of craft. The nozzle fittings for both types of fuels were therefore the same. A tragic accident occurred near Atlanta where jet fuel was mistakenly loaded into a Martin 404 propeller craft. The engines failed on takeoff, resulting in fatalities. A similar accident occurred in 1974 in Botswana with a DC-4 and again near Fairbanks, Alaska, with a DC-6.[17] Engineers should have recognized that no amount of signs or training could prevent such tragedies. They had to modify the fuel delivery systems so that it was impossible to put jet fuel into propeller-driven airplanes and *vice versa*. An example of how this can be done is the modification of the nozzles used in gasoline stations. The orifice in the gas tank is too small now to take the nozzles used for either leaded fuel or diesel fuel. (Drivers of diesel-engine cars can still mistakenly pump gasoline into their cars, however.)

Figure 7.3 (Continued)

Note: ACGIH TLV is the American Conference of Governmental and Industrial Hygienists' threshold limit value expressed as a time-weighted average, i.e., the concentration of a substance to which most workers can be exposed without adverse effects. LOAEL is the lowest observed adverse effect level. NIOSH REL is the National Institute of Occupational Safety and Health's recommended exposure limit; NIOSH-recommended exposure limit for an 8- or 10-h time-weighted average exposure and/or ceiling. NIOSH IDLH is NIOSH's suggested concentration immediately dangerous to life or health; NIOSH-recommended exposure limit should ensure that a worker can escape from an exposure condition that is likely to cause death or immediate or delayed permanent adverse health effects or prevent escape from the environment. OSHA PEL is the Occupational Safety and Health Administration's permissible exposure limit expressed as a time-weighted average, i.e., the concentration of a substance to which most workers can be exposed without adverse effect, averaged over a normal 8-h workday or a 40-h workweek. OSHA ceiling is OSHA's short-term exposure limit; 15-min time-weighted average exposure should not be exceeded at any time during a workday even if the 8-h time-weighted average is within the threshold limit value. The lethal concentration-50% (LC_{50}) is the concentration at which half of the test animals die; thus it reflects acute toxicity. The NOAEL is the concentration at which studies indicate a threshold, i.e., the no observable adverse effect level. There is always uncertainty, so when factors of safety are considered, the safe concentration is lower than the NOAEL and is referred to as the reference concentration (RfC).

[a]Health numbers are toxicological numbers from animal testing or risk assessment values developed by the EPA.

[b]Regulatory numbers are values incorporated in Government regulations, while advisory numbers are nonregulatory values provided by the Government or other groups as advice. OSHA numbers are regulatory, whereas NIOSH and ACGIH numbers are advisory.

[c]This LOAEL is from the critical study used as the basis for the RfC set by the EPA.

Conversion factors: To convert concentrations in air (at 25°C) from parts per million (ppm) to mg m^{-3}:

$$\frac{(ppm) \times (molecular\ weight\ of\ MIC = 2.34\ mg\ m^{-3})}{24.45}.$$

Are these lessons here for medical devices? Indeed, if a patient or user can reasonably defeat a device's intended use (e.g., reverse the flow of inlet and outlet tubes), the designer must prevent this action. Such prevention should balance the device's intended use with undesired uses without sacrificing the overall efficiency and effectiveness of the device. This can be a tall order, but human factors are a reality. Ignoring them is a form of negligence.

FAILURE TYPE 5: LACK OF IMAGINATION

Every time something fails, whether a manufactured product (such as a pump or valve leakage) or a system (such as a protocol for using a medical instrument) it is viewed as an engineering failure. The job of engineers historically has been to predict the problems that can occur, and to design so as to minimize these events, protecting people from design errors, natural variability, unforeseen events, and ignorance/carelessness.

Today this mandate has become even more complicated and difficult. The engineer is now required to protect the health, safety, and welfare of the public from acts of terrorism. Until recently, it had never occurred to most engineers that they have a responsibility of protecting people from those who would want to intentionally harm other people or to destroy public facilities intentionally. This is a totally new failure mode in engineering. Sarah Pfatteicher calls such failures "intentional accidents," or failures resulting from intentional actions.[18]

Engineers now find themselves in the position of having to address these "intentional accidents." Military engineers of course have had to design against such destructive actions since the days of moats and castles but those were all structures built explicitly to withstand attack. Civilian engineers have never had to think in these terms, but are now asked to design structures for this contingency. We must ask: "How might my new technology be misused or abused?" and "What must I do to prevent such misuse and abuse?" Engineers have a new challenge – to prevent such "accidents" on civilian targets by terrorists bent on harm to the public. The response to this threat requires a two-pronged approach – technical and social.

BIOTERRORISM: THE ENGINEER'S RESPONSE

There is a distinct likelihood that a number of the designs intended for beneficence will be converted and adapted for malevolence. Engineers have responded enthusiastically to the call for better technology to fight terrorism. This effort has included better sensing devices for explosives and weapons, better communication systems to warn and predict when terrorist activity might occur, better identification systems for suspect individuals, better ways of identifying explosives and biological weapons, and many more such innovations. A recent issue of *Prism*, the magazine of the American Society of Engineering Education, had several articles summarizing technology that can be brought to bear in the fight against terrorism, and these developments no doubt will have a beneficial effect.[19] But the use of such technology also signals four potential problems.

First, the application of such technology will almost always diminish or adversely alter our sense of liberty. Witness, for example, the facial identification program used on attendees at a recent Super Bowl. Despite protests by the manufacturer and the security people to the contrary that such equipment only identifies those faces on file, many people were upset about having their picture taken. We Americans are just not used to having people or machines knowing where we are and what we do and we resent such an intrusion on our privacy. We believe, and rightly so, that equipment and procedures such as face identification have the potential for abuse and loss of our liberty. This conflict between security and liberty is an example of competing values in ethics. For example, security can be seen as a necessary end, but some of the means to achieve this end will detract from liberty. This points to one of the problems with utilitarian ethics. Likewise, the United States was founded on principles espoused by rights ethicists like John Locke (1632–1704). In fact, three of these rights, life, liberty, and the pursuit of happiness, are foundations to the US Constitution. Recently, however, the words of Benjamin Franklin have taken on added currency:

They that can give up essential liberty to obtain a little temporary safety deserve neither liberty nor safety.[20]

Thus, engineers may see the balance between security and liberty as an "optimization problem" between the two variables. Even Franklin qualified liberty to that which is "essential" and safety to that which is "temporary."

The effect of security at airports and public places has been traumatic and costly, as well as bothersome. However, most people recognize that some security, even loss of certain previously held privacy, is part of the post-9/11 world. The threshold of where security has eroded liberty is ill defined. A personal, recent example points to this uncertainty. In 2006, because of a specific threat, U.S. airports instituted rigorous controls on the amount of fluids allowed to be carried onto aircrafts. However, by early 2007, the rules were relaxed to allow most small containers of fluids on board if they were placed in a plastic, sealed bag (i.e., a "baggie"). The cynic might ask: Has one "thin blue line" been drawn at the baggie?

Second, a malicious person with money and knowledge can get around any technology we implement. Witness the scourge of viruses in our computers. As soon as the antivirus programs are modified to take care of the latest virus, unknown persons with devious minds and malicious intent can develop new viruses that get around the defenses. So it will be with antiterrorist technology.

Third, terrorists can use our own technology to create havoc in our country, the prime example being the use of fully loaded airplanes on 11 September. The terrorist organizations have used Internet banking to move money around, defense communication networks to keep in touch, and sophisticated explosives to kill innocent people – technologies we have developed as part of our national defense. The more technology we produce, the more it has the potential to harm us.

Fourth, the problem with the development of antiterrorist technology is that it is emergent and it is not something for which most engineers have yet to develop an aptitude. As evidence, few resources that would protect us from low technology assault have been developed to date. Traditionally, our research funds have been spent either to enhance our own health or to develop sophisticated weapons for countering threats from similarly technically sophisticated enemies. For example, the use-inspired and applied research budgets of the National Institutes of Health and the Department of Defense greatly surpass the budgets for basic science research efforts of the National Science Foundation. We have done very little research on how to protect the safety of the public from terrorist threats. But then even if we had, the terrorists would simply use that technology for their own purpose, thus negating its original intent.

DUAL USE AND PRIMACY OF SCIENCE

INQUISITOR: I don't know anyone who could get through the day without two or three juicy rationalizations. They're more important than sex.

RESPONDENT: Ah, come on. Nothing's more important than sex.

INQUISITOR: Oh yeah? Ever gone a week without a rationalization?

Dialogue from the movie, *The Big Chill*, 1983

Human beings have proficiency for rationalizing what we want to do. And, human institutions, like corporations and governmental agencies, scale up such rationales. For example, research that is clearly basic is often supported for its possible, yet tenuous, public benefits. Interestingly, the term "dual use" has two different connotations. The first, which grew in popularity during the Cold War and space missions, is any science, engineering and technology designed to provide both military and civilian benefits. Better pots and pans, microwave ovens, and DVD players can be touted as having been procreated out of huge, publicly funded military programs. The second definition, which in many instances seems to challenge the first, is any research or technology that simultaneously benefits and places society at risk.

Recently, concerns about terrorism and national security have piqued the public's interest in the research and technology that possibly fits the second definition. For example, in July 2006 the Congressional Research Service reported:

> An issue garnering increased attention is the potential for life sciences research intended to enhance scientific understanding and public health to generate results that could be misused to advance biological weapon effectiveness.
>
> Such research has been called "dual-use" research because of its applicability to both biological countermeasures and biological weapons. The federal government is a major source of life sciences research funding. Tension over the need to maintain homeland security and support scientific endeavor has led to renewed consideration of federal policies of scientific oversight. Balancing effective support of the research enterprise with security risks generated by such research has proven a complex challenge. Policies considered to address science and security generate tensions between federal funding agencies and federal funding recipients. To minimize these tensions while maximizing effective oversight of research, insight and advice from disparate stakeholders is generally considered essential.[21]

A real-life case of the conflict in dual use recently came to light when a study pointed out specific deficiencies in the protection of the milk supply and its vulnerability to widespread contamination by the *botulinum* toxin. The manuscript from the study was submitted to, and ultimately published in, *Proceedings of the National Academy of Sciences*.[22] The US Department of Health and Human Services had strongly opposed the publication on the grounds that it could encourage and instruct would-be bioterrorists (a dramatic example of dual use's second definition).[23] One mediating aspect of this debate is that the authors of the article are not funded by the federal government. Had they received federal assistance, the US government would likely have had a greater onus and stronger position to veto the text.

In fact, the authors and their advocates made two arguments that favored publication. First, the information could be helpful to agricultural and food security decision makers charged with protecting the milk supply (analogous to interviewing a burglar and publishing vulnerabilities in homes with a hope that this information will drive homeowners to make the necessary changes to improve home safety). Second, the information had already been made readily available via the Internet and other sources.

The case illustrates the quandary of risk avoidance. Like many other engineering writers, I recently was confronted with a decision about just how much I should say about certain vulnerabilities, even though the information being shared was indeed readily available and not confidential or secret in any way. However, engineers are trained "to connect the dots" in ways that many are not, so even if the source information is readily available to anyone interested, we know where to look and how to assimilate the information into new knowledge.

This is the bottom-up design process. We do this all the time. So, there is an additional onus on professionals to take care how we pull together information in order to avoid giving new knowledge to those who intend to use it nefariously. And, we must not fall victim to our own rationalizations that we are doing it for the public good or the advancement of the state of the science, when our real intentions are to improve our own lot (e.g., a gold standard publication, a happy client, or public recognition). There is nothing wrong and much right about improving one's own lot, but engineers are in a position of trust and must hold paramount the public's safety, health, and welfare. And, as the famous physicist Richard Feynman reminded us:

Science is a long history of learning how not to fool ourselves.[24]

And:

Science is a way of trying not to fool yourself. The first principle is that you must not fool yourself, and you are the easiest person to fool.[25]

This sounds much like the stuff of the Ancient Greek dialogues. Socrates implied the need for self-scrutiny, stripping down our intellectualism before we can build up (knowing that we do not know is requisite to ever really knowing anything). Researchers and practitioners alike need a healthy dose of realism and a critical eye toward our own justifications. The tensions between advancing the state of veterinary sciences with animal welfare illustrate this point. Every veterinarian that I have had the pleasure of knowing exudes a love of animals, including those who conduct or oversee studies wherein animals are sacrificed (i.e., the term for killing animals for research ends). Like engineers, these scientists have an ability to compartmentalize their views.

It is not possible for engineers to prevent all malicious deeds, just as it is not possible for engineers to make anything 100% safe from other kinds of failure. In a manner of speaking, this is analogous to our reliance on pharmaceuticals to treat diseases. One of the fears of medical researchers is that we may destroy most of the pathogenic microbes, but the remaining strains rapidly become resistant to even the newest antibiotics. So, physicians try not to prescribe these drugs indiscriminately, and to use a holistic approach to prevent certain contagious diseases. No matter what technology we employ, the possibility always remains that terrorists can cause great harm by using the very same technology against us. We cannot be completely "inoculated" against these diseases, nor can we expect "quick technological fixes" every time. Engineering technology is therefore only partially able to protect the health, safety, and welfare of the public from intentional destruction. But, the first step of doing so is to recognize when vulnerabilities are present and to apply engineering expertise to address them.

SOCIAL RESPONSE OF ENGINEERING TO TERRORISM

There are two approaches to solving engineering problems – attack the symptoms or attack the cause – and often the best solution to a problem is to change what is causing the problem. In engineering, it is almost always cheaper and safer to travel "up the pipeline" to find out where the problem is coming from and then change the process to eliminate or minimize it. The alternative option, which is almost always more expensive, is to treat the problem once it arrives.

The first step in finding solutions to engineering challenges is to define the problem – to understand all the variables and constraints. Using this approach to combating terrorism, we first have to understand why people would cause "intentional accidents." To gain such understanding, we have to develop a greater knowledge of diverse cultures and societies and ask why people would want to cause harm, what their motives are, and try to understand what drives their actions. There is always a driving force behind every action, especially ones that result in suicide. Understanding this driving force and taking account of multiple perspectives in the solution of engineering problems (including intentional accidents) can be extremely helpful in preventing such events. This is not justifying evil, it is simply being effective observers of why the problem exists as a first step in designing possible solutions.

SUCCESS PARADIGMS

As mentioned in Chapter 3, human actions tend to be repetitive, but we do make incremental change by pushing the envelope and eventually shifting paradigms.[26]

In his book, *To Engineer Is Human*, Henry Petroski discusses success and failure in engineering as part of an ongoing process of trial and error that leads to innovation. For example, he discusses the effects of the growth of the railroads on engineering structures. As the network of railroads in the United States was quickly expanding, there were more incentives for heavier trains to travel progressively faster on more rugged terrain. But soon enough, it was obvious that the strength that was required for earlier railroads to facilitate the earlier trains was no match for the new trains. As collapses occurred, "each defective bridge resulted in demands for excess strength in the next similar bridge built, and thus the railroad bridge evolved through the compensatory process of trial and error."[27] As the processes of engineering have become progressively more scientific in nature over the decades, engineers have had to deal with "success" and "failure" more explicitly. Thus the real trial and error that has been passed down to today's engineers by their predecessors is that of "mind over matter." According to Petroski, while no one wants to learn through mistakes, they are an integral part of the process. As engineers perpetually strive to employ new concepts to create lighter, more cost-effective structures, each new structure can be seen as a trial of sorts.[28] We must learn from any successes and our failures.

Such a heuristic approach is beneficial to biomedical ethics, since modern technologies can use information technology (e.g., models that can extend the lessons learned from previous successes and failures, such as bioinformatics and computational tools). This can be done virtually, rather than actually experiencing the failures.

CHARACTERIZING SUCCESS AND FAILURE

Essentially, engineering involves design under constraint.[29] Engineers may strive to optimize the level of elegance of a given device while subject to the constraints of fixed physical and mental resources, as well as the demands of the marketplace. At the same time, engineers additionally face the greatest constraint of all: safety. As mentioned, this constraint becomes more important than the aesthetic and economic aspects because "the loss of a single life due to structural collapse can turn the most economically promising structure into the most costly and can make the most beautiful one ugly."[30]

Engineering success is somewhat different than many other human endeavors. Perhaps the best way to address what it means to have a successful design and project is to explore the ways we could fail, and to take steps to ensure these do not happen.

ACCOUNTABILITY

The level and type of accountability for success and failure of a design is affected by the setting in which the engineer works. However, in every engineering office or department there is a designated "engineer in responsible charge" whose job it is to make sure that every project is completed successfully and within budget. Biotechnological and other life science institutions are no different. This responsibility is often indicated by the fact that the engineer in responsible charge places his or her professional engineering seal on the design drawings or the final reports. By this action the engineer is telling the world that the drawings or plans or programs or whatever are correct, accurate, and that they will work. In some countries, not too many years ago, the engineer in responsible charge of building a bridge was actually required to stand under the bridge while it was tested for bearing capacity. This is not required in a literal sense today, but it is a reminder that the device or system being designed should be of a quality that we would personally trust it. Figuratively, the engineer "stands under" and "stands on" all of his or her designs. In sealing drawings or otherwise, by accepting responsibility the engineer in charge places his or her professional integrity and professional honor on the line. There is nowhere the engineer in charge can hide if something goes wrong. If something does go wrong, "One of my junior engineers screwed up," is not a defense because it is the engineer in charge who is supposed to have overseen the calculations or the design.

For very large projects where the responsible engineer may not even know all of the engineers working on the project, much less be able to oversee their calculations, this is clearly impossible. In a typical engineering office the responsible engineer depends on a team of senior engineers who oversee other engineers, who oversee others, and so on down the line. How can the responsible engineer at the top of the pyramid be confident that the product of collective engineering skills meets the client's requirements?

The rules governing this activity are fairly simple. Central is the concept of truthfulness in engineering communication. Such technical communication up and down the organization requires an uncompromising commitment to tell the truth no matter what the next level of engineering wants to hear.

What we value influences the attention we give to detail. For example, an engineer in the lower ranks may develop spurious data, lie about test results, or generally manipulate the basic design components. Such information might not be readily detected by supervisory engineers if the bogus information is beneficial to the completion of the project. If the information is not beneficial, on the other hand, everyone along the chain of engineering responsibility will give it a hard critical look. Therefore the inaccurate information, if it is the desired information, can steadily move up the engineering ladder because at every level the tendency is not to question good news. The superiors at the next level also want good news, and want to know that everything is going well with the project. They do not want to know that things may have gone wrong somewhere at the basic level. Knowing this, and fearing being shot as the messenger, the temptation can induce engineers to accept good news and to question bad news. In short, the axiom that

good news travels up the organization – bad new travels down

holds for engineering as well. And bad news will travel down very quickly (e.g., "You're fired!").

This inherent vulnerability can lead to devastating and tragic results. The only correcting mechanism in engineering exists at the very end of the project if failure occurs: the software crashes, the valve leaks, stock prices drop, or people die. And then the search begins for what went wrong. Eventually the truth emerges, and often the problems can be traced to the initial level of engineering design, the development of data, and the interpretation of test results. It is for this reason that engineers, young and old, must be extremely careful of the details of their work.

It is one thing to make a mistake (we all do), but it is another thing totally to use misinformation in the design. Fabricated or spurious test results can lead to catastrophic failures because there is often an absence of a failure detection mechanism in engineering until the project is completed. Without trust and truthfulness in engineering, the system will fail. To paraphrase Joseph Bronowski's quote in Chapter 4,[31] honesty is the glue that holds our profession together. If we violate this tenet, we lose trust with our clients and with our colleagues. The engineering advances of the past two centuries have been allowed to occur because the profession earned the trust of the public. In return, we must hold ourselves accountable and we must safeguard the trust bestowed upon us by our principal client, the public.

VALUE

The various definitions of success and failure require a means for determining success. We need ways to measure the level of success of an engineering design in terms of effectiveness. For instance, let's assume that pump Y has been designed to administer drug X to the liver of a sick patient's body. We can measure the efficiency of drug delivery, by a typical mass balance, such as the dose of X administered compared to the mass reaching the liver. Thus, efficiency is relatively straightforward. Effectiveness, however, may not be. For example, even if the total efficiency is high (e.g., 99.99% of the dose of X reaches the liver), it may be ineffective (nonefficacious), because metabolism changes X to an oxidized metabolite X′ before it reaches the liver that is not effective in treating the level. We place a high value on the purity of the drug that reaches the target organ. Thus, effectiveness assumes some definition of value.

CASE ANALYSIS

There are numerous ways to identify, characterize, and analyze the success and failure of cases. The key to any approach is that it should accurately describe the events, people, and decisions that were made; strive for objectivity; and be fair and consistent in assessing and ascribing moral intent and consequences. Following is a step-wise approach for analyzing a decision leading to an engineering action, whether the decision is being considered or has already been made.

Step 1 – Factual Description: Key characters and events pertinent to the case are identified. This description includes narrative, tables, figures, maps, organization charts, critical path diagrams, and photographs that are needed to place the case in an ethical context. As in any case analysis, the first step is descriptive. Who are involved and what is the role of each person in the case? What happened or is happening and where did each key step that led to the ultimate result occur? In what manner did the steps occur? Were they sequential, coincidental, or combinations of these? Describing an ethical issue makes use of photos, maps, profiles, and any other descriptive materials that will help show what happened and/or what could happen. During the descriptive stage, we are not yet ready to discuss the ethics or to ascribe blame or credit. For now, what we need is a complete and accurate telling of the case.

Step 2 – Logical Arguments and Syllogisms: Based upon the findings in Step 1, the validity of the decisions or lack thereof is analyzed. The syllogism includes a factual premise, a connecting fact–value premise, and an evaluative premise to reach an evaluative conclusion. Many moral (and scientific) arguments fail because of weaknesses in any of these components of the syllogism. Depending on the case, numerous arguments must be evaluated.

Step 3 – Ethical Problem-Solving Analysis: Once the facts and ethical problems are sufficiently identified and explained, the issues must be classified as to whether they are factual, conceptual, or moral.[32]

From the descriptions in Step 1, the depth of each type of issue can be assessed. Factual issues are those that are known. This can sometimes be apparent just by reading the events, but in certain cases the facts may not be so clear (e.g., you and I may agree on the "fact" that carbon dioxide is a radiant gas, but we may disagree on whether the build-up of CO_2 in the troposphere will lead to increased global warming). This may mean that we agree on first principles of science and even the data being used but may disagree on the relative weightings in indices and models. This leads to a need to ascribe causality, a very difficult problem indeed.

Step 3a – Identify Possible Causal Relationships: To begin to evaluate whether a model is valid, oftentimes the best that science usually can do in this regard is to provide enough weight of evidence between a cause or causes and an effect. The medical research and epidemiological communities use a number of criteria to determine the strength of an argument for causality, but the first well-articulated criteria were Hill's Causal Criteria[33] (Chapter 3). Depending on the case, some of Hill's criteria are more important than are others.

Conceptual issues involve different ways that the meaning may be understood. For example, what you and I consider "contamination", "disease" or "good lab practices" may vary (although the scientific community strives to bring consensus to such definitions).

Many engineers and scientists believe it is the job of technical societies and other collectives to try to eliminate factual and conceptual disagreements. Most of us agree on first principles (e.g., fundamental physical concepts like the definitions of matter and energy), but unanimity fades as the concepts drift from first principles. Scientists should agree about the facts.[34] The progress of research and knowledge helps to resolve factual issues (eventually), and the consensus of experts aids in resolving conceptual issues. But since complete agreements are generally not possible even for the factual and conceptual aspects of a case, the moral or ethical issues are further complicated.

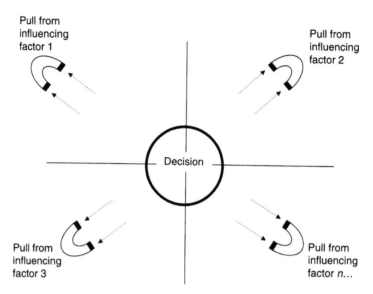

Figure 7.4   Decision force field.

Step 3b – Apply Decision Force Field Analysis: Whether scientists and engineers want to believe it, few, if any, decisions can be made exclusively from physical scientific principles. We can visualize these decision-forcing functions within a force field, where a magnet is placed in each sector at a point equidistant from the center of the diagram (Figure 7.4). If we assume that the decision being considered will be pulled in the direction of the strongest magnetic force, the stronger the magnet the more the decision that will actually be made will be pulled in that direction. For example, lawyers may proceed in one direction, while engineers in another, and the business clients in another, all applying different forces that "deform" the decision.

A decision that is almost entirely influenced by strong science will appear something like the force field in Figure 7.5. However, if there are other factors, such as a plant closing down or the possibility of attracting new jobs at the expense of environmental, geological, or other scientific influences, the decision would migrate toward these stronger influences, as shown in Figure 7.6.

The engineering risk management process is informed by the quantitative results of the risk assessment process. The shape and size of the resulting decision force field diagram give an idea of what are the principal driving factors that lead to decisions. A force field diagram can be a useful, albeit subjective, tool to assess each decision.

Step 3c – Net Goodness Analysis: This is a subjective analysis of whether a decision will be moral or less than moral. It puts the case into perspective, by looking at each factor driving a decision from three perspectives:

1. How good or bad would the consequence be.
2. How important is the decision.
3. How likely is it that the consequence will occur.

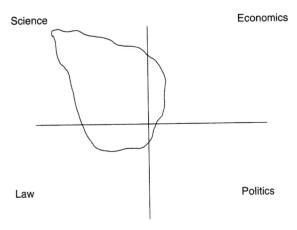

Figure 7.5 Decision force field for a scientifically based decision, although other factors (legal, financial, and political) are influencing the final outcome.

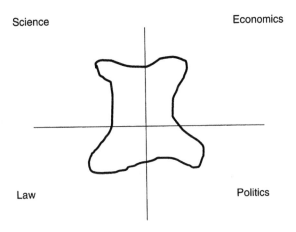

Figure 7.6 Decision force field for complex decisions with multiple influences with about equal influence on the outcome.

These factors are then summed to give the overall net goodness of the decision:

$$\text{NG} = \Sigma \ (\text{goodness of a consequence}) \times (\text{importance}) \times (\text{likelihood}) \qquad (7.1)$$

Thus, this can be valuable in decisions that have not yet been made, as well as in what decisions "should" have been made in a case. For example, these analyses sometimes use ordinal scales, such as 0 through 3, where 0 is nonexistence (e.g., zero likelihood or zero importance) and 1, 2, and 3 are low, medium, and high, respectively. Thus, there may be many small consequences that are near zero in importance and, since NG is a product, the overall net goodness of the decision is driven almost entirely by one or a few important and likely consequences. The scale of goodness may range from −3 to +3 (immoral to very moral).

An example is to compare two options: A. Make a device that helps 90% of the general population, harms 9% and kills 1% *versus* B. A device that does not work as well (helps only 15% of the population), but harms no one. If one of our moral principles is to protect life, the 1% mortality rate of Device A may render the goodness component to be negative. If the importance and likelihood of outcomes are about the same, the comparative scores are:

$$A = (-3) \times (1) \times (1) = -3$$
$$B = (1) \times (1) \times (1) = +1$$

So of the two options that are both somewhat important and somewhat likely, option B is clearly more ethical. Note that a goodness value of 0 can either mean there is disagreement about its moral standing (e.g., a bimodal distribution) or it is not really a moral decision (i.e., an amoral question).

There are two cautions in using this approach. First, although it appears to be quantitative, the approach is very subjective. Second, as we have seen many times in cases involving health and safety, even a very unlikely but negative consequence is unacceptable.

Step 3d – Line Drawing: Graphical techniques like line drawing, flow charting, and event trees are very valuable in assessing a case. Line drawing is most useful when there is little disagreement on what the moral principles are, but when there is no consensus about how to apply them. The approach calls for a need to compare several well-understood cases for which there is general agreement about right and wrong and to show the relative location of the case being analyzed. This is a type of ethical extrapolation from the known to the unknown. Two of the cases are extreme cases of right and wrong, respectively. That is, the positive paradigm is very close to being unambiguously moral and the negative paradigm unambiguously immoral as shown in Figure 7.7.

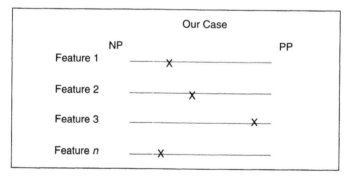

Figure 7.7 First steps in line drawing analysis: comparing features of a decision. The number and types of features depend on the case or decision being considered. For example, a conflict of interest decision is influenced by the size and the level of funding of the project being affected, the reasons for the conflict, and the effects or outcomes likely to result from the conflict (negative and positive). For example, if the outcome results in diminished product quality or safety, the decision is clearly unethical. However, if the outcome results in no product quality or even improved quality (e.g., the spouse's company is really better than the one that would have received the bid without the conflict), the decision is still unethical because it violates the principles of fairness and honesty.

Adapted from: C.E. Harris, 2004, "Methodologies for Case Studies in Engineering Ethics," in: National Academy of Engineering. *Emerging Technologies and Ethical Issues in Engineering: Papers from a Workshop*, 14–15 October 2003, pp. 77–94.

Figure 7.8 Relative location of text case (T) to other similar cases in terms of ethical acceptability.

Next, our case (T) is put on a scale showing the positive paradigm (PP) and the negative paradigm (NP), as well as other cases that are generally agreed to be less positive than PP but more positive than NP. This shows the relative position of our case T (Figure 7.8). The proximity of each feature to the negative and positive paradigms, and the importance (i.e., weight) determine where the whole project or decision compares to an overall NP or PP. This tells us how our proposed approach stacks up. For example, if Feature 3 is very important (e.g., product safety) compared to other features (e.g., reason for conflict, such as the spouse's employer, or timing of disclosure), then based on the outcome, the overall decision is rather positive. Of course, this varies, such as when timing of disclosure is crucial – i.e., the engineer waited until after others identified the conflict to disclose it or disclosed only selective information that put others at risk. In fact, the engineer is held to professional standards that do not usually allow unfettered "risk–benefit analyses" for each decision (e.g., even if I can show that it was better for everyone in this case to give a bribe or gratuity, it is still an ethical breach). So, before dismissing a seemingly "small" ethical matter (only a few dollars involved), it may still constitute a violation of an engineering ethical canon. If something is prohibited in the code of ethics, it must be followed. There is no threshold. Thus, even a positive outcome does not preclude professional duty.

This gives us a sense that our case is more positive than negative, but still short of being unambiguously positive. In fact, two other actual, comparable cases (2 and 3) are much more morally acceptable. This may indicate that we should consider taking an approach similar to these if the decision has not yet been made. If the decision has been made, we will want to determine why the case being reviewed was so different from these.

Although being right of center means that our case is closer to the most moral than to the most immoral approach, other factors must be considered, such as feasibility and public acceptance. Like risk assessment, ethical analysis must account for trade-offs and balancing ethical principles (e.g., security *versus* liberty). We will analyze a case using line drawing in Chapter 8.

Step 3e – Flow Charting: Critical paths, PERT charts, and other flow charts are commonly used in design and engineering, especially computing and circuit design. They are also useful in ethical analysis if sequences and contingencies are involved in reaching a decision, or if a series of events and ethical and factual decisions lead to the consequence of interest. Thus, each consequence and the decisions that were made along the way can be seen and analyzed individually and collectively. A flow chart may show only one of the decisions involved in a case, such as selecting materials to fabricate a device. Other charts need to be developed for safety training, the need for fail-safe measures, and proper operation and maintenance. Thus, a "master flow chart" can be developed for all of the decisions and subconsequences that ultimately led to a failure.[35]

Step 3f – Event Tree Analysis: Event trees or fault trees allow you to look at the possible consequences of each decision. Figure 7.9 provides a simple example. In fact, an actual event tree would be much more complicated. For example, the suboptions would have to consider more than a "yes-no" dichotomy. The

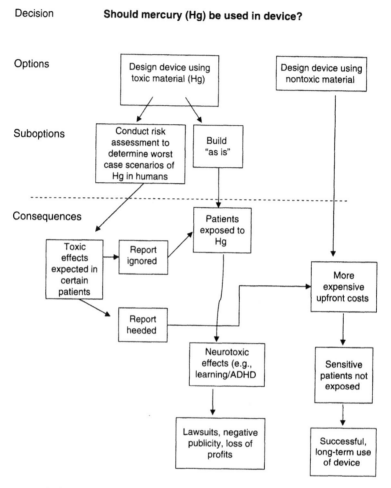

Figure 7.9 Event tree on whether to use mercury in a medical device.

consequences of not using mercury (Hg) might also include having to resort to another element with less electrical conductivity, and a less reliable device. The neurotoxicity effects are being traded for a less reliable design unless the nontoxic substitute is at least as efficient as Hg.

The event tree can build from all of the other analytical tools, starting with the timeline of key events and list of key actors. What are their interests and why were the decisions made? The event tree allows you to visualize a number of different paths that could have been taken that would have led to better or worse decisions. This is done for every option and suboption that should have been considered in the case, comparing each consequence. It may be, for example, that even in a disaster there may have been worse consequences than what actually occurred. Conversely, even though something did not necessarily turn out all that badly, the event tree could point out that you are just fortunate or that an even better, more moral approach is possible. Each consequence leading to the ultimate outcomes has a probability

Table 7.2
Hypothetical performance of a medical device (fictitious data for illustrative purposes only)

Consequence	Probability of occurrence
Performs as designed	0.9900
Fails due to material fatigue	0.0090
Fails due to variability in fabrication	0.0009
Fails due to designer error	0.0001
Sum of all outcomes	1.0000

Further, the consequences can be subdivided as to the type of failure. For example, if the failures are divided into nuisance (e.g., irritation or noise, but with no effect on the health of the patient), required replacement of device (but patient survives without serious effects), and death, each of the failures will have their own profile for these outcomes. So the table can be reconstructed as follows:

Performs as designed	0.990000
90% material failures are nuisances	0.008100
10% material failures need to be replaced	0.000900
0% material failures are fatal	0.000000
50% fabrication failures are nuisances	0.000450
30% fabrication failures need replacement	0.000330
20% fabrication failures are fatal	0.000120
80% designer error failures are nuisances	0.000080
15% designer error failures need replacement	0.000015
5% designer error failures are fatal	0.000005
Sum of all outcomes	1.000000

of occurance. In a failure analysis where the outcome is known, the sum of all consequences below the suboptions (dashed line in Figure 7.9) must equal unity (100%). In fact, the fault tree approach applies a probability to each option and suboption. (Table 7.2)

Thus, the design fatality rate is $1.2 \times 10^{-4} + 5.0 \times 10^{-6} = 1.25 \times 10^{-4}$. Five per million of the patients using the device are at risk of death from the designer's errors. This is one of the features (product safety) that are factored into the ethical analysis. The engineering community and regulators must decide the extent to which these failures are weighted in deciding how to uphold the health, safety and welfare of the public and whether the engineer has performed competently and has served as a faithful agent to the users of the device.

Step 3g – Risk Trade-off Analysis: The event tree mentioned above depicts an unusually dichotomous decision (e.g., whether or not to use a toxic substance). Often, however, decisions are fraught with advantages and disadvantages no matter what is decided. For example, not using Hg in the device could result in premature failure that may lead to complications, such as risks associated with reimplantation. There are also opportunity risks when choosing the "no action" alternative.

Depending on the case to be analyzed, any or all of these tools are useful. However, always keep in mind that ethical issues are seldom quantitative. Some aspects (such as risk and reliability calculations) are, but many aspects of an ethical decision are qualitative.

NOTES AND COMMENTARY

[1] H.A.L. Fisher, "Preface," in *A History of Europe* (1936).

[2] The derivation of the mythical Pandora's box is found in Chapter 2.

[3] George Santayana, *The Life of Reason*, vol. 1 (1905).

[4] Thomas Bayes, English preacher and mathematician, argued that knowledge of prior events is needed to predict future events. Thus Bayes, like Santayana for political thought, advocated for the role of memory in statistics. Bayes' theorem, which was published two years after his death in 1761 in *An Essay towards Solving a Problem in the Doctrine of Chances*, introduced the mathematical approach to predict, based on logic and history, the probability of an uncertain outcome. This is very valuable in science, i.e., it allows uncertainty to be quantified.

[5] Human Factors and Ergonomics Society, "HFES History," http://www.hfes.org/WEB/AboutHFES/history.html (accessed 31 July 2006).

[6] US Department of Health and Human Services, Food and Drug Administration, Guidance for Industry and FDA Premarket and Design Control Reviewers – Medical Device Use–Safety: Incorporating Human Factors Engineering into Risk Management, July 18, 2000.

[7] Ibid.

[8] Ibid.

[9] Ibid.

[10] *The American Heritage Dictionary of the English Language*, 2004, 4th ed. Houghton Mifflin Company, New York, NY (accessed at *Answers.com* on 8 July 2006).

[11] J.-K. Hammitt, E.S. Belsky, J.I. Levy, and J.D. Graham, "Residential building codes, affordability, and health protection: A risk-tradeoff approach," *Risk Analysis* 19, no. 6 (1999): 1037–58.

[12] N. Augustine, "Ethics and the Second Law of Thermodynamics," *The Bridge*, 32(3). Fall 2002.

[13] J. Song, P.P. Lee, D.L. Epstein, S.S. Stinnett, L.W. Herdon Jr., S.G. Asrani, R.R. Allingham, and P. Challa, "High failure rate associated with 180 degrees selective laser trabeculoplasty," *Journal of Glaucoma* 14, no. 5 (2005): 400–8.

[14] L.J. Morse, J.A. Bryan, J.P. Hurley, J.F. Murphy, T.F. O'Brien, and T.F. Wacker, "The Holy Cross Football Team Hepatitis Outbreak," *Journal of the American Medical Association* 219 (1972): 706–8.

[15] The principal sources for this case are M.W. Martin and R. Schinzinger, *Ethics in Engineering*, 3rd ed. (New York: McGraw-Hill, 1996); and C.B. Fledderman, *Engineering Ethics* (Upper Saddle River, NJ: Prentice Hall, 1999).

[16] Many would consider the nuclear meltdown at the Chernobyl, Ukraine power facility to be the worst industrial accident in history.

[17] Aviation Safety Network, 2002, http://aviation-safety.net/database/index.html.

[18] S. Pfatteicher, "Learning From Failure: Terrorism and Ethics in Engineering Education," *Technology and Society Magazine, IEEE* 21, no. 2 (2002): 8–12, 21.

[19] American Society for Engineering Education, *Prism* 11, no. 6 (2002).

[20] B. Franklin, 1759, *Historical Review of Pennsylvania*.

[21] D.E. Shea, Congressional Research Service, 2006, "Oversight of Dual-Use Biological Research: The National Science Advisory Board for Biosecurity," Updated July 10, 2006, Order Code RL33342.

[22] L.M. Wein and Y. Liu, "Analyzing a Bioterror Attack on the Food Supply: The Case of *Botulinum* Toxin in Milk," *Proceedings of the National Academy of Sciences of the United States of America*, 102 (2005), 9984.

23 J. Kaiser, "ScienceScope," *Science*, 309 (2005), 31; and A. McCook, 2005, "PNAS Publishes Bioterror Paper, After All," *The Scientist* (2005).

24 In K.C. Cole, *The Universe and the Teacup: The Mathematics of Truth and Beauty* (New York, NY: Harcourt Brace, 1999).

25 F.P. Feynman and J. Robbins, *The Pleasure of Finding Things Out* (New York, NY: Perseus Publishing Co., 1999).

26 T.S. Kuhn, *The Structure of Scientific Revolutions* (Chicago, IL: University of Chicago Press, 1962).

27 Ibid., 58.

28 Ibid., 62–3.

29 National Academy of Engineering, *The Engineer of 2020: Visions of Engineering in the New Century* (Washington, DC: National Academy Press, 2004).

30 Petroski, *To Engineer Is Human*, 41.

31 From Ian Jackson, *Honor in Science* (Research Triangle Park, NC: Sigma Xi, 1956), 7; and from J. Bronowski, *Science and Human Values* (New York/Messner, 1894), 73.

32 C.E. Harris Jr., M.S. Pritchard, and M.J. Rabins, *Engineering Ethics, Concept and Cases* (Belmont, CA: Wadsworth Publishing, 2000).

33 A. Bradford Hill, "The Environment and Disease: Association or Causation?," Proceedings of the Royal Society of Medicine, *Occupational Medicine* 58 (1965): 295.

34 John Ahearne's comments on this topic were made at the National Academy of Engineers' workshop on emerging technologies and ethics held in Washington, DC, in November 2003.

35 C.B. Fleddermann, *Engineering Ethics*, 2nd ed. (Upper Saddle River, NJ: Pearson Education, 2004).

Chapter 8

Justice and Fairness as Biomedical and Biosystem Engineering Concepts

Man's capacity for justice makes democracy possible, but man's inclination to injustice makes democracy necessary.

Reinhold Neibuhr (1944)[1]

Are engineers just and fair?

The reality articulated by Neibuhr indicates the complexities and failings of the human condition, which necessitate vigilance and hard work in providing the public benefit of high-quality health care along with science, engineering, and technologies needed to support it. Justice and fairness comprise the vital aspect of the "bio" part of bioethics, as in "life," liberty, and the pursuit of happiness. As evidence, the second paragraph of the United States' Declaration of Independence states:

We hold these truths to be self-evident, that all men are created equal, that they are endowed by their Creator with certain unalienable Rights that among these are Life, Liberty and the pursuit of Happiness.... That whenever any Form of Government becomes destructive of these ends, it is the Right of the People to alter or to abolish it, and to institute new Government, laying its foundation on such principles and organizing its powers in such form, as to them shall seem most likely to effect their Safety and Happiness.

These unalienable rights of life, liberty, and the pursuit of happiness depend on a modicum of a mutually assured healthy lifestyle. As is often the case, there is an optimal range of governmental and private provisions. The Declaration aptly warns against a destructive government. Democracy and freedom are at the core of achieving fairness and Americans rightfully take great pride in these foundations of our Republic. The framers of our Constitution wanted to make sure that life, liberty, and the pursuit of happiness were available to all: first with the protection of property rights and later with the Bill of Rights by granting human and civil rights to all the people.

Certainly, a modern connotation of "safety and happiness" requires some level of risk reduction, optimized with a concomitant amount of personal responsibility (freedom). Western society often appears parabolic in its distribution of benefits and optimization scheme for various risks. For example, people are not allowed to smoke tobacco in many private venues, yet in many places, relatively untrained drivers are allowed on most roads. The result is, to a dispassionate observer, an inordinate amount of societal costs (deaths, injuries, insurance, and property damage) resulting in one relatively unchecked behavior (driving) *versus* those of another (exposure to environmental tobacco smoke). Such decisions are reached in various ways. Sometimes, it is purely historic inertia (people have always done this, so much energy is needed to overcome the *status quo*). Other times, it is based on rational, or at least "scientific" premises (e.g., health studies, epidemiology, actuarial analyses). Sometimes, there is no clear anthropology on why certain laws, regulations, and social norms and mores exist. No matter how injustices or other societal needs have come about, they call on any responsible engineer to be an agent of justice.

Ethical principles are "general norms that leave considerable room for judgment."[2] Applications of such principles are formally codified into professional codes of practice. They are also stipulated informally by societal norming, such as by religious, educational, and community standards. In fact, most principles of professional practice are derivatives from a small core of moral principles,[3] such as:

1. *Respect for autonomy* – Allowance for meaningful choices to be made. Autonomous actions generally should be taken intentionally, with understanding and without controlling influences or duress.
2. *Beneficence* – Promotion of good for others and contribution to their welfare.
3. *Nonmaleficence* – Affirmation of doing no harm or evil.
4. *Justice* – The fair and equal treatment of people.

Three of these moral principles were codified in the 1979 release of *The Belmont Report: Ethical Principles and Guidelines for the Protection of Human Subjects of Research.*[4] As discussed in Chapter 4, the *Belmont Report* resulted from the abuses of Nazi science and ultimately, those of scientists in the US Public Health Service sponsored Tuskegee Syphilis trials. These travesties led to a consensus among the scientific community for the need to regulate research more diligently and to codify regulations to ensure that researchers abide by these principles. The other principle, nonmaleficence, follows ethical precepts required in many ethical frameworks, including harm principles, empathy, and consideration of special populations, such as the infirm and the children.

Ethical principles also follow basic tenets of virtue. For example, "cardinal virtues" are virtues on which morality hinges (Latin: *cardo*, hinge): justice, prudence, temperance, and fortitude. Justice is the empathic view and is basic to many faith traditions, notably the Christian's "Golden Rule" and the Native American's and Eastern monks' axiom to "walk a mile in another's shoes." Actually, one of the commonalities among the great faith traditions is that they share the empathetic precept; for example:[5]

- Judaism, Shabbat 31a, Rabbi Hillel: "Do not do to others what you would not want them to do to you."
- Christianity, Matthew 7, 12: "Whatever you want people to do to you, do also to them."
- Hinduism, Mahabharata XII 114, 8: "One should not behave towards others in a way which is unpleasant for oneself; that is the essence of morality."

- Buddhism, Samyutta Nikaya V: "A state which is not pleasant or enjoyable for me will also not be so for him; and how can I impose on another a state that is not pleasant or enjoyable for me?"
- Islam, Forty Hadith of an-Nawawi, 13: "None of you is a believer as long as he does not wish his brother what he wishes himself."
- Confucianism, Sayings 15:23: "What you yourself do not want, do not do to another person."

Justice is a universal human value crucial to bioethics. It is a concept that is built into every professional code of practice and behavior, including the codes of ethics of all engineering and other technical and design disciplines. And, it is at the heart of assuring quality of life, health care, and environmental protection. It is the linchpin of social responsibility. In ending the Pledge of Allegiance with the phrase ". . . with liberty and justice for all," Americans couple the cherished value of freedom with the principle of fairness, which is also articulated throughout the Constitution of the United States.

Justice must be universalized and applied to everyone. We may be tempted to assume that systems are fair simply because "most" are satisfied with the current situation. In the words of Reverend Martin Luther King, "Injustice anywhere is a threat to justice everywhere."[6] By extension, if any group is disparately denied a healthy life, then the whole nation is subjected to inequity and injustice. Put in a more positive way, we can work to provide a healthful life by including everyone, leaving no one behind.

Engineers are trained to be technical experts. Yes, we practice in a milieu of law, politics and the social sciences, but our forte is within the realm of the physical sciences and engineering principles. However, the contemporary engineering profession is simultaneously demanding that we be better equipped technically and technologically, as well as in the social and human sciences. This calls for a systematic approach to engineering education and practice, which is consistent with elements defined by the National Academy of Engineering to be included in the guiding strategies for the engineer of the future:

- Applying engineering processes to define and to solve problems using scientific, technical, and professional knowledge bases
- Engaging engineers and other professionals in team-based problem solving
- Using technical tools
- Interacting with clients and managers to achieve goals
- Setting boundary conditions from economic, political, ethical, and social constraints to define the range of engineering solutions and to establish interactions with the public[7]

The combination of technical skill and ethical character that defines an engineer is not sequential. Rather, it is integral (Figure 8.1). Ethics is built into the process. The "best" engineering solution, therefore, has to be fair and just.

FAIRNESS AND DISTRIBUTIVE JUSTICE[8]

Part of the difficulty in defining justice is that justice and injustice are not often distinguished by the "what" so much as the "how." For example, currently, bioethical debates center around how much information about a microbe, a rabbit, or you and me can be patented. Can such property be held and, if so, can it be transferred? Does the recent controversy over eminent domain provide lessons?

Rights, property, freedom, and other actions and behaviors are routinely given up or actively taken in the society as part of the social contract. The act of "taking" is morally neutral (i.e., either good

A: Bioethics deferred

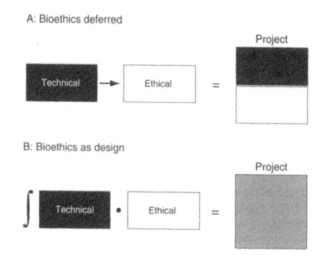

B: Bioethics as design

Figure 8.1 Two approaches to designing bioethics into a project: sequential and retrofit (A) *versus* integral (B).

or bad depending on the conditions of the taking). Taking, in fact, has been highly visible recently, culminating in the United States Supreme Court ruling in July 2005 that certain private concerns may use eminent domain ostensibly to take private property for the public good. Some would say that this is good, since the private enterprise is improving things (e.g., enhancing the local tax base). This is a utilitarian perspective; that is, the proponents perceive a greater good, with the end (larger tax revenues) achieved by defined means (taking of private property that would have yielded much less tax revenues). Others consider such taking to be immoral and a violation of the intent of the United States Constitution, since such powers are granted only to public entities and the "public good" is strictly defined. Also, they see the takings as an encroachment or even an outright assault on individual freedoms. At a more basic level, "stealing" is an immoral form of taking.

Fairness, the equal application of morality to all people, is a much more sophisticated concept. Fairness is a relatively new idea in human history and is considered by many ethicists and philosophers to be at some higher level than most basic moral rules such as prohibitions against lying, stealing, and the like.

Fairness is built into many value systems, and has been throughout recorded history. For example, during the time of the Roman Empire, tax collectors were local citizens in the remote provinces who were required to collect a certain sum from citizens. In addition, they were allowed to collect monies beyond what was due to the empire as personal commission. Thus, the tax collector was despised by the local people because he was seen as disloyal and his methods of collection were deemed unfair.[9] It is also one of the first examples of professional ethics, or more correctly, of the public's dissatisfaction with the ethics of a profession. And, it represents another example of how justice is defined by the "how" *versus* the "what" in a matter. Many reasonable persons at the time may not have begrudged the local tax collector his rightful wages. The injustice was the inflated amounts taken, as well as the extortion and the tactics used in the gain. Fast forward to contemporary times and there is similar discomfort with unfairness, such as insider trading in the stock market, exorbitant interest rates, price gouging, corporate cheating, excessive CEO salaries, "big box" department stores replacing local downtown businesses, political chicanery, and even "legacy" college admissions. Fundamentally, these are perceived as unfair practices (the "big guy" exploiting the "little guy").

The idea of fairness as a moral vehicle for individual and professional ethics, however, was not exploited until John Rawls wrote his highly influential book, *A Theory of Justice*, in 1958 in which he proposed that justice is fairness.[10] For Rawls, justice emerges when there is a fair compromise among the members of a true community. If individuals are fairly situated and have the liberty to move and to better their position by their own industry, justice results when they agree to a mutually beneficial arrangement. Fairness is the right ordering of distributed goods or "bads," and fair persons are those who, when they control distributive processes, make those processes fair.

This is a circular definition, of course, defining fairness as the fair distribution of goods. So, let us try another approach, keeping with the concept of distribution, which is familiar to the engineer. For a moment let us think of justice as a commodity that can and should be distributed fairly throughout society, i.e., "distributive justice." Allocating things of value that are limited in supply relative to demand has various dimensions. Justice, in this regard, depends on what is being allocated, such as wealth and opportunities. For example, one often hears arguments about what is a "fair redistribution of wealth" in discussions on fair taxation. Other variables include the diversity of the subjects of the distribution (e.g., people in general, people of a certain country or national origin, citizenship status, socioeconomic status, or even people *versus* nonhuman species). The basis of how the goods should be distributed also varies. For example, some philosophies call for equal distribution to every member of society (known as "strict egalitarianism"), others for the characteristics of individuals comprising a population (varying by age, handicaps, or historical biases), and still others are based purely on market forces (e.g., "strict libertarianism").[11]

While the idea of fairness is tied to many ethical ideas such as justice, reciprocity, and impartiality, the word "fair" can have other meanings as well. For example, there is the problem of the "free rider," a person who uses the contributions of others in society in order to better his or her position but does not participate in the cost of the society. A person who does not pay tax for religions reasons still uses the roads and public services for which others pay. Some, and arguably most, we would deem such actions as "unfair" since that person would be taking social goods without contributing to social welfare.[12]

Another meaning of "fair" is the receipt of good or bad events beyond the control of society. For example, a person whose mobile home is destroyed by a tornado while other mobile homes in the vicinity are spared would call this "unfair," although there is nothing unfair (in moral terms) about a random event of nature. However, if the random occurrence is followed by a willful act, such as increasing the costs of needed supplies following a natural disaster, i.e., "gouging," such an act would be considered unfair. Since it places inequitable burdens on those suffering from the natural disaster.

A popular use of the word "fair" relates to how events, beyond the control of society, treat the person. For example, a person might get a debilitating disease such as multiple sclerosis, a neurological illness that strikes only young. Contracting multiple sclerosis, while it is a tragedy for that individual and his or her family and friends, is not a case of unfairness. It is a difficult event, but it is not unfair.

On the other hand, if human suffering is caused by premeditated human actions, such as decisions to allow an unsafe drug or device to be marketed and thereby increasing the risk of human illness, then such unethical decisions would constitute unfairness. Such unfairness is exacerbated if one group of society is disparately affected (e.g. certain ethnic groups are more sensitive to the allergies elicited by the unsafe drug).

Thus, we are looking for a connotation of fairness that distinguishes such unfortunate confluences of events (e.g., genetic expression of chronic diseases) from those where human decisions have caused unfairness or have not properly accounted for certain groups and which have led to adverse consequences. Thus, fairness occurs with the honoring of appropriate and just claims. Another way of saying it is that fairness is a process where the legitimate claims of each person are respected.[13]

The ancient Greeks considered fairness to exist when equals were treated equally; and unequals were treated unequally; that is, fairness occurs when identically situated people are treated identically. When there are no substantial differences among various people, then they all ought to be treated equally.[14]

What characteristics are sufficient to cause unfair treatment? For example, two people, A and B, need a drug supplied at a private company. The first one is offered the drug for $1000 per year and the second one $500 per year. Since the needs are identical, there now has to be a justification for the difference in costs. Suppose A is a man and B is a woman. Clearly, the difference in gender ought not to apply and B would have a legitimate cause for claiming unfair treatment. However, what if person A earns 100 times more than person B? Would a difference in costs be justified? In this case, the ability to pay is offered as a justification for the lower costs.

We often approve of such differences if they are significant, but the term "significant" is confusing to scientists and engineers because, for us, it has a very distinct and clear definition. It is the probability that an observed result is not due to chance alone. For example, an experimental finding is statistically significant if there is a less than 1% probability that the difference observed would occur by chance alone if the treatments being compared were equally effective (i.e., a p-value of less than 0.01). So, for the engineer, significance is an expression of the probability of a hypothesis given the data. But, how does this help us to get a handle on fairness?

Often, since engineering involves numerous elements of uncertainty, we are willing to accept significance levels much less restrictive than those of our colleagues in the basic sciences. Physicists, for instance, may require a p-value several orders of magnitude more restrictive (e.g., $p < 0.000001$). In this instance, the physicist will not accept a hypothesis if there is more than one in a million probability that the outcome was due to chance. On the other hand, biomedical engineers, who deal with people with all their uncertainties, may be quite happy to accept a 95% confidence or $p < 0.05$. People are hard to predict and difficult to study.

The equal treatment of equals is also one of the conundrums of "affirmative action." Does fairness demand retribution for past wrongs committed to an identifiable social group? When is fairness the same as "equity" (equal treatment), such as equal housing and employment opportunities, and when does fairness require a more affirmative approach to repair past and ongoing injustices (e.g., lingering effects of generations of uneven educational achievements, union membership, and career opportunities due to intentional and even sanctioned biases). Equal opportunity seems to imply "equals treated equally," while affirmative action calls for some effort to treat "unequals unequally."

One way to resolve this might be to define fairness as a lack of envy, or when no participant envies the lot of any other. However, this is not necessarily fair, since the claims of some people might be exaggerated. For example, suppose a farmer is retiring and wants to distribute his farm of 300 acres among his three sons. If the sons are equal in all significant (there is that word again!) ways, the farmer would divide his farm into three 100 acre plots. But suppose one son claims to be a better farmer than the other two, and insists that this ought to result in his having a larger share of the 300 acres. A second son might need 120 acres because he wants to sell the land for a new airport, and thus stakes his claim for the larger lot. A third might say that since he has more children than the first two, he needs a larger share because he will have to eventually subdivide his plot.

Are any of these claims legitimate enough to change the initial distribution of 100 acres each? It would be unlikely that a disinterested arbitration board would respect any of these claims, and thus the different claims should not result in a division different from the 100/100/100 distribution. Each of the three sons might go away unhappy, but nevertheless the process has resulted in a fair division of the goods. Further, the quality of the acreage, not simply the amount, is also a component of fair distribution. The injustices done to the Native Americans in moving them to reservations was not necessarily the land area, but the loss of sacred and lands deemed more useful (e.g., for fishing and hunting) than the reservations. Even if some bureaucrat counts up all the allotted acreages and finds that they are not so different from the evacuated lands, then this is in no way a measure of justice. Western culture, and especially we engineers, likes to quantify. But, the essence of most things are more than the sum of their parts. Even very concrete commodities have subjective and abstract meaning and value that can easily be missed in an actuarial report. Thus, human values are seldom completely fungible.

Another problem with the "envy-free" approach to fairness is that it depends on each person having a similar personality. Suppose, of the three sons in the above example, one is a very generous person, and would not object if his brothers took much more than their share. At the conclusion of the division, with one brother getting 60 acres and the other two sharing the remaining 240 acres might result in an envy-free division of the goods, but this would be eminently unfair to the generous brother. Thus, defining fairness as a lack of envy does not seem to be useful. And, at its worst, it can be a tool for unfair distributions. After all, there is no shortage of those who live by the maxim, "Never give a sucker an even break."[15] Unfortunately, there is also no shortage of those who would take advantage of other's ignorance, naïveté, and sense of fair play.

Perhaps we can go to other professions for an acceptable definition of fairness. One means of determining fairness in the legal profession is the "reasonable person standard." A fair distribution of goods occurs when an objective outsider, taking into account all the claims of the participants, renders a decision that would be agreed upon by most rational, impartial people, as equitable to all, regardless of each individual claim. In common law, the reasonable person standard is a "legal fiction" since there really is no such person. It provides an objective means to analyze a situation for its moral content. Creating a hypothetical individual whose view is based solely on reason provides a means of looking at the situation in a less biased way. Bias is another of those terms with a distinct engineering meaning; that is, it is a systematic error. Thus, the reasonable person standard helps to recalibrate our sense of fairness in the same way as we calibrate our scientific apparatus against a known (i.e., rational) standard. We would expect an arbitration board to apply such a rational approach to determining fairness.

Another way of describing fairness is to define what we mean by its opposite, or unfairness. Rescher[16] identifies three types of claims of unfairness that might be valid:

- Inequity
- Favoritism
- Non-uniformity

Inequity. Giving people goods not in proportion to their claim is inequity. The opposite would be "equity," or a condition where people's shares are proportional to their just and appropriate claims.

For example, suppose a business goes broke and creditors are lining up for their share. Say the business owed $100 but three creditors are owed $50, $100, and $250. An equitable distribution of the available funds would pay each one 25 cents on the dollar, so the three claimants would get $12.50, $25, and $62.50 each. Of course, the claims have to proven to be just claims.

Favoritism. Some conditions that have nothing to do with the issue at hand, for example one's relatives or one's religion ought to have nothing to do with the situation or claim. The opposite would be "impartiality," the evenhanded distribution of goods without favoritism.

Suppose in the above bankruptcy example the executor decides to pay out $50, $50, and $0 to the three creditors because the first one is a local merchant and the last one an international bank, believing that it is more important to support the local merchant than some far-off impersonal bank. This distribution is unfair since the type of business ought not to be germane to the distribution of the available funds. The problem here is in the evaluative premise. That is, some would indeed believe the type of business indeed *is* morally germane. This can be problematic, however, in biomedical decisions. Granting special status for certain segments of society is difficult and can be immoral. For example, if the only beneficiaries of a medical device are users of elicit drugs, ought that condition influence the decision to proceed?

Non-uniformity. "Equal treatment under the law" means that the law is to be applied to all people regardless of their status or wealth. The opposite is "uniformity," or the uniform application of the rules.

Suppose a dinnertime rule in a family is that all vegetables have to be eaten before dessert is served. Some children will not always kill off the last leaf of spinach, hoping to get away with the small transgression. If the parents allow this for all their children save one, the ones who are held to the strict rule can rightfully claim to have been treated unfairly. The rule was not evenly applied.

However, returning to the Greek definition, fairness is not equalitarianism, i.e., the treatment of all people equally. Society, in order to function, occasionally has to impose unequal treatment of some. In our non-uniformity example, what if the child allowed to avoid eating her spinach has a rare liver illness that makes eating spinach dangerous? The military draft was patently inequalitarian. Only some people were to be drafted and others were not. The ones drafted may have ended up in harms way, and certainly lost time out of their lives. But it is not possible to send partial persons into the army. If the need is for 100 000 soldiers out of an eligible population of 10 000 000, everyone has a 1% chance of being drafted. The key here is that the draft, or the process by which the 100 000 will be chosen, has to be fair. Everyone ought to have an equal chance of being drafted unless they can show some significant reason why they ought to be exempted from the draft. And, if you recall the Vietnam era, such exemptions as college deferments and conscientious objections were the stuff of controversy and moral debates.

One of the principles of our society is that all persons are to be treated equally under the law. But this does not mean unqualifiedly equal. Some identifiable groups of people, such as professionals, are treated differently under the law. All professional pharmacists, for example, are allowed to dispense drugs, while this activity would be illegal for the nonprofessional. All people in the category "pharmacists" then are being treated differently from other people. Unfairness occurs when a pharmacist, because of some irrelevant differences such as gender, religion, or shoe size, is not allowed to dispense drugs. Similarly, while we want to treat all people in the same way when they have committed a crime, this is seldom done. A first offender might receive a different sentence than a repeat offender for the same crime. And, the jurisdiction prosecuting the crime often determines the severity of the consequences.

Equality before the state is also important, in that the goods distributed by the state (and the goods taken by the state) are not equal, but should be equitable. The progressive income tax requires rich people to pay more on a per person basis than poor people, and welfare recipients need to show that they are destitute before they can receive assistance. The important objective of fairness is that each person should be treated equitably within the process. So, a rich woman ought not to have to pay more taxes than a rich man, all other things being equal.

Perhaps the best definition of fairness that is useful in our discussions of bioethics is to say that fairness is treating each person the same according to morally accepted and agreed on rules, and whenever these rules result in unequal treatment, there has to be a good and acceptable reason for this inequality.[17]

What makes the unfair treatment of individuals immoral? Such treatment becomes immoral when the claims of individuals are not respected. This points to a more important aspect of fairness: systematic exclusion. For example, it has only been a few decades since job selection criteria included questions about one's intentions to have children, whether one owned an automobile or whether one can lift 50 pounds. While some of these factors are important in certain careers, for example, a firefighter needing to lift an unconscious person or the likelihood that the person may be exposed to chemicals that can harm babies in the womb, they must not be used to ensure that only a European–American male should be employed. The job selection process should be fair. For example, if an analytical chemist is needed in a biomedical engineering laboratory, there are few if any situations where more than a few pounds needs to be lifted and it is really no one's business how a person gets to work (private automobile, bike, public transportation, or a Jetsons' flying car!) and thus an irrelevant question can only result in unfair treatment. A general rule is that if the job that needs to be performed is not affected by the answer given to a question in an interview, such questions should never have been asked in the first place.

A fair division of goods may not be democratically popular. Consider a country with two primary religious sects, one with 40 and the other with 60% of the population. An election is held, and the majority of people (60%) decide to prevent any and all goods from going to the minority (40%). This is obviously unfair, even though the result has been arrived at democratically.

Fairness also has a time component; that is, like other aspects of engineering, justice is constrained by time and space. The example of siting the noxious facility unfairly is a spatial injustice (i.e., where we put the medical waste landfill determines the injustice). Sometimes, it is not so much what or where, but when an action takes place, that determines its fairness. In the words of the British statesman, William E. Gladstone (1809–1898), "Justice delayed is justice denied." Excluding people from key decision points or waiting to involve them until enough momentum for the project has been gathered is a manner of injustice. Not sufficiently accounting for future generations (e.g., nuclear waste dumps, mine subsidence, or future declines in property values) is arguably an unfair practice.

Finally, a different connotation of the word "fair" is to denote that something is neither very good nor very bad. For example, in the old adjective grading systems, a "C" grade was often described as fair. This was to let the guardians of the student know that while the student was not failing, he or she was not doing an excellent job in the course. At first blush, this connotation may be seen as very similar to "equal opportunity." Rather, fair in this use is actually a utilitarian or statistical concept. A fair decision from a utilitarian view is the one that provides the greatest amount of benefit for the greatest number of people. So, if a benefit-to-cost ratio (B/C) is calculated from different segments of the society, the option that provides the best overall B/C ratio for the largest segment of society would be chosen. This is akin to the statistical concept of a normal distribution (Figure 8.2). In other words, if we assume that the benefits and costs (or risks) in a given situation are normally distributed within a population, we would be able to select the most fair option as the one where most of the population receives, on average, the largest benefits *versus* costs or risks. This assumes that the very well-off and the very poor receive the

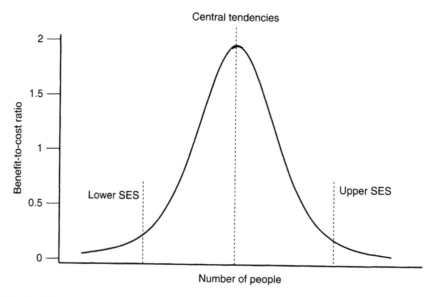

Figure 8.2 Hypothetical selection of a "fair" option from a utilitarian perspective when the benefits are normally distributed.

least benefits, and most of the population receives the most benefit. This can occur, for example, in economics, where the very poor receive a stipend from the government or low-paying jobs and the very rich are taxed at an increasingly high marginal rate (i.e., a "progressive" tax), but the majority of the population receives the most goods and services from the government. The highest B/C ratios are those near the statistical measures of central tendency (i.e., the mean, median, and mode).

This sounds plausible until other aspects of fairness are considered. In fact, this can be one of the most unfair ways to decide on bioethical issues, since it places an undue and disparate burden on a few groups. And, these are often the ones least likely to be heard in terms of pointing out costs and risks. This is known as the tyranny of the majority.[18] Often, the curve is not normally distributed, but is skewed in favor of the higher socioeconomic strata (Figure 8.3). Sometimes, it is doubly unfair because the people assuming most of the costs and risks are those that receive the fewest benefits from this particular decision (e.g., less access to state-of-the-science biomedical technologies).

These curves demonstrate the importance of the concept of "harm" in fairness. No group should have to bear an "unfair" amount of costs and risks relative to the benefits. This is why John Stuart Mill added the "harm principle" to utilitarianism and why John Rawls argues that one must empathize with the weakest members of the society (see Discussion Box: Harm and the Hippocratic Oath). Rawls argues that the only fair way to make a moral decision is to eschew personal knowledge about the situation that can tempt a person to select principles of justice that will allow them an unfair advantage. This is known as the "veil of ignorance," but it is really a way to implement Mill's harm principle. So, fairness also implies more than utility and more than a good B/C ratio, it requires virtue.

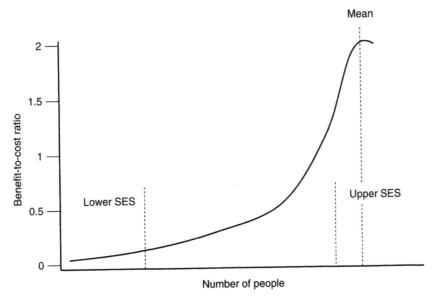

Figure 8.3 Conceptual model for selecting a "fair" option from a utilitarian perspective when the benefits are skewed in favor of the higher socioeconomic strata.

Discussion Box: Harm and the Hippocratic Oath

The Hippocratic oath for physicians is an example of a precautionary principle, when it states "First do no harm...." The traditional text of Hippocrates of Cos (c. 460–377 BC) is:

I swear by Apollo the physician, by Æsculapius, Hygeia, and Panacea, and I take to witness all the gods, all the goddesses, to keep according to my ability and my judgment, the following Oath.

To consider dear to me as my parents him who taught me this art; to live in common with him and if necessary to share my goods with him; to look upon his children as my own brothers, to teach them this art if they so desire without fee or written promise; to impart to my sons and the sons of the master who taught me and the disciples who have enrolled themselves and have agreed to the rules of the profession, but to these alone the precepts and the instruction. I will prescribe regimens for the good of my patients according to my ability and my judgment and never do harm to anyone. To please no one will I prescribe a deadly drug nor give advice which may cause his death. Nor will I give a woman a pessary to procure abortion. But I will preserve the purity of my life and my art. I will not cut for stone, even for patients in whom the disease is manifest; I will leave this operation to be performed by practitioners, specialists in this art. In every house where I come I will enter only for the good of my patients, keeping myself far from all intentional ill-doing and all seduction and especially from the pleasures of love with women or with men, be they free or slaves. All that may come to my knowledge in the exercise of my profession or in daily commerce with men, which ought not to be spread abroad, I will keep secret and will never reveal. If I keep this oath faithfully, may

I enjoy my life and practice my art, respected by all men and in all times; but if I swerve from it or violate it, may the reverse be my lot.

Many contemporary physicians continue to take a reworded oath based on the Hippocratic oath. The oath actually includes a number of precautionary elements instructive to the engineer. The closest parallel is the first engineering canon that "hold paramount the health, safety, and welfare of the public," another example of a precautionary principle. It is a call to empathize with those who may be affected by bioethical decisions and actions.

Weaknesses in this connotation of fairness can be demonstrated using an ethical analytical tool introduced in Chapter 7, i.e., line drawing.[19]

Consider the following example, adapted from Fledderman:[20]

Teachable Moment: Disposal of a Slightly Hazardous Waste

A company is trying to decide how and to what extent it should dispose of a "slightly hazardous" waste. The company's current waste stream contains about 5 parts per million (5 ppm) of contaminant A. The State Environmental Department allows up to 10 ppm of contaminant A in effluent drained into its sanitary sewers. The company has no reason to suspect that at 5 ppm any health effects would result. In fact, most consumers would not detect the presence of contaminant A until relatively high concentrations (e.g., 100 ppm). The city's wastewater treatment plant discharges to a stream that runs into a lake that is used as a drinking water supply. So is it ethical and fair for the company to dispose of slightly hazardous waste directly into the city sewers?

Solution and Discussion

The positive paradigm in this case is that the company would do what is possible to enhance the health of people using the lake as a drinking water source. The negative paradigm is an action that causes the drinking water to be unhealthy. After a literature review, we chose some hypothetical cases that the company can pursue:

1. Company dumps contaminant A at the regulatory limit (100 ppm). No harm, but unusual taste is detected by a few sensitive consumers.
2. The company lowers its concentration of contaminant A so that it is effectively removed by the town's existing drinking water system.
3. Company discharges contaminant A into the sewer at 5 ppm or below, but ensures that it is effectively removed by the town with new equipment bought by the company and donated to the town.
4. Contaminant A can be removed by equipment paid for by taxpayers.
5. Seldom, but occasionally, contaminant A concentrations in water will make people feel sick, but the feeling only lasts for an hour.

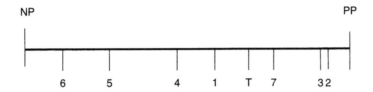

Figure 8.4   Line drawing for decision to dispose of hazardous waste.

6. Contaminant A passes untreated (i.e., 5 ppm) causing sensitive people to become acutely ill, but only for a week with no long-term harm.
7. Equipment is installed at the company that reduces the loading of contaminant A to 1 ppm.

Drawing each case and our proposed case (T) gives us a relative location with respect to the negative and positive paradigms (Figure 8.4).

Cases 2 and 3 are closest to the positive paradigm, so they appear to be the most ethical. In these cases, the amount of contaminant A that reaches the public is kept well below the regulatory limit and any health threshold. Case 7 is also relatively close to the positive paradigm, since it is well below the regulatory limit, but even at these levels some sensitive people (e.g., newborns, immuno-compromised, and elderly) could experience effects. Cases 6 and 7 are less ethical because they resemble the negative paradigm, i.e., actions that make the water less safe to drink. The key, though, is that in the middle of the diagram (Case 4), the burden of the problem caused by the company is shifted from the private company to the public. This is not the fairest option by any means.

Although being right of center means that this case is closer to the most moral approach than to the most immoral approach, other factors, such as feasibility and public acceptance, must be considered. The location on the line indicates that being fair is different from receiving a "C" grade. Fairness implies that we need to search for options that move us closer to the positive paradigm, i.e., the ideal. As we migrate toward options in the negative direction, we give up a modicum of fairness. This is the nature of balancing benefits and costs, but the engineer must be fully aware that these balances are taking place. So, like risk assessment, professional judgment in selecting the most fair designs and projects must account for tradeoffs (e.g., cost effectiveness *versus* fairness, security *versus* liberty, and short-term needs *versus* long-term effects).

Questions

What three steps can be taken to move our case closer to the positive paradigm?

Care should always be taken when trying to apply objective and quantitative tools to concepts like ethics and justice. The social sciences and philosophical principles are often highly subjective. While the natural sciences and engineering strive for objectivity, they also must deal with subjectivity from time to time.

Bioethics Question: What characterizes a morally commendable engineer?

Subjectivity is common not only to justice, but to all moral principles. Subjectivity is different from situational ethics. It simply means that every moral principle cannot be completely explained in empirical and measurable ways. External facts are crucial, but they are only part of ethical decision making. Impressions and feelings also come into play.

Philosophers argue about the sources of morality, such as whether moral principles are learned or innate. For example, moral objectivism holds that certain acts are inherently right or wrong, and their morality does not depend on the opinions of humans. The basis for their rightness or wrongness can be reason, divine revelation, or both. This is why it can be difficult to discuss ethics without, at some level, discussing religion also. The moral relativists hold an opposite view, i.e., they believe that there are not really inherent moral principles, such as honesty or courage, but that these are always and completely the product of a culture. Notwithstanding the source, deciding the degree and extent of morality is often quite subjective in the sense that the normal means of testing them do not work well, at least the way engineers empirically measure things. Perhaps, another thought experiment can help to elucidate this challenge.

Thought Experiment: Who Is More Ethical?

The saintly Mother Theresa is walking down the street when suddenly a bag of money falls out of the back of the truck, without the driver or anyone else aware of the event. Written on the side of the truck is "Boys and Girls Club of America." Theresa, a moral exemplar, becomes overjoyed that she can recover the money and return it to its rightful owner. She does this out of the habit of moral excellence (as counseled by Aristotle). In fact, she is not even tempted to keep the money.

Meanwhile, the evil Roman Emperor Caligula is walking down the next street and another bag of money falls out of the same truck, again unbeknownst to the driver or any other witnesses. Caligula's first instinct is to take the money, but is suddenly struck that this is an opportunity to behave morally. He anguishes over the decision, but ultimately decides to return the money to the charity; clearly contrary to his normal behavior.

Is Theresa or is Caligula more morally commendable for the decision? Who of the two is behaving most morally?

Some would say that Caligula is more morally commendable because it is so hard for such a person to behave morally. In fact, the ease with which Theresa makes this moral decision, in this way of thinking, diminishes the commendation she deserves. She may even enjoy the good act, while the emperor suffers dreadfully the pain of giving up the money. So, this line of argument is that the one who suffers the most from the pain of making the decision is more moral.

We tend to honor those who overcome weaknesses and frailties. In fact, we give awards to those who overcome adversity, even if it is moral adversity and even if it is the result of one's own immoralities. For example, when my daughter was in middle school, she wondered why her fellow student received a special award. For most of the year he had received in-school suspensions and many other disciplinary actions, but for the final grading period he was able to avoid them. Thus, he "merited" an award. My daughter wondered why she received no award even though she never had to be disciplined.

This is akin to the Biblical parable of the prodigal son,[21] who left his home and committed many immoral (and stupid) acts. He was truly a poor decision maker, ending up sharing a home with the swine (literally). However, when he came to his senses (morally and intellectually), his father welcomed him back home and even threw a party to celebrate his return. The parable illustrates the goodness and mercy of the Creator, but I believe it has other lessons relevant to engineering ethics. Notably, why does the parable include another son? We tend to remember that he is put off by the special treatment of his brother:

> Now the older son had been out in the field and, on his way back, as he neared the house, he heard the sound of music and dancing.
> He called one of the servants and asked what this might mean.
> The servant said to him, "Your brother has returned and your father has slaughtered the fattened calf because he has him back safe and sound."
> He became angry, and when he refused to enter the house, his father came out and pleaded with him.
> He said to his father in reply, "Look, all these years I served you and not once did I disobey your orders; yet you never gave me even a young goat to feast on with my friends.
> But when your son returns who swallowed up your property with prostitutes, for him you slaughter the fattened calf."
> He said to him, "My son, you are here with me always; everything I have is yours.
> But now we must celebrate and rejoice, because your brother was dead and has come to life again; he was lost and has been found."[22]

Arguably the father's comment to the older son carries some important truths for the engineering profession. First, "you are here with me always" indicates trustworthiness, which is the keystone of professionalism. "Here" and "always" are crucial reminders that our success is tied to diligence and constancy; or as our codes put it, we are "faithful agents." In return for this, "everything I have is yours" indicates the reciprocation of our clients giving us this trust. Finally, the fact that the father tells the older brother ("the engineer") that we *must* celebrate is because of the goodness. So, even if our designs and projects are not always appreciated for their real worth, we have the joy of knowing how they are benefiting the society. Even if the prodigal son never returns, the older son is a success so long as he is a faithful agent.

An element of ethical decision making is the focus of the decision. In our thought experiment, both Theresa and Caligula are concerned about the Boys and Girls Club. However, Theresa's habit of excellence moved her focus toward others. This is her default. It would be out of character for her to be selfish. Caligula's norm is to be selfish. Even this act is somewhat selfish, since he has to weigh the good against the bad. In a sense, he is being utilitarian and quickly calculating a benefit–cost ratio.

Ethical egoists and hedonists (hedonism grounds utilitarianism) might argue that all actions are selfish, even those that are altruistic. In professional development, there is some evidence of that. "Good" students make A's. These students get the best jobs. Much of this is driven by selfishness. Recall that Kohlberg's preconditional stages of moral development are very selfish. We shed this selfishness only after external forces (e.g., mentorship, codes of ethics, learning from mistakes, guidance, and team memberships) help us progressively to internalize moral principles (e.g., spiritual growth and maturity). Theresa undoubtedly started off selfish and was trained and developed into a moral exemplar. Conversely, Caligula never advanced beyond the preconditional phase. Recall that engineering is an active endeavor. So is moral development. If we do not invest moral energy into ourselves we will not grow.

Thus, the person who habitually behaves morally is the one who should be commended. This applies to engineering ethics also. This means that engineers have to be Theresa every day, rather than Caligula on a good day. If you think about it, do you really want to be a client of an engineer (or any other professional, for that matter) who is only occasionally honest, even if being good causes the engineer great pain? Engineering is not middle school, nor it is professional sports. Society does not reward engineers with "comeback of the year" or "most improved" awards. Professional trust is not merited intermittently, but earned continuously.

Some of the subjectivity is the result of the steady march of inquiry and the iterative nature of science. As we learn more, key scientific concepts are refined. Some are even abandoned, but their remnants remain in the lay literature (e.g., watching the "sun rise") or even in scientific circles (e.g., "nature abhors a vacuum"). Frequently, the changes are subtle.

A certain amount of subjectivity can present itself in all sciences, and we must take care not to be condescending to our colleagues in the social sciences and the humanities. Two books by Sheldon Rampton and John Stauber provide ample cases of how the public has been manipulated by experts: S. Rampton and J. Stauber, 2001, *Trust Us, We're Experts: How Industry Manipulates Science and Gambles with Your Future*, Penguin Putnam, Inc., New York, NY; and S. Rampton and J. Stauber, 1995, *Toxic Sludge is Good for You*, Common Courage Press, Monroe, ME. These books may be better classified as "muckraking" than as scholastic endeavors, but they do point out some of the ways that "spin" is used to justify decisions. As such, they provide cautionary tales to engineers on how they may, unwittingly, be party to deception. The lesson is that the perception of uncertainty can be amplified when engineers communicate with diverse audiences. We may well know that the physical principles hold and that we are applying them appropriately in a particular project, but certain segments of the society may perceive that we are not being completely honest with them. This is further exacerbated in neighborhoods that have traditionally been excluded from decision making or with whom the track record of "experts" has been tainted with prevarications and unjustifiable "bills of clean health."

At the risk of stating the obvious, good risk communication is not an invitation to compromise the quality of science or to introduce "pseudoscientific" methods, or junk science to keep everyone happy. On the contrary, it is a reminder that we must be just as open and honest about what we do not know as we are about what we know. On more than a few occasions, engineers so strongly believe in the merit of a project that they become advocates to the point that they begin to compromise the actual scientific rationale for the project.

Some of this is the result of "sunken costs," that is, they are so committed and so far down the road that rethinking the projects' design and approach is not, in their minds, a viable option. Another factor is the "us and them" problem, wherein the engineer begins to see those who complain less as clients and those who complain more as obstacles that must be overcome. Whatever be the reasons, an unwillingness to examine and reexamine a project in terms of its scientific credibility is dangerous.

PROFESSIONAL VIRTUE AND EMPATHY

Being fair and advancing the cause of justice is morally admirable. People who devote their lives to doing the right thing are said to behave virtuously. The classical works of Aristotle, Aquinas, Kant, et al. make the case for life being a mix of virtues and vices available to humans. Virtue can be defined as the power to do good or a habit of doing good. In fact, one of Aristotle's most memorable lines is that "Excellence is habit." If we do good, we are more likely to keep doing good. Conversely, vice is the power and habit of doing evil.

Aristotle tried to clarify the dichotomy of good and evil by devising lists of virtues and vices, which amount to a taxonomy of good and evil. One of the many achievements of Aristotle was his keen insight as to the similarities of various kinds of living things. He categorized organisms into two kingdoms, plants and animals. Others no doubt made such observations, but Aristotle documented them. He formalized and systematized this taxonomy. Such a taxonomic perspective also found its way into Aristotle's moral philosophy.

Few engineering tragedies have resulted from wanton and nefarious acts, although the attacks of 9/11 and the Oklahoma City federal building bombing are noteworthy examples that the growing likelihood of such acts must be considered. Most engineering failures have been in large part due to "sins of omission" rather than from "sins of commission." Proper safeguards must be extended to include the externalities and contingencies of our decisions. Is it possible that some future chain of events and contingencies will result in severe hardship or malady as a result of an action that has eluded us today?

Teachable Moment: Albert Schweitzer and the Reverence for Life

A man is truly ethical only when he obeys the compulsion to help all life which he is able to assist, and shrinks from injuring anything that lives.

Albert Schweitzer (1875–1965) was a "missionary surgeon," but that characterization falls short. He was also a leader in modern bioethics. Born in Alsace, he followed the calling of his father and grandfather and entered into theological studies in 1893 at the University of Strasbourg where he obtained a doctorate in philosophy in 1899, with a dissertation on religious philosophy. He began preaching at St Nicholas Church in Strasbourg in 1899 and served in various high ranking administrative posts. In 1906, he published *The Quest of the Historical Jesus*, a book on which much of his fame as a theological scholar rests.

Schweitzer had a parallel career as an organist. He had begun his studies in music at an early age and performed in his father's church when he was nine-years old. He eventually become an internationally known interpreter of the organ works of Johann Sebastian Bach. From his professional engagements, he earned funds for his education, particularly his later medical schooling.

He decided to embark on a third career, as a physician, and to go to Africa as a medical missionary. After obtaining his MD at Strasbourg in 1913, he founded his hospital at Lambaréné in French Equatorial Africa. In 1917, however, the first world war intervened and he and his wife spent the year in a French internment camp as prisoners of war. Returning to Europe after the war, Schweitzer spent the next six years preaching in his old church, and giving lectures and concerts to raise money for the hospital.

Schweitzer returned to Lambaréné in 1924, and except for relatively short periods of time, spent the remainder of his life there. With the funds earned from his own royalties and personal appearance fees and with those donated from all parts of the world, he expanded the hospital to seventy buildings, which, by the early 1960s, could take care of over 500 patients in residence at any one time.

On one of his trips up the Congo to his hospital, Schweitzer saw a group of hippopotamuses along the shore, and had a sudden inspiration for a new philosophical concept that he called "reverence for life," which has had wide influence in Western bioethical thought.

He was awarded the Nobel Peace Prize in 1953. Among the many bioethical lessons from Schweitzer's life is the need to consider the entirety of complex moral issues. We are tempted to work on our specific tasks, sometimes forgetting why we are doing them. Our lives are more than our jobs. To quote Schweitzer, "Do something for somebody everyday for which you do not get paid."

Questions

1. How does Schweitzer's concept of reverence compare to that of respect articulated in the "Belmont Report" and the prototypical institutional review board (IRB) requirements for research using human subjects?
2. Does this respect extend to embryos and fetuses? Explain.
3. How does Schweitzer's concept relate to animal subjects of research?
4. Does this respect extend to genetic modifications? Explain.
5. How does this reverence apply to animals as food source? (For example, does the respect require everyone to be a vegetarian?)

Engineers are well prepared for our fields, for some of us beginning long ago, by first grasping mathematics and the physical principles of science. We are often put off by philosophy and its ilk, but these disciplines can be really valuable to us. Most of us do not consider daily gluons, quarks, and imaginary numbers to do our jobs, but we have considered such theoretical concepts along the way as part of our academic preparation. By analogy, an understanding of bioethics must also be steeped in an appreciation of what we do. The moral principles and canons espoused in our codes of ethics are practical manifestations of deeper moral and philosophical justifications, in much the same way that our designs and calculations are rooted in mathematical and scientific foundations. Thus, it is worthwhile to consider the moral rationale for our daily practice in what makes us not only competent, but moral professionals. Since we are pragmatists a good place to start is with reason.

REASON

Bioethics Question: Is ethics rational?

Bioethics requires reason, but is not wholly rational. Reason must be informed by practical experience and a set of values. For engineers, these values are, to a limited extent, codified in our standards of practice (i.e., codes of ethics). But, here is the challenge. Ethics cannot be formulaic. We cannot plug in some values, set certain initial and boundary conditions, and expect some proven and general principle to yield a quantitatively supportable result. If bioethical principles were universally upheld by everyone at all times, then there would be no need to justify them. For example, some would say that morality should be based on suffering. Many rationales for and against bioethical issues are based on pain and suffering: we do not need to justify that pain is, by definition, unpleasant. But, does unpleasantness always translate into a negative? No. Indeed, pain is an inherent defense mechanism. Without it animals would repeat many harmful behaviors. However, we can agree that unnecessary pain is bad. The point here is that the response itself is not pleasant and, if given the choice, most of us would choose not to have pain. Furthermore, psychopathologists may note that masochists "enjoy" pain. This is probably a learned activity, overcoming the innate avoidance response. It points out that even supposedly human perspectives that enjoy consensus seldom have unanimity of thought. Thus, even though pain has value, and is still something to be avoided if possible, it does not inherently support or reject bioethical principles.

We cannot rely completely on scientific reasoning in ethical decision making because unethical behavior manifests itself a number of ways, many of which are difficult to quantify. One of the challenges of equal opportunity and affirmative action, for instance, is that the metrics appear to be quantitative, but often are not (at least using mathematics that are better suited for the physical sciences). Thus, there are ethical disagreements about whether an act or a policy can be classified as unfair. One way to start at least to address ethical disagreement is to be honest about the facts. To do so, factual information needs to be separated from opinion and from moral evaluation (see Teachable Moment: Abortion, Fairness, and Justice). For example, Lawrence Hinman, Director of the Values Center at the University of California – San Diego, has recommended that to disentangle ethical disagreements, we must first ask five questions:[23]

1. What is the present state?
2. What is the ideal state?
3. What is the minimally acceptable state?
4. How do we get from the present to the minimally acceptable state?
5. How do we get from the minimum to the ideal state?

Moving from a present, unjust state to the minimally acceptable state often requires direct coercion (e.g., by legislation, law enforcement, government action, etc.). However, moving from the minimal state to the ideal state usually entails non-coercive approaches, especially incentives, such as recognition, media attention, and delegation of professional trustworthiness. Nevertheless, all of these questions and steps are dependent on reliable and honest exposition of the pertinent facts.

Teachable Moment: Abortion, Fairness, and Justice

Constitutional expert, Nat Hentoff recently wrote of a long-time friend, whose husband is a physician who performs abortions:

> At the dinner table one recent evening, their 9-year old son – having head a word whose meaning he didn't know – asked, "What is an abortion?" His mother, choosing her words carefully, described the procedure in simple terms. "But," said her son, "that means killing the baby." The mother then explained that there are certain months during which an abortion cannot be performed, with very few exceptions. The 9-year old shook his head. "But," he said, "it doesn't matter what month. It still means killing the babies."[24]

In the United States, during the past three decades, about 1.8 million abortions were performed, compared to approximately 4 million live births. Most of the abortions were legal. However, as we have seen in other macroethical issues, like slavery, legality does not mean morality. One of the key scientific factors involved in the decision of whether an abortion is moral has to do with whether and when the embryo and fetus are human. Next, if the answer is "yes," the factual premises forming the moral conclusion must address whether the human entity is a person. The logical syllogism then results:

- If the entity (embryo or fetus) is a person, then that person is entitled to specific rights. (Primarily, the right to life and the commensurate right not to be harmed unjustly.) What would right ethics (John Locke), utilitarians (John Stuart Mill), and Rawlsians (John Rawls) say about this condition? And, why do their positions differ?
- In addition, justice requires that all persons are entitled to every right granted by the society to its members. What would the social contractarians (Thomas Hobbes) say about this condition?

The connecting premise is that personhood grounds our rights.

John T. Noonan Jr. in his famous essay, "An Almost Absolute Value in History,"[25] argues that "the most fundamental question involved in the long history of thought on abortion is: How do you determine the humanity of a being?"

His argument and syllogism is:

	1	It is morally wrong to harm another human being without sufficient reason.	
	2	Except in cases of cancerous uterous and ectopic (tubal) pregnancy, abortion harms another human being without sufficient reason.	
Therefore	3	Except in cases of cancerous uterous and ectopic pregnancy, abortion is morally wrong.	**1 & 2**

Adding Noonan's concern that, from conception, the embryo is a human person, the syllogism can be extended as follows:

	1	If x is an act of harming a human being and x has no sufficient reason, then x is morally wrong.	
	2	If x is an abortion then x is an act of harming a human conceptus (embryo, fetus).	
	3	A human conceptus is a human being.	
Therefore	4	If x is an abortion then x is an act of harming a human being.	2 & 3
	5	If x is an abortion then, if x is not a case of cancerous uterous or ectopic pregnancy, then x has no sufficient reason.	
Therefore	6	If x is an abortion and x is not a case of cancerous uterous or ectopic pregnancy, then x is an act of harming a human being and x has no sufficient reason.	4 & 5
Therefore	7	If x is an abortion and x is not a case of cancerous uterous or ectopic pregnancy, then x is morally wrong.	

In the article mentioned earlier, Hentoff further states:

As time goes on, my deepening concern with the consequences of abortion is that its validation by the Supreme Court, as a constitutional practice, helps support the convictions of those who, in other controversies – euthanasia, assisted suicide and the "futility doctrine" by certain hospital ethics committees – believe that there are lives not worth continuing.[26]

Questions

1. Consider Noonan's and Hentoff's factual premises and moral conclusions. Apply Mill's harm principle, Rawls' veil of ignorance, and Kant's categorical imperative to this argument.
2. Do the moral conclusions you reach differ from those of Noonan and Hentoff? Explain (factually and from a social justice perspective) why you agree or disagree.

UTILITY

Utility is an economic term that expresses the amount of satisfaction with goods and services. Utility is often the basis for rational decisions and forms the foundation for all consequentialist ethical theories, most notably utilitarianism.

Engineers apply utility somewhat differently than do ethicists, at least those other than the empiricists. If there is a bias within the modern intellectual community it is toward the scientific "rationalism." However, it should be noted that ethics should not be "force-fitted" into a convenient model, especially one that limits all utility to what can be physically measured. And, this is what the empiricists do.

As mentioned in Chapter 3, resistance against shifting paradigms results from groupthink.[27] For biotechnologies and other interests where bioethics is paramount, finding the right amount of shift is a challenge. For example, refusing to change in the face of compelling evidence is succumbing to groupthink, but changing the paradigm beyond what needs to be changed can be unprincipled, lacking in scientific rigor. Groupthink is a complicated concept. An undergraduate team in one of my recent courses thought of groupthink as a positive concept. While they either did not read or disagreed with the assigned text's discussion of the matter, they made some good points about the value of group thinking in similar ways. One major value is that when a group works together, they have synergies of ideas and economies of scale (Figure 8.5). Their point is well taken. Pluralistic views are often very valuable, but a group can also stifle differing opinions.

The utilitarian view, when carried to the extreme, can be ethically dangerous. The President's Council on Bioethics included this warning in its consideration of the utility of stem cell, cloning, and other "human enhancement" research areas:

> We have tried to conduct our inquiry into human cloning unblinkered, with our eyes open not only to the benefits of the new biotechnologies but also to their challenges—moral, social, and political. We have not suppressed differences but sought rather to illuminate them, that all might better appreciate what is at stake. We have eschewed a thin utilitarian calculus of costs and benefits, or a narrow analysis based only on individual "rights." Rather, we have tried to ground our reflections on the broader plane of human procreation and human healing, with their deeper meanings. Seen in this way, we find that the power to clone human beings is not just another in a series of powerful tools for overcoming unwanted infertility or treating disease.[28]

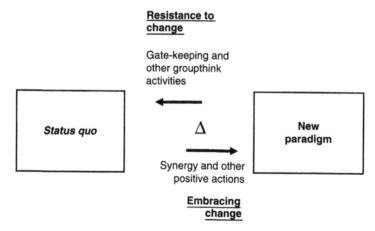

Figure 8.5 Resistance and openness to change: the difference between groupthink and synergy.

Teachable Moment: Utility and Futility

Recently, medical practitioners were asked whether removing a feeding tube is an act of killing. The results are shown in Table 8.1. The respondents generally disagree with the statement, meaning that they believe that they are not killing the patient when they remove the tube. On its face, this appears to be enough to fail them in an introductory biology course, wherein we teach that thermodynamics and energy balances require a sustained food source. If we withhold the energy source (food), we kill the organism. How can language be modified so drastically? Is it to mitigate a "tough choice" or is it simply an example of "groupthink" where gatekeepers ensure that the common ethos is maintained? If we use such strong, albeit accurate terms, like "killing," practitioners may feel obliged to refuse such medical procedures.

One of the factors in the medical caregivers' definition may be that they perceive the patient to be beyond help. So, they tend to label certain patients as such. This is a decision. If we were to draw an event tree, somewhere in the patient's critical path, the patient's treatment changed from efficacious to futile. Since this label change affects the very essence of the treatment, great care must be taken in ascribing it. For example, many disagreed that the care of Terri Schiavo was futile and that withholding food and water was immoral. Reporter, Diane Alden put it this way:

A young woman named Terri Schiavo is on a deathwatch through court-ordered starvation and dehydration. Her death comes courtesy of the efforts of her husband, Michael Schiavo; right-to-die activists like lawyer George Felos and Dr. Ronald Cranford; judges George Greer and Richard Lazzara; and the Florida Supreme Court and the U.S. Supreme Court, which have refused to hear the case.

Terri's parents want her to live, and after viewing the videos and talking to people, so do I.[29]

Table 8.1
Practitioner agreement with the statement "Disconnecting a feeding tube is killing a patient" by nationality and profession*

UK nurses	US medical attendings	US surgical attendings	US house officers	US nurses	US overall
18 (2.22)	11 (1.84)	12 (1.89)	9 (1.62)	12 (1.97)	12 (1.89)

*Values are percentage (with mean score in parentheses) of those agreeing or strongly agreeing, on a scale of 1–5 (1, strongly disagree and 5, strongly agree), with the statement.
Source: D.L. Dickenson, 2001, Practitioner Attitudes in the United States and United Kingdom toward Decisions at the End of Life: Are Medical Ethicists out of Touch? *Western Journal of Medicine*, 174 (2), pp. 103–9, http://www.pubmedcentral.gov/articlerender.fcgi?artid=1071268.

What we call things is important. It literally can be a matter of life and death. Alden goes on to say:

> The various "right to die" organizations apparently are setting us up to accept the notion that it is our responsibility to end our agony at some point – particularly if we become a burden to the system, depressed, old or inconvenient. This particular right will be given to us by a doctor, nurse, medical institution, judge or relative who makes it their mission in life to dispose of us for a host of "humane" reasons.
>
> The list of "quality of life" standards is growing. It is a list consisting of elements on the slippery slope from being humane to being rid of the unwanted. It has evolved from simply withholding heroic efforts to save a brain-dead or terminal human to a laundry list of whether or not we are costly to maintain, unproductive, dependent, burdensome, poor or depressed.

Euphemisms like "quality of life" can be dangerous. For engineers, it could translate into perceiving whole groups or classes of fitness or wellness as not worthy of the effort to save them. They simply become something other than members of the public whose health we must hold paramount.

Engineers by nature hate to give up on a project. Physicians by nature hate to give up on a patient. The point at which it is no longer justifiable to continue treatment is a hard choice. Beyond ethics are legal considerations. Doctors fear patients and their advocates may perceive that insufficient treatment is being provided, even if medicine is no longer efficacious. Treatment may have some effect, but indeed may not be beneficial to the patient. Still the physician may be tempted to continue as a form of "defensive medicine." R.S. Downie, Professor of Philosophy at the University of Glasgow has attempted to characterize the term "medical futility."[30] This is an important bioethical threshold. Among the challenges put forth by Downie is that the definition is influenced by what we value, ethically and economically.[31] One type of definition of futility removes control from the attending physician, and another type is a threat to patients' autonomy. Medicine has to balance these two sometimes conflicting values. Another definition could lead to greater costs to hospitals.

So, then, whose definition should be used? Downie decided on the following definition of a futile action to be one "that cannot achieve the goals of the action, no matter how often it is repeated."[32] This definition is value-laden. What are the goals? Who has final say on which goals have priority? What if the hospital had a goal to move people out of beds as quickly as possible? Would this not create an ethical conflict of interest between the patient and the caregiver; that is, pressure the physician to hasten the status of treatment from efficacious to futile? Downie identifies six reasons that are advanced to justify futile treatment of the hopelessly ill:

1. Fear of the law
2. Conceptual confusion over doctor's beneficence
3. Sanctity of life argument
4. Confusion between treatment and palliation
5. Medical ethos that encourages the development and use of expertise in the name of Samaritan compassion
6. Argument based on the consumer conception of patient autonomy

Questions:

1. Is this an exhaustive list of the reasons to continue treatment in the face of the hopelessly ill? If not, what should be added?
2. Which of these reasons do you support? Explain.

3. In commenting on Downie's article, Vilhjálmur Árnason[33] states that:

> There are at least two other relevant constraints on exercise of choice in Mill's theory of freedom. The first has to with the agent's *competence*. Although the provision of relevant information is a necessary condition for a free choice, it is not sufficient. The agent must as well be able to understand and reflect upon the information. This means that he must not be in a state of mind which is "incompatible with the full use of the reflecting faculty."

Árnason further argues that:

> Downie may well be correct when he argues that for Kant the decision to persist with futile treatment would be irrational and therefore not autonomous. Nevertheless, a Kantian argument could be made for respecting such a decision. The cardinal moral demand of Kantian ethics is that we respect people as persons.

Consider these dissents from utilitarian and deontological perspectives as to whether Mill and Kant would indeed endorse the concept of futile treatment and its use to justify removal of a feeding tube from certain patients.

4. Although Downie's focus on medical futility and much of that of the media has been on the physician, does it apply to engineers as well? For example, a biomedical engineer in a clinical setting may have the ability to adapt a device to be more palliative, but which will not benefit the treatment. Is this worth doing? Is it morally obligatory? Can fear of the law or the engineer's beneficence factor into end-of-life ethics? How?

PRECAUTION AS A BIOETHICAL CONCEPT

Design assumptions may range from paranoid to simply judicious. Unfortunately, the design engineer often does not know that an assumption is not sufficiently protective until after the design phase, so we are expected to err on the side of caution, especially in such situations where the stakes are very high and uncertainties are large. Thus, the prudent course of action may be to take preventative and precautionary measures. The so-called precautionary principle is applied when an activity threatens harm to human health, so that precautionary measures are taken even if some cause and effect relationships are not established scientifically. The concept was first articulated in 1992 as an outcome of the Earth Summit in Rio de Janeiro, Brazil.[34] The proponent of an activity (such as a pharmaceutical company's development of a new chemical or a biotechnology company's research in genetic engineering), rather than the public, bears the burden of proof in these cases.[35] The precautionary principle provides a margin of safety beyond what may be directly construed from science. This is a shift in onus from having to prove that a harm exists to proving that the harm does not exist at the outset. However, some have argued that if the principle is carried to an extreme, it could severely reduce technological advancement because it could severely limit the type of risk taking that has led to many scientific and medical breakthroughs.[36] Perhaps, one way to balance risks is to consider any harm that can result from even very positive outcomes (see Discussion Box: The Tragedy of the Commons).

Discussion box: The Tragedy of the Commons

In his classic work "Tragedy of the Commons," Garrett Hardin gives an example of the individual herder and the utility of a single cow and what is best for the pasture.[37] If everyone takes the egocentric view, the pasture will surely be overgrazed. So, the farmer who will gain immediate financial gain by adding a cow to the herd must decide the utility of the cow *versus* the collective utility of pasture. The utilities, for the herder, are not equal. The individual utility is 1, but the collective utility is less than 1. In other words, the farmer may be aware that the collective cost to each herder on the pasture adding a cow is that overgrazing will cause the pasture to be unproductive for all herders at some threshold. So, the utility becomes inelastic at some point. The damage may even be permanent, or at least it may take a very long time to recover to the point where it may "sustain" any cows including those of the individual herder.

Hardin's parable demonstrates that even though the individual sees the utility of preservation (no new cows) in a collective sense, the ethical egoistic view may well push the decision toward the immediate gratification of the individual at the expense of the collective good. The benefits differ in kind, time, and space from the risks.

Libertarians argue that the overall collective good will come as a result of the social contract. Utilitarianism determines that a moral act should produce the greatest amount of good consequences for the greatest number of beings. Even Mill, however, saw the need for the "harm principle" to counterbalance that temptation to use good ends to rationalize immoral methods, i.e., "ends justifying the means." The harm principle states:

> [T]he sole end for which mankind are warranted, individually or collectively, in interfering with the liberty of action of any of their number, is self-protection. That the only purpose for which power can be rightfully exercised over any member of a civilised community, against his will, is to prevent harm to others. His own good, either physical or moral, is not sufficient warrant. He cannot rightfully be compelled to do or forbear because it will be better for him to do so, because it will make him happier, because, in the opinion of others, to do so would be wise, or even right.... The only part of the conduct of anyone, for which he is amenable to society, is that which concerns others. In the part which merely concerns himself, his independence is, of right, absolute. Over himself, over his own body and mind, the individual is sovereign.[38]

More recently, Rawls conceptualized the "veil of ignorance." Rawls argues that rational people will adopt principles of justice when they reason from general considerations, rather than from their own personal situation.[39] Reasoning without personal perspective and considering how to protect the weakest members comprise Rawls' veil of ignorance. Both the harm principle and the veil of ignorance are buffers against pure ethical egoism; that is, the utilitarian view requires that one not be so self-centered that a personal decision causes harm to another, and the Rawlsian view requires that the weakest members of society be protected from the expressions of free will of others. So, the need to "protect the pasture" must be balanced against decisions based on individual utilities. The tragedy of the commons is that even when we know we are creating harm to others and ourselves, we may still likely decide to take that action if the consequences are to some amophorons society.

Justice and fairness are difficult to define, but are often easier to understand in their absence. They are important design constraints. They even operate at the planetary scale, as we will see in the next chapter.

NOTES AND COMMENTARY

[1] Foreword to *The Children of Light and the Children of Darkness* (New York: Charles Scribner's Sons, 1944).

[2] N. Naurato and T.J. Smith, "Ethical Considerations in Bioengineering Research, " *Biomedical Sciences Instrumentation* 39(2003): 573–8.

[3] These core principles are articulated by T.L. Beauchamp and J.F. Childress, "Moral Norms," in *Principles of Biomedical Ethics*, 5th ed. (New York, NY: Oxford University Press, 2001).

[4] US Department of Health, Education and Welfare, National Commission for the Protection of Human Subjects of Biomedical and Behavioral Research, 1979, *The Belmont Report: Ethical Principles and Guidelines for the Protection of Human Subjects of Research*, 18 April 1979. The notice states:

> The Belmont Report attempts to summarize the basic ethical principles identified by the Commission in the course of its deliberations. It is the outgrowth of an intensive four-day period of discussions that were held in February 1976 at the Smithsonian Institution's Belmont Conference Center supplemented by the monthly deliberations of the Commission that were held over a period of nearly four years. It is a statement of basic ethical principles and guidelines that should assist in resolving the ethical problems that surround the conduct of research with human subjects. By publishing the Report in the Federal Register, and providing reprints upon request, the Secretary intends that it may be made readily available to scientists, members of Institutional Review Boards, and Federal employees.

[5] B. Allenby, 2003, Presentation of this collection as well as the later discussions regarding macro- and micro-ethics, National Academy of Engineering Workshop, "Emerging Technologies and Ethical Issues," Washington, DC, 14–15 October 2003.

[6] "Letter from Birmingham Jail" in Martin Luther King, *Why We Can't Wait* (New York, NY: HarperCollins, 1963).

[7] National Academy of Engineering, *Educating the Engineer of 2020: Adapting Engineering Education to the New Century* (Washington, DC: National Academies Press, 2005).

[8] I would like to thank P. Aarne Vesilind for his insights regarding distributive justice. Some of the foregoing discussion is the result of our conversations and our recent collaboration: D.A. Vallero and P.A. Vesilind, *Socially Responsible Engineering: Justice in Risk Management* (Hoboken, NJ: John Wiley & Sons, 2006).

[9] One early example of distributive justice is the advice of John the Baptizer to the tax collectors and soldiers who asked how to follow the precepts he had laid out. Luke 3:10–14 states:

> "What should we do then?" the crowd asked.
>
> John answered, "The man with two tunics should share with him who has none, and the one who has food should do the same." Tax collectors also came to be baptized. "Teacher," they asked, "what should we do?"
>
> "Don't collect any more than you are required to," he told them.
>
> Then some soldiers asked him, "And what should we do?"
>
> He replied, "Don't extort money and don't accuse people falsely – be content with your pay."

[10] John Rawls, *A Theory of Justice* (1971; repr., Cambridge, MA: Belknap Press, 1999).

[11] Interestingly, John Rawls mentioned earlier is considered to be a "contractarian," as are the libertarians, since both subscribe to a form of social contract theory as posited by Thomas Hobbes. Rawls modulated strict contractarianism by adding the "veil of ignorance" as a protection for weaker members of society.

[12] The concept of free rider shows up on some supply–demand curves in economics. In fact, in a pure supply–demand relationship, the free rider would not exist; that is, since the supply of all goods and services are provided according to their demand (the more the demand, the greater the cost), no person would receive any goods or services without payment.

[13] Nicholas Rescher, *Fairness: Theory and Practice of Distributive Justice* (New Brunswick, CT: Transaction Publishers, 2002).

[14] The concept of treating equals equally and unequals unequally shows up in theological and religious discourses. For example, the late Harry Werner, a Jesuit missionary who lived among native Americans on reservations, noted that the concepts of borrowing and lending are empathic. For example, Werner noted that a person may ask to borrow $10.00, but never return the money. Werner observed this behavior regularly and was informed by the locals that it makes little sense for someone who has little to return money to those who have much. Werner had a car and lived in a decent abode, so until the people had such stations in life, there was no moral obligation to "repay" the money.

    Another instance of treating unequals unequally is the concept of usury. This is mentioned in the Bible, for example, but draws seemingly little attention in Western society. While the contemporary definition is more akin to "gouging," the term "usury" originally meant the charge of any interest on a loan. The practice is prohibited in Islam. In the strictest sense, Jews are prohibited from charging interest on loans to other Jews. St Thomas Aquinas considered the act of charging of interest to be immoral since it charges doubly, i.e., both the item (money) borrowed and the use of the thing. The lender charges for the loan by insisting that the loan be repaid. Thus the repayment is the charge for the loan. Any further charge is a charge for using the loan. To Aquinas, charging interest on a loan is likened to selling a person a bottle of wine and adding another charge if the person actually drinks the wine! Of course, this is moral reasoning and not business acumen; that is, lending can be a moral good, but this goodness is diminished if interest is required.

    These examples illustrate that fairness is a social construction, at least in part.

[15] This is the title of a 1941 film by the comedian W.C. Field. It is similar to the sentiments in the phrase, "There's a sucker born every minute." This phrase, which in its simplest connotation means that people are easy to manipulate, has been erroneously attributed to the circus entrepreneur P.T. Barnum. In fact, the quote was made by his competitor David Hannum. In the late 1860s, it seems Hannum and Barnum were both bidding on what they thought was a petrified giant that was "found" (actually planted by a hoaxer). When Barnum allegedly had his own giant made, Hannum was quoted as lamenting that these people were being fooled, not knowing that he was the original brunt of the hoax. This account was shared by R.J. Brown online at http://www.historybuff.com/library/refbarnum.html (accessed 13 July 2005).

[16] N. Rescher, *Fairness: Theory and Practice of Distributive Justice* (New Brunswick, CT: Transaction Publishers, 2002).

[17] There is, of course, the problem of the majority in a democracy choosing to act immorally. Racial discrimination in the South (and much of the North, for that matter) was for years supported by

the majority, but this did not make it morally right or fair to the African-Americans. We have to assume in this definition that the decisions of fairness are based on defensible moral principles by the popular majority.

[18] The phrase was coined by Alexis de Tocqueville and considered at some length by John Stuart Mills. In a democracy, the majority is very powerful. It can influence and even control the entire people, as what happened in Germany where elected officials, including Hitler, gained and abused power. This was much on the mind of the framers of the United States Constitution, so that the duly elected do not become the tyrants and so that those with little power are not crushed.

[19] C.B. Fleddermann, *Engineering Ethics* (Upper Saddle River, NJ: Prentice-Hall, 1999).

[20] Ibid.

[21] Luke 15: 11–32.

[22] Luke 15: 25–32.

[23] L.M. Hinman, "Understanding Moral Disagreements," 2005, http://ethics.sandiego.edu/presentations/AppliedEthics/Disagreement/Moral%20Disagreement_files/frame.htm (accessed 29 November 2006).

[24] N. Hentoff, *Chapel Hill (NC) Herald*, 15 June 2006, 2.

[25] From John T. Noonan Jr., ed., *The Morality of Abortion: Legal and Historical Perspectives* (Cambridge, MA: Harvard University Press, 1970), 51–9.

[26] Hentoff, 2.

[27] I. Janus, *Groupthink: Psychological Studies of Policy Decisions and Fiascoes*, 2nd ed.(Boston, MA: Houghton Mifflin Company, 1982).

[28] L.R. Kass, The President's Council on Bioethics, 2002, Transmittal Memo to "Human Cloning and Human Dignity: An Ethical Inquiry," Washington, DC.

[29] D. Alden, "Futile Care: The Terri Schiavo Case," *NewsMax.com*, 17 October 2003.

[30] R.S. Downie, "Medical Technology and Medical Futility," *Ends and Means: Journal of the University of Aberdeen, Center for Philosophy, Technology and Society*, 2, no. 2 (1998).

[31] R. Halliday, "Medical Futility and the Social Context," *Journal of Medical Ethics*, 23 (1997), 148–53.

[32] Taken from: L. Schneiderman, M. Jecker, and A. Jonsen, Medical Futility: Its Meaning and Ethical Implications," *Annals of Internal Medicine*, 112 (1990), 949–54.

[33] V. Árnason, "Kant, Mill and Consumer Autonomy: A Response to R.S. Downie," *Ends and Means: Journal of the University of Aberdeen, Center for Philosophy, Technology and Society*, 3, no. 2 (1999).

[34] United Nations Conference on Environment and Development, 1992, Principle 15, Declaration on Environment and Development, Rio de Janeiro, Brazil. The principle reads:

> In order to protect the environment, the precautionary approach shall be widely applied by States according to their capabilities. Where there are threats of serious or irreversible damage, lack of full scientific certainty shall not be used as a reason for postponing cost-effective measures to prevent environmental degradation.

[35] I. Goklany, *The Precautionary Principle* (Washington, DC: Cato Institute, 2001). In Goklany's reference the definition by Carolyn Raffensperger and Joel Tickner is paraphrased.

[36] See J. Morris, *Rethinking Risk and the Precautionary Principle* (Burlington, MA: Butterworth-Heinemann, 2000).

[37] Garrett Hardin, "Tragedy of the Commons," *Science* 162 (13 December 1968).

[38] J.S. Mill, *On Liberty* (London, UK: Longman, Roberts & Green, 1869).

[39] J. Rawls, *A Theory of Justice* (1971; repr., Cambridge, MA: Belknap Press Reprint, 1999).

Chapter 9

Sustainable Bioethics

It is our aspiration that engineers will continue to be leaders in the movement toward the use of wise, informed, and economical sustainable development. This should begin in our educational institutions and be founded in the basic tenets of the engineering profession and its actions.

National Academy of Engineering (2004)[1]

GREEN IS GOOD

Ethical engineering is socially responsible engineering.

Harkening back to the advice of the founder of bioethics, Van Rensselaer Potter II, ensuring a livable and healthy environment is a moral imperative. Environmental conscientiousness evolved in the twentieth century from a peculiar interest of a few committed advocates to an integral part of every engineering disciple. In fact, one of the most important macroethical challenges for engineers is to provide sustainable designs. The US Environmental Protection Agency defines "green engineering" as:

> [T]he design, commercialization and use of processes and products that are feasible and economical while reducing the generation of pollution at the source and minimizing the risk to human health and the environment.[2]

Green engineering is also a bioethical tenet, wherein "environmentally conscious attitudes, values, and principles, combined with science, technology, and engineering practice, all directed toward improving local and global environmental quality."[3]

Green has become recognized as a code for sustainable programs. So, the "green engineer" is no longer a term for a neophyte to the profession (opposite of a "grey beard"), it is now more likely to mean an environmentally oriented engineer.

One of the principles of green engineering is the recognition of the importance of "sustainability."

Bioethics Question: What is the role of the engineer in sustaining the environment?

SUSTAINABILITY

Their recognition of an impending and assured global disaster led the World Commission on Environment and Development, sponsored by the United Nations, to conduct a study of the world's resources. Also known as the Brundtland Commission, their 1987 report, *Our Common Future*, introduced the term "sustainable development" and defined it as "development that meets the needs of the present without compromising the ability of future generations to meet their own needs."[4] The United Nations Conference on Environment and Development (UNCED), i.e., the Earth Summit held in Rio de Janeiro in 1992, communicated the idea that sustainable development is both a scientific concept and a philosophical ideal. The document, *Agenda 21*, was endorsed by 178 governments (not including the United States) and was hailed as a blueprint for sustainable development. In 2002, the World Summit on Sustainable Development (WSSD) identified the major areas that are considered key to moving sustainable development plans forward.

The underlying purpose of sustainable development is to help developing nations manage their resources, such as rain forests, without depleting these resources and making them unusable for future generations. In short, the objective is to prevent the collapse of the global ecosystems. The *Brundtland Report* presumes that we have a core ethic of intergenerational equity, and that future generations should have an equal opportunity to achieve a high quality of life. The report is silent, however, on just why we should embrace the ideal of intergenerational equity, or why one should be concerned about the survival of the human species. As such, sustainability is not only a response to biological threats at the planetary scale, it is a statement of an inherent macroethical principle: Human life is valuable and is worthy of protection. The goal is a sustainable global ecological and economic system, achieved in part by the wise use of available resources.

We are creatures that have different needs. Maslow[5] articulated this as a hierarchy of needs consisting of two classes of needs: basic and growth (Figure 9.1). The basic needs must first be satisfied before a person can progress toward higher level growth needs. Within the basic needs of classification, Maslow separated the most basic physiological needs, such as water, food, and oxygen, from the need for safety. Therefore, one must first avoid starvation and thirst, satisfying minimum caloric and water intake, before being concerned about the quality of air, food, and water. The latter is the province of environmental protection, yet it is also clearly a statement of biosystematics and bioethics. The most basic of needs must first be satisfied before we can strive for more advanced needs. Thus, we need to ensure adequate quantities and certain ranges of the quality of air, water, and food. Providing food requires ranges of quantity and quality of soil and water for agriculture. Thus, any person and any culture that is unable to satisfy their most basic needs cannot be expected to "advance" toward higher order values, such as free markets and peaceful societies. In fact, the inability to provide basic needs militates against a peaceful society. This means that when basic needs go unmet, societies are frustrated even if they strive toward freedom and peace. And, even those that begin may enter into vicious cycles wherein any progress is

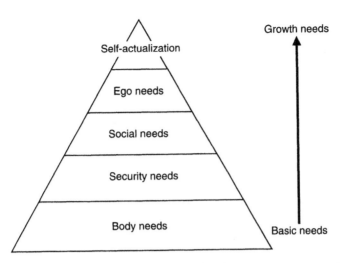

Figure 9.1 Maslow's hierarchy of needs. The lower part of the hierarchy (i.e., basic needs) must first be satisfied before a person can advance to the next growth levels. Bioengineering directly addresses body, security, and social needs.

undone by episodes of scarcity. We generally think of peace and justice as the province of religion and theology, but engineers will increasingly be called upon to "build a better world."

Even mechanical engineers, whom we may at first blush think of as being mainly concerned about nonliving things, are embracing sustainable design in a large way. In many ways the mechanical engineering profession is out in front on sustainable design. For example, the American Society of Mechanical Engineers (ASME) website draws a systematic example from ecology:

> To an engineer, a sustainable system is one that is in equilibrium or changing at a tolerably slow rate. In the food chain, for example, plants are fed by sunlight, moisture and nutrients, and then become food themselves for insects and herbivores, which in turn act as food for larger animals. The waste from these animals replenishes the soil, which nourishes plants, and the cycle begins again.[6]

Therefore, sustainability is a systematic phenomenon; so, it is not surprising that engineers have embraced the concept of "sustainable design." At the largest scale, manufacturing, transportation, commerce, and other human activities that promote high consumption and wastefulness of finite resources cannot be sustained. At the individual designer scale, the products and the processes that engineers design must be considered for their entire lifetime and beyond.

Teachable Moment: Rational Ethics and Thermodynamics

As mentioned in Chapter 8, in 1968 Garrett Hardin wrote the hugely influential article entitled *The Tragedy of the Commons*, which has become a must-read in every ecology course. Recall that Hardin imagines an English village with a common area where everyone's cow may graze.

The common area is able to sustain the cows and village life is stable, until one of the villagers figures out that if he gets two cows instead of one, the cost of the extra cow will be shared by everyone, while the profit will be his alone. So he gets two cows and prospers, but others see this and similarly want two cows. If two, why not three – and so on – until the village common area is no longer able to support the large number of cows, and everyone suffers. Hardin applies this to the problem of birth control, arguing that the environmental cost of a child born today (air, water, and other resources) is shared by everyone, while the benefit is only to the parents. Hence, every family will want more children, until the ability of the earth to support these populations is exceeded, resulting in global disaster. The flaw in such thinking is the same as that expressed by Malthus and Ehrlich discussed earlier. It does not account for human ingenuity. Nor does it respect the morality articulated in various belief systems. However, many of these faith traditions call for "stewardship" of the environment. Many of the anti-religious perspectives show unease with stewardship, possibly because it places primacy on human values (the anthropocentric view) rather than the biodiversity of other species (the biocentric view). Nonetheless, even if we humans put ourselves first among creatures, the systematic utilitarianism forces us to sustain a rich and biodiverse planet. This is indeed the "spaceship earth" or closed and isolated thermodynamic system perspective.

If we treat diminishing resources, such as oil and minerals, as capital gains, we will soon find ourselves in the "common" difficulty of having too many people and not enough resources. This is the macroethical challenge of what it means to be good stewards.

A thread running all through Hardin's books is that ethics has to be based on rational argument and not on emotion. He argues that for ethics to be useful, people have to be literate; they must use words correctly, and they must appreciate the power of numbers. His interesting book *Stalking the Wild Taboo* takes on any number of social misconceptions that demand rational reasoning. The problem with his application of absolute rationalism is that ethics does not completely concur with the scientific method. Ethical principles cannot usually be derived from measurable data.

Questions

1. What is your personal ecological footprint?
2. What steps can you take to lighten it?

Hardin's parable is useful in that it demonstrates that even though the individual sees the utility of preservation (no new cows) in a collective sense, the ethical egoistic view may well push the decision toward the immediate gratification of the individual at the expense of the collective good. Ironically, yet predictably, this view can result in large-scale harm (i.e., pollution and waste of resources).

Sustainability requires adopting new and better means of using materials and energy. The operationalizing of the quest for sustainability is "green engineering," which recognizes that engineers are central to the practical application of the principles of sustainability to everyday life. The relationship between sustainable development, sustainability, and green engineering is progressive:

$$\text{Sustainable development} \rightarrow \text{Green engineering} \rightarrow \text{Sustainability} \qquad (9.1)$$

Sustainable development is an ideal that can lead to sustainability, but this can only be done through green engineering. Systematics require that we close the loop:

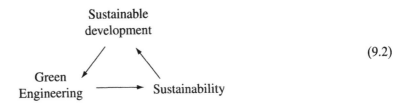

$$(9.2)$$

Green engineering[7] treats environmental quality as an end in itself. The US EPA definition given earlier includes another element, the systematic view:

> The discipline embraces the concept that decisions to protect human health and the environment can have the greatest impact and cost effectiveness when applied early to the design and development phase of a process or product.[8]

Thus, green engineering must be built into the product, building, or any system being designed. Green engineering approaches are being linked to improved computational abilities (Table 9.1) and other tools that were not available at the outset of the environmental movement. Increasingly, decision makers have come to recognize that improved efficiencies save time, money, and other resources in the long run. This includes decision making at biomedical facilities (e.g., "green hospitals"). Hence, companies are thinking systematically about the entire product stream in numerous ways:

- Applying sustainable development concepts, including the framework and foundations of "green" design and engineering models
- Applying the design process within the context of a sustainable framework, including considerations of commercial and institutional influences
- Considering practical problems and solutions from a comprehensive standpoint to achieve sustainable products and processes
- Characterizing waste streams resulting from designs
- Understanding how first principles of science, including thermodynamics, must be integral to sustainable designs in terms of mass and energy relationships, including reactors, heat exchangers, and separation processes
- Applying creativity and originality in group product and building design projects.

There are numerous industrial, commercial, and governmental green initiatives, including design for the environment (DFE), design for disassembly (DFD), and design for recycling (DFR).[9] These are replacing or at least changing pollution control paradigms. For example, the concept of a "cap and trade" has been tested and works well for some pollutants. This is a system where companies are allowed to place a "bubble" over a whole manufacturing complex or trade pollution credits with other companies in their industry instead of a "stack-by-stack" and "pipe-by-pipe" approach, i.e., the so-called command and control approach. Such policy and regulatory innovations call for some improved technology-based approaches as well as better quality-based approaches, such as leveling out the pollutant loadings and

Table 9.1

Biosystem and biomedical aspects of principles of green engineering programs

Principle	Description	Example	Biosystem and biomedical aspects
Waste prevention	Design chemical syntheses and select processes to prevent waste, leaving no waste to treat or clean up	Use a water-based process instead of an organic solvent-based process	Bioinformatics and data mining can provide candidate syntheses and processes
Safe design	Design products to be fully effective, yet have little or no toxicity	Using microstructures, instead of toxic pigments, to give color to products. Microstructures bend, reflect, and absorb light in ways that allow for a full range of colors	Systems biology and "omics" technologies (i.e., genomics, proteomics, metabanonics) can support predictions of cumulative risk from products used in various scenarios
Low hazard chemical synthesis	Design syntheses to use and generate substances with little or no toxicity to humans and the environment	Select chemical synthesis with toxicity of the reagents in mind upfront. If a reagent ordinarily required in the synthesis is acutely or chronically toxic, find another reagent or new reaction with less toxic reagents	Computational chemistry and biology can help to predict unintended product formation and reaction rates of optional reactions
Renewable material use	Use raw materials and feedstocks that are renewable rather than those that deplete nonrenewable natural resources. Renewable feedstocks are often made from agricultural products or are the wastes of other processes; depleting feedstocks are made from fossil fuels (petroleum, natural gas, or coal) or that must be extracted by mining	Construction materials can be from renewable and depleting sources. Linoleum flooring, e.g., is highly durable, can be maintained with nontoxic cleaning products, and is manufactured from renewable resources amenable to being recycled. Upon demolition or reflooring, the linoleum can be composted	Systems biology, informatics, and "omics" (e.g., genomics and proteomics) technologies can provide insights into the possible chemical reactions and toxicity of the compounds produced when switching from depleting to renewable materials

Catalysis	Minimize waste by using catalytic reactions. Catalysts are used in small amounts and can carry out a single reaction many times. They are preferable to stoichiometric reagents, which are used in excess and work only once	The Brookhaven National Laboratory recently reported that it has found a "green catalyst" that works by removing one stage of the reaction, eliminating the need to use solvents in the process by which many organic compounds are synthesized. The catalyst dissolves into the reactants. Also, the catalyst has the unique ability of being easily removed and recycled, because, at the end of the reaction, the catalyst precipitates out of the products as a solid material, allowing it to be separated from the products without using additional chemical solvents[a]	Computation chemistry can help to compare rates of chemical reactions using various catalysts
Avoiding chemical derivatives	Avoid using blocking or protecting groups or any temporary modifications if possible. Derivatives use additional reagents and generate waste	Derivativization is a common analytical method in environmental chemistry, i.e., forming new compounds that can be detected by chromatography. However, chemists must be aware of possible toxic compounds formed, including leftover reagents that are inherently dangerous	Computational methods and natural products chemistry can help scientists start with a better synthetic framework
Atom economy	Design syntheses so that the final product contains the maximum proportion of the starting materials. There should be few, if any, wasted atoms	Single atomic- and molecular-scale logic used to develop electronic devices that incorporate design for disassembly, design for recycling, and design for safe and environmentally optimized use	The same amount of value, e.g., information storage and application, is available on a much smaller scale. Thus, devices are smarter and smaller, and more economical in the long term. Computational toxicology enhances the ability to make product decisions with better predictions of possible adverse effects, based on the logic

(Continued)

Table 9.1
Biosystem and biomedical aspects of principles of green engineering programs—cont'd

Principle	Description	Example	Biosystem and biomedical aspects
Nano-materials	Tailor-made materials and processes for specific designs and intent at the nanometer scale (≤ 100 nanometer)	Emissions, effluent, and other environmental controls; design for extremely long life cycles. Limits and provides better control of production and avoids overproduction (i.e., "throwaway economy")	Improved, systematic catalysis in emission reductions, e.g., large sources like power plants and small sources like automobile exhaust systems. Zeolite and other sorbing materials used in hazardous waste and emergency-response situations can be better designed by taking advantage of surface effects; this decreases the volume of material used
Selection of safer solvents and reaction conditions	Avoid using solvents, separation agents, or other auxiliary chemicals. If these chemicals are necessary, use innocuous chemicals	Supercritical chemistry and physics, especially that of carbon dioxide and other safer alternatives to halogenated solvents, are finding their way into the more mainstream processes, most notably dry cleaning	To date, most of the progress has been the result of wet chemistry and bench research. Computational methods will streamline the process, including quicker "scale-up"
Improved energy efficien-cies	Run chemical reactions and other processes at ambient temperature and pressure whenever possible	To date, chemical engineering and other reactor-based systems have relied on "cheap" fuels and, thus, have optimized on the basis of thermodynamics. Other factors, e.g., pressure, catalysis, photovoltaics, and fusion, should also be emphasized in reactor optimization protocols	Heat will always be important in reactions, but computational methods can help with relative economies of scale. Computational models can test feasibility of new energy-efficient systems, including intrinsic and extrinsic hazards, e.g., to test certain scale-ups of hydrogen and other economies. Energy behaviors are scale dependent, for example, recent measurements of H_2SO_4 bubbles when reacting with water have temperatures in the range of those found on the surface of the sun[b]

Design for degradation	Design chemical products to break down to innocuous substances after use so that they do not accumulate in the environment	Biopolymers, e.g., starch-based polymers, can replace styrene and other halogen-based polymers in many uses. Geopolymers, e.g., silane-based polymers, can provide inorganic alternatives to organic polymers in pigments, paints, etc. These substances, when returned to the environment, retain their original parent form	Computation approaches can simulate the degradation of substances as they enter various components of the environment. Computational science can be used to calculate the interplanar spaces within the polymer framework. This will help to predict persistence and to build environment friendly products, e.g., those where space is adequate for microbes to fit and biodegrade the substances
Real-time analysis to prevent pollution and concurrent engineering	Include in-process real-time monitoring and control during syntheses to minimize or eliminate the formation of byproducts	Remote sensing and satellite techniques can provide link to real-time data repositories to determine problems. The application to terrorism using nanoscale sensors is promising	Real-time environmental mass spectrometry can be used to analyze whole products, obviating the need for any further sample preparation and analytical steps. Transgenic species, while controversial, can also serve as biological sentries, e.g., fish that changes colors in the presence of toxic substances
Accident prevention	Design processes using chemicals and their forms (solid, liquid, or gas) to minimize the potential for chemical accidents, including explosions, fires, and releases to the environment	Scenarios that increase the probability of accidents can be tested	Rather than waiting for an accident to occur and conducting failure analyses, computational methods can be applied in prospective and predictive mode; that is, the conditions conducive to an accident can be characterized computationally

First two columns, except "Nanomaterials" adapted from: US Environmental Protection Agency, 2005, "Green Chemistry," http://www. epa.gov/greenchemistry/principles.html; accessed 12 April 2005. Other information from discussions with Michael Hays, US EPA, National Risk Management Research Laboratory, 28 April 2005.

[a] US Department of Energy, *Research News*, http://www.eurekalert.org/features/doe/2004-05/dnl-brc050604.php; accessed 22 March 2005.
[b] D.J. Flannigan and K.S. Suslick, 2005, Plasma formation and temperature measurement during single-bubble cavitation, *Nature*, 434, pp. 52–5.

using less expensive technologies to remove the first large bulk of pollutants, followed by higher operation and maintenance (O&M) technologies for the more difficult to treat stacks and pipes. But, the net effect can be a greater reduction of pollutant emissions and effluents than treating each stack or pipe as an independent entity. This is a foundation for most sustainable design approaches, i.e., conducting a life cycle analysis, prioritizing the most important problems, and matching the technologies and operations to address them. The problems will vary by size (e.g., pollutant loading), difficulty in treating, and feasibility. The easiest ones are the big ones that are easy to treat (so-called low hanging fruit). You can do these first with immediate gratification! However, the most intractable problems are often those that are small but very expensive and difficult to treat, i.e., less feasible. Of course, as with all paradigm shifts, expectations must be managed from both a technical and an operational perspective. Not the least, the expectations of the client, the government, and those of the individual engineer must be realistic in how rapidly the new approaches can be incorporated.

Historically, environmental considerations have been approached by engineers as constraints on their designs. For example, hazardous substances generated by a manufacturing process were dealt with as a waste stream that must be contained and treated. The hazardous waste production had to be constrained by selecting certain manufacturing types, increasing waste handling facilities, and if these did not entirely do the job, limiting rates of production. Green engineering emphasizes the fact that these processes are often inefficient economically and environmentally, calling for a comprehensive, systematic life cycle approach. Green engineering attempts to achieve four goals:

1. Waste reduction
2. Materials management
3. Pollution prevention
4. Product enhancement

This can be particularly challenging in biomedical venues, where patient and provider health and safety hold primacy. Green techniques must not detract or diminish these primary missions. Thus, sustainable designs must be incorporated into medical systems, not the other way around. A tool to do this and achieve all four green engineering goals is to take a long-term life cycle point of view.

LIFE CYCLES AND CONCURRENT ENGINEERING

A life cycle analysis is a holistic approach to consider the entirety of a product, process or activity, encompassing raw materials, manufacturing, transportation, distribution, use, maintenance, recycling, and final disposal. In other words, assessing its "life cycle" should yield a complete picture of the product. It is a closed system wherein each stage affects and is affected by the subsequent and previous stages.

The first step in a life cycle assessment is to gather data on the flow of a material through an identifiable society. Once the quantities of various components of such a flow are known, the important functions and impacts of each step in the production, manufacture, use, and recovery/disposal are estimated.

A relatively new approach to engineering is to design and manufacture a product simultaneously rather than sequentially, known as "concurrent engineering." This approach may allow environmental improvements under real-life manufacturing conditions. However, changes made in any step must consider possible effects on the rest of the design and implementation.

Case Study Box: SIDS, A Concurrent Engineering Failure

One of the most perplexing and tragic medical mysteries of the past fifty years has been sudden infant death syndrome (SIDS). The syndrome was first identified in the early 1950s.

Numerous etiologies have been proposed for SIDS, including a number of environmental causes. A recent study, for example, found a statistically significant link between exposure of newborn infants to fine aerosols and SIDS.[10] The study found that approximately 500 of the 3800 SIDS cases in 1994 were associated with elevated concentrations of particle matter with aerodynamic diameters less than 10 microns (PM_{10}) in the United States. This estimate is based only on metropolitan areas in counties with standard PM_{10} monitors. Based on the metropolitan area with the lowest particle concentrations, there appears to be a threshold; that is, particulate-related infant deaths occurred at PM_{10} levels below $11.9\,\mu g\ m^{-3}$.

Extrapolations from these data show that almost 20% of all SIDS cases each year in the top twelve most polluted metro areas in the United States are associated with PM_{10} pollution. The number of annual SIDS cases associated with PM_{10} in Los Angeles, New York, Chicago, Philadelphia, and Detroit metropolitan areas ranges from 20 to 44. The study found that 10 states accounted for more than 60% of the particle-related SIDS cases, with 93 in California, 37 in Texas, and 32 in Illinois.

Since particle matter has been linked to SIDS cases, a logical extension would be to suspect the role of environmental tobacco smoke (i.e., "side stream" exposure) in some cases, since this smoke contains both particulate- and gas-phase contaminants that are released into the infant's breathing zone. Also, when a pregnant woman smokes *in utero* exposures to toxic substances (e.g., nicotine and other organic and inorganic toxins) may make the baby more vulnerable.

Another suspected etiology for SIDS is the exposure to pollutants via consumer products. For example, polyvinyl chloride (PVC) products have been indirectly linked to SIDS. The most interesting link is not the PVC itself, but the result of an engineering "solution."

Plastics came into their own in the 1950s, replacing many other substances, because of their ease of fabrication, light weight and durability. However, being a polymer, physical and chemical conditions affect the ability of PVC to stay "hooked together." This can be a big problem for plastics used for protection, such as waterproofing. One such use was as a tent material.

Serendipity often plays a role in linking harmful effects to possible causes. In 1988, Barry Richardson was in the process of renting a tent for his daughter's wedding. Richardson, an expert in material science and deterioration, while renting a tent from proprietor Peter Mitchell inquired about its durability and found that PVC tents tend to break down. Richardson surmised that the rapid degradation was microbial and in fact due to fungi. The tent manufacturers decided to correct the PVC durability by changing the manufacturing process, i.e., by concurrent engineering. In this case, they decided to increase the amount of fungicide, 10,10'-oxybis(phenoxarsine) (OBPA).

10-10'-Oxybis (phenoxarsine)

A quick glance at the OBPA structure shows that when it breaks down it is likely to release arsenic compounds. In this case, it is arsine (AsH_3), a toxic gas (vapor pressure = 11mm Hg at 20 °C). It is rapidly absorbed when inhaled, and easily crosses the alveolo-capillary membrane and enters red blood cells. Arsine depletes the reduced glutathione content of red blood cells, leading to the oxidation of sulfhydryl groups in hemoglobin and, possibly, red cell membranes. These effects produce membrane instability with rapid and massive intravascular hemolysis. It also binds to hemoglobin, forming a metalloid–hemoglobin complex.[11] These can lead to acute cardiovascular, neurotoxic, and respiratory effects.

Increasing the OBPA to address the problem of PVC disintegration is an example of the problem of ignoring the life cycle and systematic aspects of most engineering problems. In this case, production and marketing would greatly benefit from the type of PVC that does not break down readily under ambient conditions. In fact, if that problem cannot be solved, the entire camping market might be lost, since fungi are ubiquitous in the places where these products are used. But was this decision ethical?

Had the engineers and planners considered the chemical structure and the possible uses, however, they at least might have restricted the PVC treated with high concentrations of OBPA to certain uses, such as only on tent materials, and not in materials that come in contact or near humans (bedding materials, toys, etc.). On the contrary, the PVC manufacturers blatantly disregarded the underpinning science. Richardson, the expert, from the outset had warned that increasing the amount of fungicide would not only increase the hazard and risk, but would make the product less efficacious (even more vulnerable to fungal attack). He stated, "The biocide won't kill this fungus – instead, the fungus will consume the biocide as well as the plasticizer. Since the biocide contains arsenic, the fungus will generate a very poisonous gas which would be harmful to your staff working with the marquees." Plasticizers are semivolatile organic compounds (e.g., phthalates) that can serve as a food source for microbes, once they become acclimated. The engineers should have known this, since it is one of the biological principles upon which much wastewater treatment is based. But, the manufacturers wanted to approach the situation as a linear problem with a simple solution, i.e., increase fungicide and decrease fungus. Further, the PVC manufacturer argued that the fungicide was even approved for use in baby mattresses.

The extent to which arsine gas released by the degradation of OBPA was a causative agent in SIDS cases is a matter of debate. But, the fact that a toxic gas could be released leading to exposures of a highly susceptible population (babies) is not debatable. The decision was clearly unethical and would not likely stand against the "reasonable engineer standard."

Pollution and consumer products are only some of the possible causes of SIDS. Others include breathing position (probably increased carbon dioxide inhalation), poor nutrition, and physiological stress (e.g., overheating).[12]

The overall lesson is that there are many advantages to concurrent engineering, such as real-time feedback between design and build stages, adaptive approaches, and continuous improvement. However, concurrent engineering works best when the entire life cycle is considered. The designer must ask how even a small change to improve one element in the process can affect other steps and systems within the design and build process.

Life cycle analyses are performed for several reasons, including the comparison of products for purchasing and the comparison of products by industry. In a biomedical setting, the total effect of reusable probe, for example, could be compared to the effect of single-use probe. If all of the factors going into the manufacture, distribution, and disposal of both types of sensors are considered, one probe might be shown to be clearly superior. However, this requires a "sensitivity analysis." For example, certain single-use devices are absolutely necessary in medical practice because of health and safety consideration (e.g., communicable disease controls involving sharps and other hazardous devices). While the other factors may support a reuse device, the design specifications are so sensitive to safety that this factor overrides all the others.

Thus, life cycle analyses, at least in biomedical venues, are akin to an index which is a mathematical means of combining a number of variables into a single indicator. Indices can give information about larger systems, such as an air quality index for a metropolitan area or an ecosystem condition index.

It is not my intent to try to make engineers into ecologists. Rather, ecologists are quite adept at forming indices that allow for sensitivity analyses of a wide array of factors. For example, the estuarine health index in Table 9.2 actually embeds a number of other indices, which themselves digest numerous variables. One of the weaknesses of an index is that it may lose information in the process of consolidating data. For example, even if chlorophyll, phosphorous, and nitrogen variables are considered adequate, but if the dissolved oxygen concentrations fall below a certain value (e.g., $DO < 2 \text{ mg } l^{-1}$), the estuary would be very unhealthy. Likewise, in a biomedical analogy, if one's heart rate, glucose, and blood pressure are normal, but breathing is well below a healthy rate, that person's health is threatened, even if the index "washes out" or dilutes the impact of poor respiration. In other words, the index is not sufficiently sensitive to the health effect of respiration. Likewise, life cycle assessments not only need to choose the important variables carefully, but also include an understanding of how these variables interrelate. Again, the sensitivity analysis can be very useful in this regard.

To illustrate an informal green index, we might determine if the use of phosphate builders in detergents is more detrimental than the use of substitutes that have their own problems in treatment and disposal. The variables would be weighted according to their importance and the best option selected.

One problem with such studies is that they are often conducted by industry groups or individual corporations, and (surprise!) the results often promote their own product. For example, Proctor & Gamble, the manufacturer of a popular brand of disposable baby diapers, found in a study conducted for them that the cloth diapers consume three times more energy than the disposable kind. But a study by the National Association of Diaper Services found disposable diapers consume 70% more energy than cloth diapers. The difference was in the accounting procedure. If one uses the energy contained in the

Table 9.2
Estuarine health index results, based on raw values

GOOD ESTUARINE HEALTH $\longrightarrow$ POOR ESTUARINE HEALTH

	Sinepuxent Bay	Chincoteague Bay	Assawoman Bay	Isle of Wight Bay	Newport Bay	St Martin River
Water quality						
Water quality index[1]	0.85	0.74	0.33	0.53	0.35	0.33
Chlorophyll a (μg l^{-1})[2]	5	5	15	11	15	16
Total nitrogen (mg l^{-1})[2]	0.35	0.54	1.19	0.84	2.08	1.93
Total phosphorus (mg l^{-1})[2]	0.04	0.04	0.05	0.05	0.07	0.09
Dissolved oxygen (mg l^{-1})[2]	6.1	6.1	6.1	5.6	6.0	5.5
Brown tide (max. cells μl^{-1})[3]	35–200	>200	35–200	35–200	>200	35–200
Macroalgal biomass (max. g m^{-2})[4]	50	320	100	250	10	390
Living resources						
Benthic index[5]	3.5	3.6	3.4	3.1	3.4	2.2
Hard clam density (clams m^{-2})[6]	0.32	0.27	0.16	0.28	0.14	0.04
Sediment toxicity[7]	10	8	12	11	13	19
Habitat						
Seagrass area (percentage of bay)[8]	36	32	8	5	4	<1
Wetland area (percentage of watershed)[9]	61	45	45	16	23	16
Natural shoreline (percentage of total)[10]	81	98	72	35	96	52

[1] Ranges from 0 (no reference criteria met) to 1 (all criteria met). Calculated from chlorophyll a, total nitrogen & phosphorus and dissolved oxygen.

[2] Medians of monthly measurements from 2001 to 2003, from 57 sites.

[3] Maximum values, monitored since 1999 at 15 sites.

[4] Survey of 388 sites throughout the Coastal Bays in 2001 and 2003.

[5] Combines a range of benthic fauna measurements from 54 sites between 2000 and 2001. Range is from 1 (poor) to 5 (good).

[6] Averages from 1994 to 2000 from a total of 1499 sites.

[7] Apparent effect threshold-combines critical levels of a range of toxicants, measured between 1991 and 1996 from >900 sites.

[8] 2002 aerial photographic survey.

[9] Survey carried out in 1988 and 1989.

[10] Aerial photographic survey carried out in 1989.

Source: Maryland Department of Natural Resources, 2004, "Maryland's Coastal Bays Ecosystem Health Assessment," Report No. DNR-12-1202-0009, Annapolis, MD.

disposable diaper as recoverable in a waste-to-energy facility, then the disposable diaper is more energy efficient.[13]

As we have seen, bioethical decisions require reliable information, but some of the information critical to the calculations is virtually impossible to obtain. For example, something as simple as the tonnage of solid waste collected in the United States is not readily calculable or measurable. And even

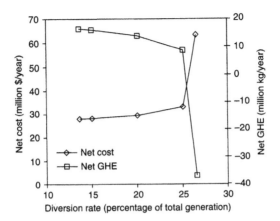

Figure 9.2 The cost in dollars and adverse environmental impact increase dramatically when the fraction of solid waste recycled exceeds 25%. An ethical decision print appears near the 25% diversion rate and the $30 million per year net cost.
Adapted from: E. Solano, R.D. Dumas, K.W. Harrison, S. Ranjithan, M.A. Barlaz, and E.D. Brill, *Integrated Sold Waste Management Using a Life-Cycle Methodology for Considering cost, Energy, and Environmental Emissions – 2. Illustrative Applications.* Department of Civil Engineering, North Carolina State University, Raleigh, NC.

if the data were there, the procedure suffers from the unavailability of a single accounting system. Is there an optimal level of pollution, or must all pollutants be removed 100% (a virtual impossibility)? If there is air pollution and water pollution, how must these be compared?

A recent study supported by the US EPA developed complex models using principles of life cycle analysis to estimate the cost of materials recycling. The models were able to calculate the dollar cost, as well as the cost in environmental damage caused at various levels of recycling. Contrary to intuition, and the stated public policy of the US EPA, it seems that there is a breakpoint at about 25% diversion; that is, as shown in Figure 9.2, the cost in dollars and adverse environmental impact start to increase at an exponential rate at about 25% diversion. Should we therefore even strive for greater diversion rates, if this results in unreasonable cost in dollars and actually does harm to the environment?

Discussion Box: The Coffee Cup Debate

A simple example of the difficulties in life cycle analysis is illustrated in a funny, yet revealing television commericial. A young man is standing at a grocery store checkout and is asked: "Paper or plastic?" We see his thoughts: Pulp mills, trees being cut down, etc. *versus* chemical plants and waste dumps. Similarly, much can be learned by one's morning decision – whether to use paper coffee cups or polystyrene coffee cups. The answer most people would give is not to use either, but instead to rely on the permanent mug. But, nevertheless, there are times when disposable cups are necessary (e.g., in hospitals), and a decision must be made as to which type to choose.[14] So let us use a crude life cycle analysis to make a decision.

The paper cup comes from trees, but the act of cutting trees results in environmental degradation. The foam cup comes from hydrocarbons such as oil and gas and this also results in adverse environmental impact, including the use of nonrenewable resources. The production of the paper cup results in significant water pollution, while the production of the foam cup contributes essentially no water pollution. The production of the paper cup results in the emission of chlorine, chlorine dioxide, reduced sulfides and particulate, while the production of the foam cup results in none of these. The paper cup does not require chlorofluorocarbons (CFCs), but neither do the newer foam cups ever since the CFCs in polystyrene were phased out. However, the foam cups results in the emission of pentane, while the paper cup contributes none. From the material separation perspective, the recyclability of the foam cup is much higher than the paper cup since the latter is made from several materials, including the plastic coating on the paper. They both burn well, although the foam cup produces 17 200 Btu/lb (40 000 kJ kg^{-1}) while the paper cup produces only 8600 Btu/lb (20 000 kJ kg^{-1}). In the landfill, the paper cup degrades into CO_2 and CH_4, both greenhouse gases, while the foam cup is much more inert. Since it is inert, it will remain in the landfill for a very long time, while the paper cup will eventually (but very slowly!) decompose. If the landfill is considered as a waste storage receptacle, then the foam cup is superior, since it does not participate in the reaction, while the paper cup produces gases and probably leachate. On the other hand, if the landfill is thought of as a treatment facility, then the foam cup is highly detrimental since it does not biodegrade.

So which cup is better for the environment? If you wanted to do the right thing, which cup would you use? This question, like so many others in this book, is not an easy one to answer. Private individuals can of course practice pollution prevention by such a simple expedient as not using either plastic or paper disposable coffee cups, but by using a refillable mug instead. Even taking into account the energy and materials needed to manufacture the mug, so long as it is used enough times the argument as to whether the plastic or paper cup is better becomes moot. It is better not to produce the waste in the first place. In addition, the coffee tastes better from a sustainable, green mug!

Once the life cycle of a material or product has been analyzed, the next engineering step is to manage the life cycle. If the objective is to use the least energy and cause the least detrimental effect to the environment, then it is clear that much of the onus is on the manufacturers of these products. The users of the products can have the best intentions for reducing adverse environmental effect, but if the products are manufactured in such a way as to make this impossible, then the fault is with the manufacturers. On the other hand, if the manufactured materials are easy to reuse, separate and recycle, then most likely energy is saved and the environment is protected. This process has become known as "pollution prevention" in industry, and there are numerous examples of how industrial firms have reduced emissions or the production of other wastes, or have made it easy to recover waste products, and in the process saved money. Some automobile manufacturers, for example, are modularizing the engines so that junked parts can be easily reconditioned and reused. Printer cartridge manufacturers have found that refilling cartridges is far cheaper than remanufacturing them, and they now offer trade-ins. There are similar opportunities in biomedical setting, although with additional constraints and complications (e.g., need for sterile devices). All the efforts to reduce waste (and save money in the process) will influence the solid waste stream in the future.

A key component of a life cycle design criterion is its reliability, a common theme of this book. The public expects an engineer to "give results, not excuses,"[15] and risk and reliability are accountability measures of their success. People using medical devices, at least intuitively, assess the risks and, when presented with solutions by physicians (who were informed by device specifications written by engineers) make the final decisions about their treatments. Next, the device user makes decisions about the reliability of the designs, again hopefully after giving informed consent to the physician. Device users, for good reason, want to be assured that they will be "safe." But, safety is a relative term. Calling something "safe" integrates a value judgment that is invariably accompanied by uncertainties. The safety of a product or process can be described in objective and quantitative terms. Factors of safety are a part of every design. The weight of evidence always includes some nonquantifiable factors. This is true for sustainable design decisions.

The sheer amount and complexity of biomedical and biosystematic data and information are enormous at present and will continue to grow. Engineers must be comfortable with ambiguity, since every professional decision is made under uncertainty, often a great deal of it. A lot of what scientists and engineers do does not always seem logical to the lay public. Thus, explaining the meaning of data can be very challenging. That is in part due to the incompleteness of our understanding of the methods used to gather data. Even well-established techniques like chromatography have built-in uncertainties. Scientific and methodological uncertainties remind us again of Plato's Allegory of the Cave. We cannot see the puppets (i.e., "reality"), only the shadows cast (i.e., measurements).[16] Thus, our instruments and other methods are indications of reality and not reality itself. Engineers have a bioethical responsibility to be "faithful agents" in the interpretations of these indicators. We are purveyors of truth. And the truth affects how we approach macroethical issues, including those associated with sustainable designs.

THE BIOETHICS OF COMBUSTION

One challenge in learning about and teaching ethics in a way that reaches a technical audience is that the subject matter can be abstract and even obtuse. But, ethics matters in all we do, so let us discuss it within the context of some "hard" sciences.

Humans have been around fire throughout their existence. Fire is dichotomous. We need it, yet we fear it. Combustion is a relatively simple phenomenon; that is, it is the oxidation of a substance in the presence of heat. Chemically, efficient combustion is:

$$(CH)_x + O_2 \rightarrow CO_2 + H_2O \qquad (9.3)$$

Of late, this simple reaction has been at the center of geopolitical and ethical debates about global climate. In light of the fact that much of the world's economy is based on combustion, this reaction takes center stage. Less CO_2 generation is supported by many to reduce the effects of greenhouse gases, but it also means that most industrial outputs, at least in their present forms, would have to be dramatically reduced.

More complex combustion reactions are shown in Table 9.3.

Most fires, however, do not reach complete combustion. They are usually oxygen-limited, so that a large variety of compounds, many that are toxic, are released. Decomposition of a substance in the absence of oxygen is known as pyrolysis. So, a fire consists of both combustion and pyrolytic processes.

Table 9.3
Balanced combustion reactions for selected organic compounds

Chlorobenzene	$C_6H_5Cl + 7O_2 \rightarrow 6CO_2 + HCl + 2H_2O$
TCE (tetrachloroethene)	$C_2Cl_4 + O_2 + 2H_2O \rightarrow 2CO_2 + HCl$
HCE (hexachloroethane)	$C_2Cl_6 + 1/2O_2 + 3H_2O \rightarrow 2CO_2 + 6HCl$
Post-chlorinated polyvinyl chloride (CPVC)	$C_4H_5Cl_3 + 4^1/2O_2 \rightarrow 4CO_2 = 3HCl + H_2O$
Natural gas fuel (methane)	$CH_4 + 2O_2 \rightarrow CO_2 + 2H_2O$
PTFE teflon	$C_2F_4 + O_2 \rightarrow CO_2 + 4HF$
Butyl rubber	$C_9H_{16} + 13O_2 \rightarrow 9CO_2 + 8H_2O$
Polyethylene	$C_2H_4 + 3O_2 \rightarrow 2CO_2 + 2H_2O$

Wood is considered to have the composition of $C_{6.9}H_{10.6}O_{3.5}$. Therefore, the combustion reactions are simple carbon and hydrogen combustion:

$$C + O_2 \rightarrow CO_2$$
$$H + 0.25O_2 \rightarrow 0.5H_2O$$

Source: US Environmental Protection Agency.

And the fire itself is not homogeneous, with temperatures varying in both space and time. Plastic fires, for example, can release over 450 different organic compounds.[17] The relative amount of combustion and pyrolysis in a fire affects the actual amounts and the types of compounds released. Temperature is also important, but there is no direct relationship between temperature and pollutants released. For example, Figure 9.3 shows that in a plastics fire (i.e., low-density polyethylene pyrolysis), some compounds are generated at lower temperatures, while for others the optimal range is at higher temperatures. However, the aliphatic compounds in this fire (i.e., 1-dodecene, 9-nonadecane, and 1-hexacosene) are generated in higher concentrations at lower temperatures (about 800 °C), while the aromatics need higher temperatures (Figure 9.4).

All societies depend on fire, so combustion plays a major role in macroethics. Western Civilization is often criticized for its disproportionate demand for fuel, especially nonrenewable fossil fuels. The way a nation addresses burning is an important measure of how advanced it is, not only in dealing with pollution, but the level of sophistication of its economic systems. As evidence, many poorer nations are confronted with the choice of saving sensitive habitat or allowing large-scale biomass burns. And, the combustion processes in developing countries are usually much less restrictive than those in more developed nations.

Industrial processes can vary significantly between developed and underdeveloped nations. For example, in Canada and the United States, the major sources of dioxins, a chemical group comprising some of the world's most toxic and carcinogenic compounds, range from dispersed activities, such as trash burning, to concentrated heavy industry (Figure 9.5). Interestingly, the second biggest source of dioxin emissions in the United States is from medical incineration. So, engineering steps to reduce one public health threat (i.e., preventing nosocomial and other infectious agents) have increased risks in another area (i.e., exposure to airborne dioxins).

Sources of dioxin emissions in Latin America are distributed differently than those in Canada and United States. Actual emission inventories are being developed, but preliminary information indicates that much of the dioxin produced in Mexico, for example, is from backyard burning, such as neighborhood scale brick making. Refractory processes in developing nations are often small-scale neighborhood

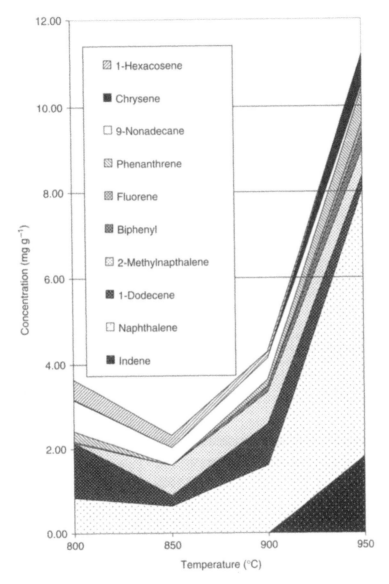

Figure 9.3 Selected hydrocarbon compounds generated in a low-density polyethylene fire (pyrolysis) in four temperature regions.
Adapted from: Data from R.A. Hawley-Fedder, M.L. Parsons, and F.W. Karasek, 1984, Products Obtained During Combustion of Polymers under Simulated Incinerator Conditions, *Journal of Chromatography*, 314, pp. 263–72.

operations. Often, the heat source used to reach refractory temperatures is a furnace with scrapped materials such as fuel, especially petroleum-derived substances like automobile tires.

The metropolitan area of El Paso, Texas, and Ciudad Juarez, Mexico, with a combined population of two million, is located in a deep canyon between two mountain ranges, which can contribute to

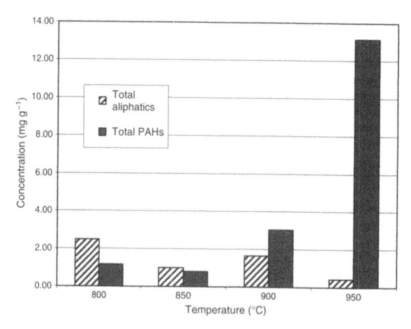

Figure 9.4 Total aliphatic (chain) hydrocarbons *versus* polycyclic aromatic hydrocarbons (PAHs) generated in a low-density polyethylene fire (pyrolysis) in four temperature regions.
Adapted from: Data from R.A. Hawley-Fedder, M.L. Parsons, and F.W. Karasek, 1984, "Products Obtained During Combustion of Polymers under Simulated Incinerator Conditions", *Journal of Chromatography*, 314, pp. 263–72.

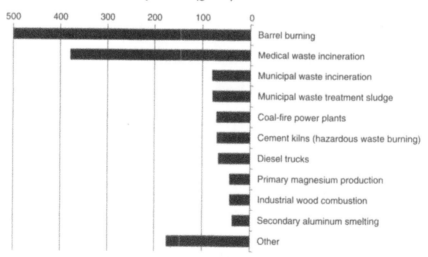

Figure 9.5 Industrial categories of dioxin emitters in the United States in 2000.
Adapted from: US Environmental Protection Agency, 2005, The Inventory of Sources of Dioxin in the United States (External Review Draft), http://cfpub.epa.gov/ncea/cfm/recordisplay.cfm?deid=132080; accessed on 12 June 2006.

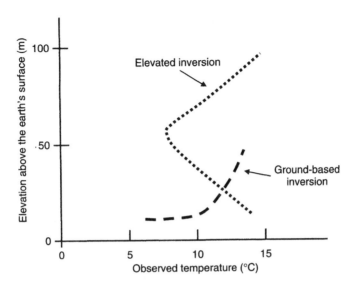

Figure 9.6   Two types of thermal inversions that contribute to air pollution.

thermal inversions in the atmosphere (Figure 9.6). The air quality has been characterized by the US Environmental Protection Agency as seriously polluted, with brick making on the Mexican side identified as a major source.[18]

In Mexico, workers who make bricks are called "ladrilleros." Many ladrilleros live in unregulated shantytowns known as "colonias" on the outskirts of Ciudad Juarez. The kilns, using the same design as those in Egypt thousands of years ago, are located within these neighborhoods, next to the small houses. The ladrilleros are not particular about the fuel, burning anything with caloric value, including scrap wood and old tires, as well as more conventional fuels like methane and butane. The dirtier fuels, like the tires, release large black plumes of smoke that contains a myriad of contaminants.

Children are at an elevated risk of health problems when exposed to these plumes, since their lungs and other organs are undergoing prolific tissue growth. Thus, the ladrilleros' families are at elevated risks particularly due to their frequent and high dose exposures. "The health impact is not only of concern to the worker but also the entire family, especially pregnant women and children who, because of their socioeconomic status, tend to be undernourished," according to Beatriz Vera, project coordinator for the US–Mexico Border Environment and Health Projects. She adds that "many times the entire family participates in the process. Sometimes children are put directly into the area where the kiln is fired."

The two nations' governments are at least somewhat cognizant of the problem, as are numerous nongovernmental organizations (NGOs). These have included Environmental Defense (ED), Physicians for Social Responsibility, the Federacion Mexicana de Asociaciones Privadas de Salud y Desarrollo Comunitario (FEMAP), and El Paso Natural Gas (EPNG). For example, FEMAP and EPNG offer courses to the ladrilleros from throughout the region on ways to use higher quality fuel, including improved safety, and business practices as well. Often, however, even if the brick makers know about cleaner fuels, they cannot afford them. For example, they have used butane, but in 1994 the Mexican government started to phase out its subsidy and about the same time the peso was devalued, leading to a sharp increase in butane costs. The ladrilleros were forced to return to using the cheaper fuels. In the meantime, the Mexican government banned the burning of tires; resulting in much of the more surreptitious tire burning done at night.

A number of solutions to the problem have been proposed, including more efficient kilns. However, arguably the best approach is to prevent the combustion in the first place. In fact, many of the traditional villages where bricks are now used had previously been constructed with adobe. A return to such a non-combustion approach could hold the key. The lesson here is that often in developing countries, the simpler, "low-tech" solutions are the most sustainable.

In the mid-1990s, the US EPA and the Texas Natural Resource Conservation Commission (TNRCC) conducted a study in the Rio Grande valley region to address concerns about the potential health impact of local air pollutants, and especially since little air quality information was available at the time. There are numerous "cottage industries," known as "maquiladoras,"[19] along both sides of the Texas–Mexico border as ascribed by the Rio Grande. In particular, the study addressed the potential for air pollution to move across the US–Mexican border into the southern part of Texas. Air pollution and weather data were collected for a year at three fixed sites near the border in and near Brownsville, TX. The study found overall levels of air pollution to be similar to or even lower than other urban and rural areas in Texas and elsewhere and that transport of air pollution across the border did not appear to adversely impact air quality across the US border. Although these "technical" findings may not be particularly surprising, the study provided some interesting results on how to work with local communities, particularly those with much cultural diversity and richness.

Scientists, engineers, and planners must be sensitive to specific challenges in addressing fairness issues. The National Human Exposure Assessment Survey (NHEXAS) and Lower Rio Grande Valley Transboundary Air Pollution Project evaluated total human exposure to multiple chemicals on community and regional scales from 1995 to 1997. Lessons learned from this research[20] include the need to:

1. Develop focused study on hypotheses/objectives and develop the linkages between the objectives, data requirements to meet the objectives, and available resources.
2. Ensure that questionnaires address study objectives and apply experiences of previous studies. Wording must apply to the study participants and reflect differences in demographics (e.g., language, race, nationality, gender, age).
3. Develop materials and scripts to address potential concerns that participants may have about the study (e.g., time burden, purpose and value of study, collection of biological samples, legitimacy of the study).
4. Before recruiting, build knowledge of the study and support in the community, including support, sponsorship, and advertising from a range of organizations.
5. Plan and conduct follow-up visits with community leaders and stakeholders for timely dissemination of information learned in the study.
6. Integrate recruitment staff into the research team for continuity with participants and to build on experience.
7. Limit field teams to two people per visit (about 1–1.5 hours) and have the same staff member work with the study participant at all visits.
8. Ensure that laboratories have the expertise and capacity for planned analyses.
9. Establish format for reporting data before any field samples are collected.

A "benchmark" of environmental measurements understandable to a highly diverse citizenry is essential for many engineering projects, especially those affecting public health. For example, people will reach their own conclusions, ranging from a failure to grasp a real health problem that exists ("false negative") to perceive a problem even when values are better than or do not differ significantly from those of the general population ("false positive").

Not long ago in the United States, the standard means of getting rid of household trash was the daily burn. Each evening, people in rural areas, small towns, and even larger cities made a trip into the backyard, dumped the trash they had accumulated into a barrel,[21] and burned the contents. Also, burning was a standard practice elsewhere, such as intentional fires to remove brush, and even "cottage industries," like backyard smelters and metal recycling operations. Beginning in the 1960s and 1970s, the public acceptance and tolerance for open burning was waning. Local governments began to restrict and eventually to ban many fires. Often, these restrictions had multiple rationales, especially public safety (fires becoming out of control, especially during dry seasons) and public health (increasing awareness of the association between particulate matter in air and diseases like asthma and even lung cancer).

This new intolerance for burning was a type of ethical paradigm shift. What one did in one's own yard was no longer the sole primacy of the homeowner. The action had an effect on a larger area; that is, wherever the plume migrated. It also had a cumulative effect. As in Hardin's "Tragedy of the Commons," a fire or two may not cause considerable harm, but when large numbers of fires occur, health thresholds could easily be crossed. In fact, the paradigm was new to environmental regulation but not to ethics, especially Kant's deontological, i.e., duty-based, view of ethics. A metric used by Kant to determine the morality of an action or decision is whether that action or decision, if universalized, would lead to a better or a worse society. This is the "categorical imperative." Somewhere in the 1960s and early 1970s, environmental science and policy shifted toward the categorical imperative. Slogans like "Think globally and act locally," and "We all live in Spaceship Earth" are steeped in the categorical imperative. Our duty and obligation to our fellow human beings, now and in the future, must drive our day-to-day decisions.

So then, how do we form our factual premises about combustion? What is so bad about open burning? Part of the answer involves understanding what happens when halogenated organic compounds are burned. Let us consider two compounds, polyvinyl chloride (PVC) and polychlorinated biphenyls (PCBs). PVC is a polymer, like polyethylene. However, rather than a series of ethylenes in the backbone chain, a chlorine atom replaces the hydrogen on each of the ethylene groups by free radical polymerization of vinyl chloride. The first thing that can happen when PVC is heated is that the polymers become unhinged and chlorine is released (Figure 9.7). Also, the highly toxic and carcinogenic dioxins and furans can be generated from thermal breakdown and molecular rearrangement of PVC in a heterogeneous process; that is, the reaction occurs in more than one phase (in this case, in the solid and gas phases). The active sorption sites on the particles allow for the chemical reactions, which are catalyzed by the presence of

Figure 9.7   Free radical polymerization of vinyl chloride to form polyvinyl chloride.

inorganic chloride compounds and ions sorbed to the particle surface. The process occurs within the temperature range of 250–450°C; so most of the dioxin formation under the precursor mechanism occurs away from the high temperature of the fire, where the gases and the smoke derived from combustion of the organic materials have cooled. Dioxins and furans may also form *de novo*, wherein moieties different from those of the molecular structure of dioxins, furans, or precursor compounds lead to reactions that generate dioxins. The process needs a chlorine donor (a molecule that "donates" a chlorine atom to the precursor molecule). This leads to the formation and chlorination of a chemical intermediate that is a precursor.

In addition, PVC is seldom in a pure form. In fact, most coated wires have to be pliable and flexible. On its own, PVC is rigid, so plasticizers must be added, especially phthalates. These compounds have been associated with chronic effects in humans including endocrine disruption. Also, since PVC catalyzes its own decomposition, metal stabilizers have been added to PVC products. These include lead, cadmium, and tin (e.g., butylated forms). For these and other reasons, "chlorine" has become a watchword in environmental debates. A chemical element has taken on ethical meaning.

Another very common class of toxic compounds released when plastics are burned are the polycyclic aromatic hydrocarbons (PAHs).

The change in the attitude and the acceptance of pollution has been dramatic, but since it has occurred incrementally, it may be easy to forget just how much the baseline has changed. For example, like the open burning that was the norm, emissions from industrial stacks were pervasive and contained myriad of toxic components. People living in the plume of coke ovens have experienced the obnoxious smelling compounds, including metallic and sulfur compounds that volatilize during the conversion of coal to coke needed for steel manufacturing. While such areas continue to be industrialized, such ambient air quality as that in the 1960s is no longer tolerated, or at least ought not to be.

Coke is produced by blending and heating bituminous coals in coke ovens to 1000–1400 °C in the absence of oxygen.[22] Lightweight oils and tars are distilled from coal, generating various gases during the heating process. Every half hour or so, the flows of gas, air, and waste gas are reversed to maintain uniform temperature distribution across the wall. In most modern coking systems, nearly half of the total coke oven gas produced from coking is returned to the heating flues for burning after having passed through various cleaning and coproduct recovery processes. Coke oven emissions are the benzene-soluble fraction of the particulate matter generated during coke production. They are known to contain human carcinogens and are truly an awful concoction.

Coke oven emissions are actually complex mixtures of gas, liquid, and solid phases, usually including a range of about 40 PAHs, as well as other products of incomplete combustion: notably formaldehyde, acrolein, aliphatic aldehydes, ammonia, carbon monoxide, nitrogen oxides, phenol, cadmium, arsenic, and mercury. More than 60 organic compounds have been collected near coke plants. A metric ton of coal yields up to 635 kilograms of coke, up to 90 kilograms of coke breeze (large coke particulates), 7–9 kilograms of ammonium sulfate, 27.5–34 liters of coke-oven gas tar, 55–135 liters of ammonia liquor, and 8–12.5 liters of light oil. Up to 35% of the initial coal charge is emitted as gases and vapors. Most of these gases and vapors are collected during byproduct coke production. Coke oven gas is comprised of hydrogen, methane, ethane, carbon monoxide, carbon dioxide, ethylene, propylene, butylene, acetylene, hydrogen sulfide, ammonia, oxygen, and nitrogen. Coke oven gas tar includes pyridine, tar acids, naphthalene, creosote oil, and coal tar pitch. Benzene, xylene, toluene, and solvent naphthas may be extracted from light oil fraction.

Coke production in the United States increased steadily between 1880 and the early 1950s, peaking at 65 million metric tons in 1951. In 1976, the United States was second in the world with 48 million metric tons of coke, i.e., 14.4% of the world production. By 1990, the United States produced 24 million metric tons, falling to fourth in the world. A gradual decline in production has continued; production has decreased from 20 million metric tons in 1997 to 15.2 million metric tons in 2002. Demand for blast furnace coke also has declined in recent years, because technological improvements have reduced the amount of coke consumed per amount of steel produced by as much as 25%.

Another example of disparate exposures to noxions combustion by-products is the junkyard. The junkyards in lower socioeconomic communities some decades ago generated their own emissions as well. In fact, the combination of fires, wet muck (comprised of soil, battery acid, radiator fluids, motor oil, corroded metal, and water), and oxidizing metals created a rather unique odor around the yards. Neurophysiologists have linked the olfactory center to the memory center of the human brain (olfactory bulb in the cerebellum; Figure 9.8). In other words, when we smell something, it evokes a strong memory response. This sensation was recounted by workers conducting air sampling in lower Manhattan shortly after the attacks on the World Trade Center on 11 September 2001. For example, the smell from the burning plastics, oxidizing metal, semivolatile and volatile organic compounds, and particulate matter

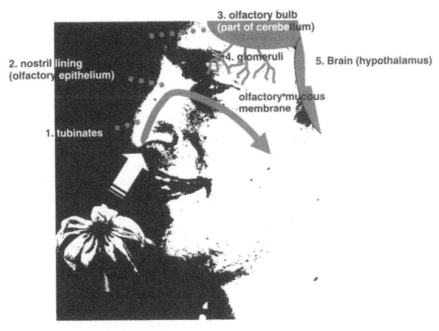

Figure 9.8 Human smell is initiated by sniffing, which transports air with concentrations of odorant molecules past curved bony structures, i.e., turbinates, in the nose. The turbinates generate turbulent airflow that mixes volatile compounds and particles and carries them to thin mucus layer that coats the olfactory epithelium. The olfactory epithelium contains odor-sensing cells.

was very similar to what I smelled as a teenager working in the East St Louis junkyards, and it brought back 35-year-old memories.

Odors have often been associated with public health nuisances. However, in addition to the link between memory and olfactory centers, the nasal–neural connection is important to environmental exposure. This goes beyond nuisance and it is an indication of potential adverse health effects. For example, nitric oxide (NO) is a neurotoxic gas released from many sources, such as confined animal feeding operations, breakdown of fertilizers after they are applied to the soil and crops, and emissions from vehicles. Besides being inhaled into the lungs, NO can reach the brain directly. The gas can pass through a thin membrane via the nose to the brain.

The nasal exposure is a different pathway from that usually used to calculate exposure. In fact, most sources do not have a means for calculating exposures other than dermal, inhalation, and ingestion. Research at Duke University, for example, linked people's emotional states (i.e., moods) to odors around confined animal feeding operations in North Carolina.[23] People who live near swine facilities are negatively affected when they smell odors from the facility. This is consistent with other research that has found that people experience adverse health symptoms more frequently when exposed to livestock odors. These symptoms include eye, nose, and throat irritation, headache, nausea, diarrhea, hoarseness, sore throat, cough, chest tightness, nasal congestion, palpitations, shortness of breath, stress, and drowsiness. There is quite a bit of diversity in response, with some people being highly sensitive to even low concentrations of odorant compounds, while others are relatively unfazed even at much higher concentrations. Actually, response to odors can be triggered by three different mechanisms. In the first mechanism, symptoms can be induced by exposure to odorant compounds at sufficiently high concentrations to cause irritation or other toxicological effects. The irritation, not the odor, evokes the health symptoms. The odor sensation is merely an exposure indicator. In the second mechanism, symptoms of adverse effects result from odorant concentrations lower than those eliciting irritation. This can be owing to genetic predisposition or conditioned aversion. In the third mechanism, symptoms can result from a coexisting pollutant, e.g., an endotoxin, which is a component of the odorant mixture.

Therefore, to address this new paradigm, better technologies will be needed. For example, the research at Duke and other findings have encouraged some innovative research, including the development of "artificial noses."[24]

During the aftermath of the World Trade Center attacks, odors played a key role in response. People were reminded daily of the traumatic episode, even when the smoke from the fire was not visible. And, the odors were evidence of exposure to potentially toxic substances. In this instance, it was not NO, but volatile, semivolatile organic, and metallic compounds. The plume from the World Trade Center fires contained elevated concentrations of PAHs, dioxins, furans, volatile organic compounds (e.g., benzene), and particles containing metals (Figure 9.9). The presence of the odors was one of the many factors that kept New Yorkers on edge, and until the odors significantly subsided, they continued to be a source of anxiety. Analysis of the samples confirmed my brain's olfactory–memory connection.

Obviously, combustion still goes on in the United States, but the "open burning" paradigm has thankfully shifted quite far toward cleaner processes and better control technologies in recent decades. In fact, the past few decades have clearly shifted toward an appreciation for the common value of air sheds, notwithstanding Hardin's laments in the "Tragedy of the Commons." This is not to say that Hardin was wrong; that is, people do tend to favor their personal utilities and there is a ranging debate about greenhouse gases, especially carbon dioxide, which is always the product of any complete combustion.

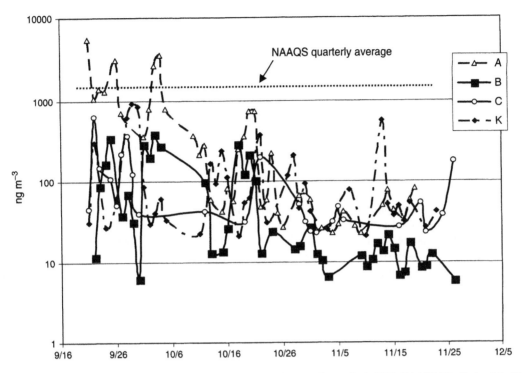

Figure 9.9 Lead concentrations (composition of $PM_{2.5}$) at the World Trade Center site in 2001. NAAQS, The National Ambient Air Quality Standard for lead, i.e., 1.55 μg m^{-3} averaged each quarter.

However, once society's members see the problems being wrought by unchecked combustion, most people have willingly accepted restrictions and regulations, so long as they are effective and perceived to be fair. In this way, the open burning paradigm shift can be seen as a codicil to Hobbes' social contract; that is, had we continued on the path of unbridled combustion and emissions, we would have reached a societal "state of nature," where the collective of "brutes" with fires would damage the entire population, including our personal and familial well-being.

SYSTEMATIC BIOETHICS

The lesson from concurrent engineering, life cycle analysis, and green engineering is that the bioethical view, like the technical view, must be systematic. Like mass and energy balances, ethical decisions must also clearly articulate what goes into the decision (e.g., assumptions, needs, goals, and underpinning science) and what will result from the decision (e.g., intended and unintended consequences, byproducts, and ancillary costs and benefits). Many biomedical failures are subtle and protracted. These are difficult to observe in real-time, so one means of understanding the complexities of intertwined factors in engineering ethics is to examine a dramatic and almost instaneous engineering failure, such as the explosion of a chemical processing facility in Seveso, Italy. This is followed by the more incrementally developed "societal" macroethical issue, poverty.

SEVESO PLANT DISASTER

This case is an example of a toxic cloud and is interesting and instructive as an example of how an engineering failure can threaten public safety. As such, it can be used as a case study of engineers not adequately attending to the first canon of the profession, i.e., to hold paramount the health, safety, and welfare of the public. It was indeed a harbinger of the toxic scares that paved the way for new environmental legislation around the world, calling for programs to address toxic substances and hazardous wastes. It was also emblematic of potential exposures to toxic substances to those millions of people living near industrial facilities and of the community "right-to-know" laws that have since emerged. But, it appears here because the disaster represents so much more than the event itself in terms of why it occurred, what were the real ethical issues, and consequently how risks from such incidents can be better managed in the future.

On 10 July 1976, an explosion at a 2,4,5-trichlorophenol (2,4,5-T) reactor at a manufacturing plant near Seveso, Italy resulted in the highest concentrations of the most toxic chlorinated dioxin (TCDD) levels known in human residential populations.[25] Up to 30 kilograms of TCDD were deposited over the surrounding area of approximately 18 square kilometer. To put this amount in perspective, scientists usually are worried about dioxin, and especially TCDD, if a gram or less is released in a year. This release of 30 000 grams occurred instantaneously.

The plant was operated by Industrie Chemiche Meda Societa, Anonima (ICMSA), an Italian subsidiary owned by the Swiss company, Givaudan, which was owned by another Swiss company, Hoffman-LaRoche.[26] The release resulted from a ruptured disk on a batch plant. The 2,4,5-T was formulated from 1,2,4,5-tetrachlorobenzene and caustic soda (NaOH) in the presence of ethylene glycol. Normally, only a few molecules of dioxins form from several trillion molecules of the batch, but in the events leading up to the explosion, the reactor temperatures rose beyond the tolerances, leading to a "runaway reaction," bursting the disk. It is likely that a large volume of hydrogen gas (H_2) was formed, which propelled the six tons of fluid in the reactor into the Seveso region. Actually, the discharge could have been worse, since upon hearing the noise being made at the beginning of the release, a plant foreman opened a valve to release cooling waters onto the heating coils of the reactor. This action likely reduced the propulsion.

The explosion is in part an example of doing the wrong thing for the right reason. The Italian legislature had passed a law requiring the plant to shut down on weekends, whether or not a batch was complete. On the weekend of the explosion, the plant was shut down after the formulation reaction, but before the complete removal of the ethylene glycol by distillation. This was the first time that the plant was shut down at this stage. Based upon chemistry alone, the operator had no reason to fear anything would go awry. The mixture at the time of shutdown was at 158 °C, but the theoretical temperature at which an exothermic reaction would be expected to occur was thought to be 230 °C. Subsequent studies have found that exothermic reactions of these reagents can begin at 180 °C, but these are very slow processes when below 230 °C. The temperature, in fact, rose mainly due to a temperature gradient from the liquid phase to the steam phase. The reactor wall in contact with the liquid was much cooler than the wall in contact with the steam, with heat moving from the upper wall to the surface of the liquid. The stirrer was switched off so that the top few centimeters of liquid rose in temperature to about 190 °C, beginning the slow exothermic reaction. In 7 hours, the runaway reaction commenced. The reactions may also have been catalyzed by chemicals in the residue that had caked on the upper wall that, with the temperature increase, was released into the liquid in the reactor. Thus, the runaway reaction and explosion could have been prevented if the plant had not had to be prematurely shut down and if the plant operators had been better trained to consider contingencies beyond the "textbook" conditions.

A continuous, slight wind dispersed the contents over the region, which included 11 towns and villages. The precipitate looked like snow. No official emergency action took place on the day of explosion. People with chemical burns checked themselves into local hospitals. The Mayor of Seveso was told of the incident only the next day

The response to the disaster and pending dioxin exposure was based on the potential risks involved. The contaminated area was divided into three zones based on the concentration of TCDD in the soil. Two hundred and eleven families in Zone A, the most heavily contaminated area, were evacuated within 20 days of the explosion and measures were taken to minimize exposure to residents in the nearby zones. In a preliminary study, US Center for Disease Control and Prevention (CDC) tested blood serum samples from five residents of Zone A who suffered from the dioxin-related skin disease known as chloracne, as well as samples from four residents from Zone A free of chloracne, and three residents from outside the contaminated area. All samples had been collected and stored shortly after the accident. Interestingly, TCDD was detected in only one sample from the unexposed group, but that one sample was comparably high, i.e., 137 part per trillion (ppt). The elevated TCDD level was thought to be due to misclassification or sample contamination. In Zone A, serum TCDD levels ranged from 1772 to 10 439 ppt for persons without chloracne and from 828 to 56 000 ppt for persons with chloracne. The TCDD concentrations were the highest ever reported in humans up to that time. The CDC is presently evaluating several hundred historical blood samples taken from Seveso residents for TCDD. There are plans to determine half-life estimates and to evaluate serum TCDD levels for participants in the Seveso cancer registry.

Scientific research has continued even after the sensational aspects of the disaster have waned. In fact, a recent study by Warner et al.[27] found that a statistically significant dose–response increased risk for breast cancer incidence with individual serum TCDD level among women in the Seveso women, i.e., a dose–response relationship of a twofold increase in the hazard rate associated with a tenfold increase in serum TCDD. This result is an early warning because the Seveso cohort is relatively young, with a mean age of less than forty-one years at the time of interview. These findings are consistent with the twenty-year follow-up study that showed that even though no increased incidence of breast cancer had been observed in the Seveso population ten and fifteen years after the incident, after twenty years breast cancer mortality emerged among women who resided in heavily contaminated areas and who were younger than fifty-five years at death (relative risk (RR),1.2; 95% confidence interval (CI), 0.6–2.2], but not in those who were older.

The findings are not statistically significant, but scary nonetheless. For example, it is very difficult to characterize a population even for a year, let alone twenty years, because people move in and out of the area. Epidemiological studies are often limited by the size of the study group in comparison to the whole population of those exposed to a contaminant. In the Seveso case, the TCDD exposure estimates have been based on the zone of residence, so the study is deficient in individual level exposure data. There is also variability within the exposure area, for example, recent analyses of individual blood serum TCDD measurements for 601 Seveso women suggest a wide range of individual TCDD exposure within zones.[28] This raises an ongoing bioethical challenge, when is it appropriate to use data, that has not yet met scientific "gold standards?" For example, in early tobacco studies, some disease was associated with exposure to environmental tobacco smoke (i.e., a non-smoker inhaling another's smoke) However, the statistical significance was only at the < 0.10 level instead of the generally accepted level of < 0.05. For engineers, the design must accomodate both scientific standards and margins of safety.

Some argue that the public's response to the real risks from Seveso was overblown, considering no one died directly from the explosion and the immediate aftermath, although hundreds were burned by the NaOH exposure and hundreds did develop skin irritations. This may be true, but part of the response is the general response to the unknown, especially when the chemical of concern is TCDD, that has been established to cause cancer, neurological disorder, endocrine dysfunction, and other chronic diseases which do not manifest themselves immediately. The medical follow-up studies may indicate that the public response was plausible, given the latent connections to chronic diseases, like breast cancer.

The toxicity, especially the very steep cancer slope for TCDD, is met with some skepticism. Exposures at Seveso were high compared to general environmental exposures. However, some high-profile cases, especially the recent suspected TCDD poisoning of the Ukrainian President Viktor Yushchenko in 2004 and the 1998 discovery of the highest blood levels of TCDD in two Austrian, have added to the controversy. In fact, both these cases were precipitated by the adverse effects, especially chloracne. Usually, dioxins, PCBs, and other persistent, bioaccumulating toxicants are studied based on the exposure, followed by investigations to see if any effects are present. This has been the case for Agent Orange, Seveso, Times Beach, and most other contamination instances. It is unusual to see such severe effects, but the very high doses (Figure 9.10) reported in the Yushchenko and Austrian cases would have come as a greater surprise had no adverse effect been seen.

This case illustrates the need to consider the likely consequences of failure early in the design process. Sufficient factors of safety are an engineering and an ethical imperative. Although, thankfully, most biomedical decisions are not at the scale and complexity of this one, there are lessons to be learned. For example, does the individual biomedical engineer always consider what goes into an autoclave? To the chemical engineer an autoclave is no less a reactor than the pesticide batch reactor of Seveso. Kant's categorical imperative compels us to think systematically about even these smaller scale decisions.

POVERTY AND POLLUTION

Although engineering decisions are often circumscribed when it comes to economics, there is little debate about whether economics and environmental quality are related. Since engineers are key agents in providing a safe, healthy, and livable environment, it behooves us to consider some of the larger bioethical issues, such as our role in reducing poverty. I would be remiss not to consider poverty in a bioethics text, since it is not only a macroethical issue for all professions, including engineering, but it is also a factor and driver of numerous other bioethical issues (e.g., human enhancement, abortion, medical care, and informed consent).

There is no consensus on what causes poverty. So, let us take a look at one major question: What is the relationship between poverty and pollution? The disagreement about economics and the environment usually comes in two forms. The most common is the degree, extent, and type of relationships between economic development and environmental condition. A commonly held perception is that greater development threatens environmental quality: more cars, more consumerism, more waste, and

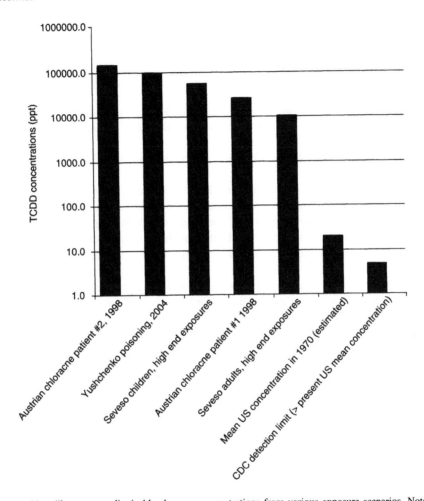

Figure 9.10 Tetrachlorodibenzo-*para*-dioxin blood serum concentrations from various exposure scenarios. Note that this is a log scale, so the measurements differ by 5 orders of magnitude.
Adapted from: Center for Disease Control and Prevention; and Dioxin facts.org: http://www.dioxinfacts.org/dioxin_health/dioxin_tissues/yushchenko.html; accessed on 24 April 2005.

more releases of toxicants. In his 2003 book, *The Real Environmental Crisis* (University of California Press, Berkeley, CA), J.M. Hollander takes an opposite view:

> People living in poverty perceive the environment very differently from the affluent. To the world's poor –
> several billion people – the principal environmental problems are local, not global. They are not the stuff of
> media headlines or complicated scientific theories. They are mundane, pervasive and painfully obvious:
>
> - Hunger – chronic undernourishment of a billion children and adults caused not only by scarcity of
>   food resources but by poverty, war, and government tyranny and imcompetence.
> - Contaminated Water Supplies – a major cause of chronic disease and mortality in the third world.

- Diseases – rampant in the poorest countries. Most could be readily eradicated by modern medicine, while others, including the AIDS epidemic in Africa, could be mitigated by effective public health programs and drug treatments available to the affluent.
- Scarcity – insufficient local supplies of fuelwood and other resources, owing not to intrinsic scarcity but to generations of overexploitation and underreplenishment as part of the constant struggle for survival.
- Lack of Education and Social Inequality, Especially of Women – lack of education resulting in high birthrates and increasing the difficulty for families to escape from the dungeons of poverty.[29]

Interestingly, the first three problems fall clearly within the domain of biomedical ethics, and a strong case can be made for scarcity, as well.

Hollander argues that environmental quality and healthy economy are highly compatible. His evidence is that most Western countries that developed economically coincidentally improved in most environmental aspects. A dramatic example is that of the island of Hispaniola. On the Dominican Republic side, there is much lush vegetation. But, on the Haiti side, the land is denuded. The Dominican Republic has embraced a more open, capitalistic marketplace, while Haiti has suffered the ravages of totalitarian regimes.

Is this anecdotal or can we expect that reducing poverty by enhancing economic development will always be associated with a cleaner environment? Perhaps the answer is not so clear or dichotomous. Perhaps, like so many issues in engineering ethics, the answer is that there is a somewhat fluid range in which to optimize both conditions, i.e., economic well being and environmental protection. Operating outside of this range, e.g., going too far toward commercialism on one end and stifling freedom on the other, may pit one against the other. The answer cannot be derived from a simple, linear formula (e.g., Capitalism = Good).

Like the life cycle analysis in green engineering, the possible societal ramifications and economic outcomes from the collective effects of all engineering projects can be addressed only systematically. Applying a chemical engineering, thermodynamic metaphor, what comes out of this "ethics mass balance" is affected by the initial conditions. These conditions are then changed as the designs are implemented. In other words, while it is difficult to predict the actual outcomes, we must consider possible effects on future and distant people.

INTERDEPENDENCE

Indeed, our actions may well improve things at a local level, which we hope will improve things for all the society, per Kant's categorical imperative, however, there are instances when the collective effects are negative (as illustrated by Garrett Hardin's "Tragedy of the Commons"). Which of these mutually exclusive outcomes applies to any specific engineering project can be answered only by:

It depends.

Jeff Peirce, my colleague at Duke, is fond of insisting that these two words are almost always the right answer to any engineering question. He says this in all of his engineering classes, including engineering economics and optimization courses. His point is that engineering is an interdependent and integral endeavor. Every factor in every design decision affects everything else. In certain scenarios, such as whether to install wells in an impoverished community to provide scarce potable water, the decision is a clear-cut "yes." However, this is the case only if it is consistent with the physical realities (e.g., that the aquifer contains high-quality drinkable water) and the social realities (e.g., the culture is not dependent on surface water, rather than ground water, since it provides an important role for those carrying water).

Thus, sustainable engineering is indeed bioethical engineering. And, since the objective is to provide and to sustain "life," engineering bioethics requires this sustainable viewpoint.

NOTES AND COMMENTARY

[1] National Academy of Engineering, *The Engineer of 2020: Visions of Engineering in the New Century* (Washington, DC: The National Academies Press, 2004), 50–1.

[2] US Environmental Protection Agency, 2006, "Green Engineering," http://www.epa.gov/oppt/green-engineering/ (accessed 13 June 2006).

[3] Virginia Polytechnic Institute and State University, 2006, http://www.eng.vt.edu/green/Program.php (accessed 13 June 2006).

[4] World Commission on Environment and Development, United Nations, *Our Common Future* (Oxford, UK: Oxford Paperbacks, 1987).

[5] Abraham Maslow, *Motivation and Personality*, 2nd ed. (New York, NY: Harper & Row, 1970).

[6] American Society of Mechanical Engineers, 2004, *Professional Practice Curriculum: Sustainability*, http://www.professionalpractice.asme.org/communications/sustainability/index.htm (accessed 2 November 2004).

[7] The source for this discussion is S.B. Billatos and N.A. Basaly, *Green Technology and Design for the Environment* (Bristol, PA: Taylor and Francis, 1997).

[8] US Environmental Protection Agency, 2004, "What Is Green Engineering?" http://www.epa.gov/oppt/greenengineering/whats_ge.html (accessed 2 November 2004).

[9] See S.B. Billatos, *Green Technology and Design for the Environment* (Washington, DC: Taylor & Francis, 1997); and V. Allada, "Preparing Engineering Students to Meet the Ecological Challenges through Sustainable Product Design," *Proceedings of the 2000 International Conference on Engineering Education* (Taipei, Taiwan, 2000).

[10] T.J. Woodruff, J. Grillo, and K.C. Schoendorf, "The Relationship between Selected Causes of Postneonatal Infant Mortality and Particulate Air Pollution in the United States," *Environmental Health Perspectives* 105, no. 6 (June 1997).

[11] R.E. Gosselin, R.P. Smith, and H.C. Hodge, *Clinical Toxicology of Commercial Products*, 5th ed. (Baltimore, MD: Williams and Wilkins, 1984).

[12] Since I brought it up, the SIDS Alliance recommends a number of risk reduction measures that should be taken to protect infants from SIDS:

> **Place your baby on his or her back to sleep.** The American Academy of Pediatrics recommends that healthy infants sleep on their backs or sides to reduce the risk for SIDS. This is considered to be most important during the first six months of age, when baby's risk of SIDS is greatest.
> **Stop smoking around the baby.** Sudden Infant Death Syndrome is long associated with women who smoke during pregnancy. A new study at Duke University warns against use of nicotine patches during

pregnancy as well. Findings from the National Center for Health Statistics now demonstrate that women who quit smoking during pregnancy, but resume after delivery, put their babies at risk for SIDS too.

Use firm bedding materials. The US Consumer Product Safety Commission has issued a series of advisories for parents regarding hazards posed to infants sleeping on top of beanbag cushions, sheepskins, sofa cushions, adult pillows, and fluffy comforters. Waterbeds have also been identified as unsafe sleep surfaces for infants. Parents are advised to use a firm, flat mattress in a safety-approved crib for their baby's sleep.

Avoid overheating, especially when your baby is ill. SIDS is associated with the presence of colds and infections, although colds are not more common among babies who die of SIDS than among babies in general. Now, research findings indicate that overheating, too much clothing, too heavy bedding, and too warm a room may greatly increase the risk of SIDS for a baby who is ill.

If possible, breast-feed. Studies by the National Institutes of Health show that babies who died of SIDS were less likely to be breast-fed. In fact, a more recent study at the University of California, San Diego, found breast milk to be protective against SIDS among nonsmokers, but not among smokers. Parents should be advised to provide nicotine-free breast milk, if breast-feeding, and to stop smoking around your baby particularly while breast-feeding.

Mother and baby need care. Maintaining good prenatal care and constant communication with your health care professional about changes in your baby's behavior and health are of the utmost importance.

[13] Dy-Dee Diaper Service, 2005, http://www.dy-dee.com/ (accessed 22 April 2005).

[14] M.S. Pritchard, *On Being Responsible* (Lawrence, KS: University Press of Kansas, 1991).

[15] C. Mitcham and R.S. Duval, "Engineering Ethics," in *Responsibility in Engineering* (Upper Saddle River, NJ: Prentice-Hall, 2000).

[16] B. Jowett, trans., *The Republic by Plato* (Oxford, UK: Oxford University Press). Another way of looking at the incompleteness of our understanding of how things work is St Paul's likening of the way that we see the world to our "... seeing through a glass, darkly ...." (1 Corinthians 13: 12). As humans, we will always be limited in our understanding of even the most fundamental aspects of the tools we use.

[17] B.C. Levin, "A Summary of the NBS Literature Reviews on the Chemical Nature and Toxicity of Pyrolysis and Combustion Products from Seven Plastics: Acrylonitrile–butadiene–styrenes; Nylons; Polyesters; Polyethylenes; Polystyrenes; Poly(vinyl chlorides) and Rigid Polyurethane Foams," *Fire Materials* 11 (1987): 143–57.

[18] The major source of information about Rio Grande brick making is *Environmental Health Perspectives*, 104, no. 5 (May 1996).

[19] The Coalition for Justice in the Maquiladoras, a cross-border group that organizes maquiladora workers, traces the term maquiladora to "maquilar," a popular form of the verb maquinar that roughly means "to submit something to the action of a machine," as when rural Mexicans speak of maquilar with regard to the grain that is transported to a mill for processing. The farmer owns the grain; yet someone else owns the mill who keeps a portion of the value of the grain for milling. So, the origin of maquiladora can be found in this division of labor. The term has more recently been applied to the small factories opened by US companies to conduct labor-intensive jobs on the Mexican side of the border. Thus, maquilar has changed to include this process of labor, especially assembling parts from various sources, and the maquiladoras are those small assembling operations along the border.

While the maquiladoras have provided opportunities to entrepreneurs along the Mexico–US border, they have also given opportunity for the workers and their families to be exploited in the interests of profit and economic gain.

[20] S. Mukerjee, "Communication Strategy of Transboundary Air Pollution Findings in a US–Mexico Border XXI Program Project," *Environmental Management* 29, no. 1 (2002): 34–56.

[21] Often, these barrels were the 55-gallon drum variety, so the first burning likely volatilized some very toxic compounds, depending on the residues remaining in the drum. These contents could have been solvents (including halogenated compounds like chlorinated aliphatics and aromatics), plastic residues (like phthalates), and petroleum distillates. They may even have contained substances with elevated concentrations of heavy metals, like mercury, lead, cadmium, and chromium. The barrels (drums) themselves were often perforated to allow for higher rates of oxidation (combustion) and to take advantage of the smokestack effect (i.e., driving the flame upward and pushing the products of incomplete combustion out of the barrel and into the plume). I recall the neighbors not being happy about burning trash while their wash was drying on the clothesline. They would complain of ash (aerosols) blackening their clothes and the odor from the incomplete combustion products on their newly washed laundry. Both of these complaints are evidence that the plume leaving the barrel contained harmful contaminants.

[22] The principal source for this section is National Toxicology Program, Eleventh Report on Carcinogens, Coke Oven Emissions, Substance Profile, 2005. http://ntp.niehs.nih.gov/ntp/roc/eleventh/profiles/s049coke.pdf (accessed 11 May 2005).

[23] S.S. Schiffman and C.M. Williams, "Science of Odor as a Potential Health Issue," *Journal of Environmental Quality* 34 (2005): 129–38.

[24] For a survey of the state of the science in electronic odor-sensing technologies, see H.T. Nagle, S.S. Schiffman, and R. Gutierrez-Osuna, "The How and Why of Electronic Noses," *IEEE Spectrum* 35, no. 9 (1998): 22–34.

[25] P. Mocarelli and F. Pocchiari, 1988, Preliminary Report: 2,3,7,8-Tetrachlorodibenzo-*p*-dioxin Exposure to Humans – Seveso, Italy, *Morbidity and Mortality Weekly Report*, 37, 733–6; and A. Di Domenico, V. Silano, G. Viviano, and G. Zappni, Accidental Release of 2,3,7,8-Tetrachlorodibenzo-*p*-dioxin (TCDD) at Seveso, Italy: V. Environmental Persistence of TCDD in Soil, *Ecotoxicology and Environmental Safety*, 4, 339–45.

[26] The principal source for the engineering aspects of the industrial accident is T. Kletz, *Learning from Accidents*, 3rd ed. (Oxford, UK: Gulf Professional Publishing, 2001).

[27] M. Warner, B. Eskenazi, P. Mocarelli, P.M. Gerthoux, S. Samuels, L. Needham, D. Patterson, and P. Brambilla, "Serum Dioxin Concentrations and Breast Cancer Risk in the Seveso Women's Health Study." *Environmental Health Perspectives* 110, no. 7 (2002): 625–8.

[28] Ibid.

[29] J.M. Hollander, *The Real Environmental Crisis* (Berkeley, CA: University of California Press, 2003).

Chapter 10

Engineering Wisdom

> By three methods we may learn wisdom: First, by reflection, which is
> noblest; second, by imitation, which is easiest; and third by experience,
> which is the bitterest.
>
> Confucius (c. 551–479 BC)

This book has discussed some of the key bioethical issues confronting the contemporary engineer. Unfortunately, the problems and uncertainties seem to outnumber the solutions. I do not expect to answer any of the questions presented to anyone's, even my own, full satisfaction. My real intention is to prime the pump and to help us to enter the "great conversation" – that has been going on for 2500 years. As Socrates would ask: How do engineers gain wisdom? The process is truly incremental. Engineers spend most of the preparation for the profession, learning the scientific principles and applying them to problems; what Confucius might have called "reflection." Next, we observe and apply lessons from our mentors. And, we hope not to experience the bitterness of direct failure by adopting practices that have worked for others in the past.

Seasoned professionals learn that texts, manuals, and handbooks are valuable; but only when experience and good listening skills are added to the mix, can they make wise (including moral) decisions. This is the sage advice offered by the great thinkers and philosophers through time.

One of the best synopses of engineering development is the counsel of Saint Peter,[1] who linked maturity with greater "self-control" or "temperance" (Greek *kratos* for "strength"). Interestingly, Saint Peter considered knowledge as a prerequisite for temperance. Thus, from a professional point of view, we could take his argument to mean that understanding and appropriately applying scientific theories and principles only becomes truly useful after practical experience. This is, in fact, the path taken toward the preparation of most engineers. For example, engineers who intend to practice must first submit to a rigorous curriculum (approved and accredited by the Accreditation Board for Engineering and Technology), then must sit for the Future Engineers (FE) examination. After some years in the profession (assuming tutelage by and ongoing intellectual osmosis with more experienced professionals), the engineer has sufficiently demonstrated *kratos* (strength), and sits for the Professional Engineers (PE) exam. Only after passing the PE exam does the state licensing authority and National Society for Professional Engineering certify that the engineer is a "professional engineer" and eligible to use the initials PE after one's name.[2] The engineer is, supposedly, now schooled beyond textbook knowledge. The professional status demarks a transition from knowing the "what" and the "how" to knowing the "why" and the "when." The engineer knows more about why technical and ethical problems require a complete understanding of the facts and possible outcomes (i.e., conditional probabilities of events). Details and timing are critical attributes of a good engineer. The wise engineer grows to appreciate that the correct answer to many engineering problems is "It depends." Engineering problems with bioethical import are even more complex and complicated than most, so the nurtured *kratos* has an added measure of importance.

Engineers are always concerned with outcomes. We want to see our ideas and designs come to fruition. When we start a project, we say, "This is what my client wants (the ends), so this is how I will make it happen (the means)." Sometimes, in the course of meeting our objectives, we may cause some harm, the well-known problem of using our ends to justify our means. Again, we are helped by the famous philosopher Immanuel Kant's prescription for moral behavior, often referred to as "duty ethics." Kant proposed a solution to the quandary of when our means become immoral and unjust, even when applied to a noble outcome. His categorical imperative[3] affirms that an act is ethical by predicting what would happen if that act were a law adopted by everyone. If the law helps humankind, the act is moral; if it hurts others on the whole, the act is immoral. Kant's categorical imperative is a societal antidote to the problem of noble engineering ends being justified by immoral means, so common to bioethical problems. For example, an argument against using embryonic stem cells (a biological means) to treat chronic maladies like Parkinson's disease (a biological end) is that the means degrades an overall respect for human life. Respect trumps medical expediency. The categorical imperative applies to all public health and global bioethical issues since if an action is not followed by a sufficient number of the members of the society, such as preventing and treating diseases, conserving energy, or properly disposing of hazardous wastes, a process or resource will not be preserved and will not be sustainable. Thus, the collective action of individuals directly affects the scale of the impact.

Recalling Elizabeth Kiss's "Six O'clock News" imperative,[4] if you are pondering whether something is ethical or not, consider how your friends and family would feel if they heard about all of its details on tonight's TV news. Put colloquially, "What would your mother think?" That may cause one to consider fully the possible consequences of one's decision. This is actually a fairly low level of moral behavior. For example, it would probably be categorized as a "conventional" type morality as defined by Kohlberg (Table 1.2).[5]

Kant's categorical imperative is but one of the tools given to us by the essential ethical thinkers. For example, empathy is not the exclusive domain of duty ethics. In teleological ethics, empathy is one of the palliative approaches to deal with the problem of "ends justifying the means." Other philosophers also incorporated the empathic viewpoint into their frameworks. John Stuart Mill's utilitarianism axiom of "greatest good for the greatest number of people" is moderated by his "harm principle," which states that, even though an act can be good for the majority, it may still be unethical if it causes undue harm to even one person. Empathy also comes into play in contractarianism, as articulated by Thomas Hobbes as social contract theory. John Rawls has moderated the social contract with the "veil of ignorance" as a way to consider the perspective of the weakest, one might say "most disenfranchised," members of the society.[6] Finally, the rational-relationship frameworks incorporate empathy into all ethical decisions when they ask the guiding question of "what is going on here?" In other words, what benefit or harm, based on reason, can I expect from actions brought about by the decision I am about to make? If we fail to identify even seemingly small outcomes that can result from our designs, we put society at risk. A warning is needed here: reason must be informed by inviolable ethical principles. One test of harm *versus* benefit is to be empathetic to all others, particularly the weakest members of the society, those with little or no "voice."

As professionals, engineers strive for excellence. This is articulated in the codes of ethics, which extend the mandate beyond avoiding actions that are clearly wrong, and move us to strive for engineering accomplishments that advance human endeavors. The engineering profession pushes us to seek ways to do what is right and to keep striving to improve. This creates tension between doing what is "safe" and what is "better." Part of the formula for ethical behavior is to know who is affected by what we do. But for whom do we strive to do what is right? Certainly, the company, agency, or holders of contracts are our clients, but our actions must not simply solve immediate problems for a specific client. Rather, they must be viewed as to how they will play out in the larger public venues and for future generations.

All design decisions are uncertain to some degree. Thus, we are seldom certain about the extent of influence of our ethical decisions. We may need to make these decisions conservatively, following a "precautionary principle," which states that if the consequences of an action, such as the application of a new technology, are unknown but the possible scenario is sufficiently devastating, then it is prudent to avoid the action. However, the precautionary approach must also be balanced against opportunity risks. In other words, by our extreme caution, are we missing opportunities that would better serve the public and future generations? Are the risks of a new technology acceptable in light of possible benefits? The key is the full and accurate characterization of risks and benefits. Unfortunately, design decisions are often not fully understood until after the fact (and viewed the prism of lawsuits and media frenzies). This text has recommended a fact-based, morally sound, logical approach.

ETHICS AND CHAOS

Scale and complexity enter into bioethical decision making. In fact, medical ethicist, Dianne Irving, argues that scientific ethical decision making is chaotic, such as morality associated with embryonic research:

> One philosopher's favorite dictum would seem quite appropriate here: "A small error in the beginning leads to a multitude of errors at the end" (paraphrased). Nowhere is this more fundamental than here. Any error in the science will have a rippling effect on the philosophical anthropology, ethics, sociology, politics, law, and theology – degrading our knowledge and understanding of the real dignity and status of the human embryo.[7]

The philosopher to whom Irving is referring is Saint Thomas Aquinas. Arguably, Aristotle had similar insights. Engineers might cast this as saying that appropriate ethical thinking must consider initial conditions, since in all likelihood they will have profound effects on the outcome. This is the teleological ethical framework, wherein the outcomes in a large way determine the ethics of a matter.

Engineering competence takes us in the right direction. We must excel in what we know and how well we do our technical work. This is a necessary requirement of the engineering experience, but it is not the only part. Engineering schools have increasingly recognized that engineers need to be both competent and socially aware. The ancient Greeks referred to this as *ethike arêtai* ("skills of character"). The competence of the professional engineer is inherently linked to character. So, being a good engineer requires both technical skill and a moral grounding. Oddly enough, the controversial and nihilistic philosopher Friedrich Nietzsche (1844–1900) foresaw the societal problems of the technical revolution and the need for engineers to be wary of complicity in harm:

> Measured by the standards of the ancient Greeks, our entire modern being, insofar as it is not weakness but power and consciousness of power, looks like sheer hubris and godlessness; for the very opposite of those things we honour today had for the longest period conscience on their side and god to guard over them. Our entire attitude to nature today, our violation of nature, with the help of machines and the unimaginable inventiveness of our technicians and engineers, is hubris . . . .[8]

Nietzsche is famous for his often misunderstood contention that the source of morality can no longer be God (note he did not capitalize God in the quote). This means that Nietzsche and the even more pessimistic Jean-Paul Sartre argued that morality cannot be traced to a cosmic moral order. Many scientists[9] and modern philosophers hold this position, but most Ancients and philosophers up to the nineteenth century would have disagreed. Also, most lay people would disagree as well. I would argue that the modernists have actually made the scientific method their "god," so their outward rationalism is betrayed by their own dogma (i.e., science is god). Aldous Huxley would be proud, but in the end scientism itself turns out to be a religion. So, there is a ready-made conflict between the so-called and self-proclaimed intellectuals and most everyone else.

MACROETHICS AND MICROETHICS

Bioethics Question: What is the bioethical responsibility of the engineering profession?

Ethical approaches can differ by scale. For example, the engineering profession has a moral responsibility to society to ensure that designs and technologies are in society's best interest. In addition, the individual engineer has a specific set of moral obligations to the public to and the client. The moral obligations of the profession as a whole are greater than the sum of the individual engineers' obligations. The profession certainly needs to ensure that each of its members adheres to a defined set of ethical expectations. This is a necessary, but insufficient, condition for the ethos of engineering. The "bottom-up" approach of ensuring an ethical engineering population does not completely ensure that many societal ills will be addressed.

Political theorist, Langdon Winner, has succinctly characterized the twofold engineering moral imperative:

> Ethical responsibility . . . involves more than leading a decent, honest, truthful life, as important as such lives certainly remain. And it involves something much more than making wise choices when such choices suddenly, unexpectedly present themselves. Our moral obligations must . . . include a willingness to engage others in the difficult work of defining what the crucial choices are that confront technological society and how intelligently to confront them.[10]

This engagement necessitates a systematic and comprehensive viewpoint that applies both the bottom-up and the top-down approaches.

Most professional ethics texts, including those addressing engineering ethics, are concerned with what has come to be known as microethics, which is "concerned with individuals and the internal relations of the engineering profession."[11] This is distinguished from macroethics, which is "concerned with the collective, social responsibility of the engineering profession, and societal decisions about technology."[12] Neither micro- nor macroethics can be ignored. Furthermore, they are interdependent in engineering.

FUTURE DIRECTIONS

The scientific research and professional community is most interested in advancing the science, with ethics just one of many constraints on this progress. Interestingly, scientific handbooks and manuals may have a separate ethics chapter or section added to the technical subject matter. Ethics is not presently "built-in." For example, bioethics does not appear in the *Oxford Dictionary of Biochemistry and Molecular Biology*,[13] nor in the *Dictionary of Cell Biology*.[14] And these are the kinds of references for which the researchers in training first reach. The challenge of Ph.D. research is that the student quickly surpasses the technical knowledge of the mentor, but the role of the mentor increases in regard to deciding the ethical paths to take. For example, the student continues to gain expertise in a specific area of engineering research, even beyond that of the Ph.D. committee members. However, the student has had less time and experience to know the difference between whether an engineer *can* and whether an engineer *should* do something.

Engineering research continues to become smaller in focus. Many research institutions have numerous nanoscale projects (within a range of a few angstroms). Nascent areas of research include ways to link protein engineering with cellular and tissue biomedical engineering applications (e.g., drug delivery and new devices); ultradense computer memory; nonlinear dynamics and the mechanisms governing emergent phenomena in complex systems; and state-of-the-art nanoscale sensors (including photonic devices). Complicating the potential societal risks, much of this research frequently employs biological materials and self-assembly devices to design and build some strikingly different kinds of devices. Among the worst case, scenarios has to do with the replication of the "nano-machines."

We are becoming increasingly separated from these process, as machines think and do more things for us. We are in an ethical quarndary. We need to advance the state of the science to improve the quality of life (e.g., treating cancer, Parkinson's disease, Alzheimer's disease, and improving life expectancies), but in so doing are we introducing new societal risks? Are we trampling on human values? In his recent book, *Catastrophe: Risk and Response*, Richard Posner, a judge of the US Court of Appeals for the Second Circuit, describes this paradox succinctly:

> Modern science and technology have enormous potential for harm. But they are also bounteous sources of social benefits. The one most pertinent . . . is the contribution technology has made to averting both natural and man-made catastrophes, including the man-made catastrophes that technology itself enables or exacerbates.[15]

Posner gives the example of the looming threat of global climate change, caused in part by technological and industrial progress (mainly the internal combustion engine and energy production tied to fossil fuels). Emergent technologies can help to assuage these problems by using alternative sources of energy, such as wind and solar, to reduce global demand for fossil fuels. We have discussed a number of other pending problems, such as the unknown territory of genetic engineering, like genetically modified organisms (GMOs) used to produce food. There is a concomitant fear that the new organisms will carry with them unforeseen ruin, such as in some way affecting living cell's natural regulatory systems. An extreme viewpoint, as articulated by the renowned physicist Martin Rees, is the growing apprehension about nanotechnology, particularly its current trend toward producing "nanomachines." For example, a recent survey by Peter D. Hart Associates, found that of persons taking a position about nanotechnology, most consider the risks to outweigh the benefits (national poll taken in August 2006 of 1014 US adults). Interestingly, the approval rate dramatically increases as people hear more about nanotechnology. Biological systems, at the subcellular and molecular levels, could very efficiently produce proteins, as they already do for their own purposes. By tweaking some genetic materials at a scale of a few angstroms, parts of the cell (e.g., the ribosome) that manufacture molecules could start producing myriad molecules designed by scientists, such as pharmaceuticals and nanoprocessors for computing. However, Rees is concerned that such assemblers could start self-replicating (like they always have), but without any "shut-off." Some have called this the "gray goo" scenario, i.e., accidentally creating an "extinction technology" from the cell's unchecked ability to exponentially replicate itself if part of their design is to be completely "omnivorous," using all organic matter as food! No other "life" on earth would exist if this "doomsday" scenario were to occur.[16]

Certainly, this is the stuff of science fiction, but it does call attention to the need for vigilance, especially since our track record for becoming aware of the dangers of technologies is so frequently tardy. Messing with genetic materials may introduce new pathogenic strains, as well as harm biodiversity, i.e., the delicate balance among species, including trophic states (producer–consumer–decomposer) and predator–prey relationships.

Engineers and scientists are expected to push the envelopes of knowledge. We are rewarded for our eagerness and boldness. The Nobel Prize, for example, is not given to the chemist or physicist who has aptly calculated important scientific phenomena, with no new paradigms. It would be rare indeed for engineering societies to bestow their most prestigious awards to the engineer who for an entire career used only proven technologies to design and build structures. This begins with our general approach to contemporary scientific research. We are rugged individualists in a quest to add new knowledge. For example, aspirants seeking Ph.D.s must endeavor to add knowledge to their specific scientific discipline. Scientific journals are unlikely to publish articles that do not contain at least some modicum of originality and newly found information.[17] We award and reward innovation. Unfortunately, there is not a lot of natural incentive for the innovators to stop what they are doing to "think about" possible ethical dilemmas propagated by their discoveries.[18]

Thus, nanoscale and other emergent research must not only include the fundamental research and possible applications but also a rigorous approach to investigate likely scenarios, along with worst-case ("doomsday") outcomes. This link between fundamental work and outcomes becomes increasingly crucial as such research reaches the marketplace relatively quickly and cannot be confined to the relative safety and rigor of the lab. Better ways are needed to ensure that laboratory and individual areas of research consider possible larger societal risks. For example, Duke University is presently performing a comparative analysis of three modes of teaching and learning macroethics:

1. Development and analysis of case studies.
2. Ongoing actual and simulated ethics education involving active learning in the student's technical context (e.g., benchside consultation and game engine use).
3. Development and/or adaptation of codes of ethics for emerging fields.

In doing so, the work will address two competing challenges: preventing future macroethical problems that may not be apparent to the researcher and encouraging the advancement of research needed to address large societal problems. The two are not mutually exclusive, so one of the evaluation criteria for the modes will be the ability to optimize utility while adding to the knowledge base within an ethically sound system.

Nanoscale and other biotechnological areas of research transport the engineer into uncomfortable venues. Whether a risk is acceptable is determined by a process of making decisions and implementing actions that flow from these decisions to reduce the adverse outcomes, or at least to lower the chance that negative consequences will occur.[19]

Risk managers can expect that whatever risk remains after their project is implemented; those potentially affected will not necessarily be satisfied with that risk. It is difficult to think of any situation where anyone would prefer a project with more risk than the one with less risk, all other things being equal. But who decides what comprises "other things being equal?" It has been said that "acceptable risk is the risk associated with the best of the available alternatives, not with the best of the alternatives which we would hope to have available."[20]

Since risk involves chance, risk calculations are inherently constrained by three conditions:

1. The actual values of all important variables cannot be known completely and thus cannot be projected into the future with complete certainty.
2. The physical science of the processes leading to the risk can never be fully understood, so the physical, chemical, and biological algorithms written into predictive models will propagate errors in the model.
3. Risk prediction using models depends on probabilistic and highly complex processes that make it infeasible to predict many outcomes.[21]

The "go or no go" decision for most biomedical and biosystem engineering designs or projects is based upon a kind of "risk-reward" paradigm, which balances benefits and costs.[22] This creates the need to have costs and risks significantly outweighed by some societal good. The adverb "significantly" reflects two problems: the uncertainty resulting from the three constraints described above and the "margin" between good *versus* bad. Significance is the province of statistics, i.e., it tells us just how certain we are that the relationship between variables cannot be attributed to chance. But, when comparing benefits to costs, we are not all that sure that any value we calculate is accurate. For example, a benefit–cost ratio of 1.3 with confidence levels at a range between 1.1 and 1.5 is very different from a benefit–cost ratio of 1.3 with a confidence range between 0.9 and 1.7. The former does not include any values less than 1, while the latter does (i.e., 0.9). This value means that even with all the uncertainties, our calculation shows that the project could be unacceptable. This situation is compounded by the second problem of not knowing the proper margin of safety; that is, we do not know the overall factor of safety to ensure that the decision is prudent. Even a benefit–cost ratio that appears to be mathematically high, i.e., well above 1, may not provide an ample margin of safety, given the risks involved.

The likelihood of unacceptable consequences can result from exposure processes, from effects processes or from both processes acting together. So, four possible permutations can exist:

1. Probabilistic exposure with a subsequent probabilistic effect
2. Deterministic exposure with a subsequent probabilistic effect
3. Probabilistic exposure with a subsequent deterministic effect
4. Deterministic exposure with a subsequent deterministic effect.[23]

A risk outcome is deterministic if the output is uniquely determined by the input. A risk outcome is probabilistic if it is generated by a statistical method, e.g., randomly. Thus, the accuracy of a deterministic model depends on choosing the correct conditions, i.e., those that will actually exist during a project's life and correctly apply the principles of physics, chemistry, and biology. The accuracy of the probabilistic model depends on choosing the right statistical tools and correctly characterizing the outcomes in terms of how closely the subpopulation being studied (e.g., a community or an ecosystem) resembles those of the population (e.g., do they have the same factors or will there be sufficient confounders to make any statistical inference incorrect?). A way of looking at the difference is that deterministic conditions depend on how well one understands the science underpinning the system, while probabilistic conditions depend on how well one understands the chance of various outcomes (Table 10.1).

Table 10.1
Examples of methods of predicting risk

	Probabilistic exposure	Deterministic exposure
Probabilistic effect	*Contracting the West Nile virus.* Although many people are bitten by mosquitoes, most mosquitoes do not carry the West Nile virus. There is a probability that a person will be bitten and another much lower probability that the bite will transmit the virus. A third probability of this bitten group may be rather high that once bitten by the West Nile virus-bearing mosquito, the bite will lead to the actual disease. Another conditional probability exists that a person will die from the disease. So, death from a mosquito bite (probabilistic exposure) leads to a very unlikely death (probabilistic effect).	*Occupational exposure to asbestos.* Exposure to asbestos from vermiculite workers is deterministic because the worker chooses to work at a plant that processes asbestos-containing substances. This is not the same as the person choosing to be exposed to asbestos, only that the exposure results from an identifiable activity. The potential health effects from the exposures are probabilistic, ranging from no effect to death from lung cancer and mesothelioma. These probabilistic effects increase with increased exposures, which can be characterized (e.g., number of years in certain jobs, availability of protective equipment, and amount of friable asbestos fibers in the air).
Deterministic effect	*Death from methylisocyanate exposure.* Exposure to a toxic cloud of high concentrations of the gas methylisocyanate (MIC) is a probabilistic exposure, which is very low for most people. But, for people in the highest MIC concentration plume, such as those in the Bhopal, India tragedy, death was 100% certain. Lower dose led to other effects, some acute (e.g., blindness) and others chronic (e.g., debilitation that led to death after months or years). The chronic deaths may well be characterized probabilistically, but the immediate poisonings were deterministic (i.e., they were completely predictable based on the physics, chemistry, and biology of MIC).	*Generating carbon dioxide from combusting methane.* The laws of thermodynamics dictate that a decision to oxidize methane, e.g., escaping from a landfill where anaerobic digestion is taking place, will lead to the production of carbon dioxide and water (i.e., the final products of complete combustion). Therefore, the engineer should never be surprised when a deterministic exposure (heat source, methane, and oxygen) lead to a deterministic effect (carbon dioxide release to the atmosphere). In other words, the production of carbon dioxide is 100% predictable from the conditions. The debate on what happens after the carbon dioxide is released (e.g., global warming) is the province of probabilistic and deterministic models of these effects, but not of the chemical reactions underpinning the effects.

Actually, the deterministic exposure/deterministic effect scenario is not really a risk scenario because there is no "chance" involved. It would be like saying that releasing a 50 kilogram steel anvil from 1 meter above the earth's surface runs the risk of falling toward the ground! The risk comes into play only when we must determine external consequences of the anvil falling. For example, if an anvil is suspended at the height of 1 meter by steel wire and used by workers to perform some task (i.e., a deterministic exposure), there is some probability that it may fall (e.g., studies have shown that the wires fail to hold one in ten thousand events, i.e., failure probability of 0.0001); so, this would be an example of a deterministic exposure followed by a probabilistic effect (wire failure), i.e., permutation number 2.

Estimating risk using a deterministic approach requires the application of various scenarios, e.g., a very likely scenario, an average scenario, or a worst-case scenario. Very likely scenarios are valuable in some situations when the outcome is not life threatening or when there are severe effects, like cancer. For example, retailers may not care so much about the worst case (a customer is not likely under most scenarios to buy their product) in designing a store, but wants to avoid the risk of losing a likely customer (e.g., the most potential Armani suit customers are not likely to want to hear loud heavy metal music when they walk in the store; but conversely loud heavy metal music may be a selling feature for most customers looking to buy tube amplifiers at a guitar store). The debate in the public health arena is often between a mean exposure and a worst-case exposure (i.e., maximally exposed and highly sensitive individuals). The latter is more protective, but almost always more expensive and difficult to attain. For example, lowering the emissions of a toxic substance from a industrial stack to protect the mean population of the state from the effects of exposures is much easier to achieve than lowering the emissions to protect an asthmatic, elderly person living just outside of the power plant property line (Figure 10.1). While the most protective standards are the best, the feasibility of achieving them can be a challenge (Figure 10.2). For many biomedical decisions, general rules will vary. For example, a device or medical protocol may have to be designed for highly unlikely conditions (e.g., two standard deviations from the mean – very sick people).

Actual or realistic values are input into the deterministic model. For example, to estimate the risk of a valve malfunction from the materials used, metabolic in a processes, the characteristics of the bodily fluids, the vulnerability of the valve materials to rupture, and the expected flow through the valve, from which the risk of failure is calculated. A probabilistic approach would require the identification of the initiating events and the metabolic states to be considered; analysis of the adverse outcome using statistical analysis tools, including event trees; application of fault trees for the systems analyzed using the event trees (i.e., reliability analyses; see Chapter 3), collection of probabilistic data (e.g., probabilities of failure and the frequencies of initiating events); and the interpretation of results.

Human beings engage in risk management decisions every day. They must decide throughout whether the risk from particular behaviors is acceptable or whether the potential benefits of a behavior do not sufficiently outweigh the hazards associated with that behavior. In engineering terms, they are optimizing their behaviors based upon a complex set of variables that lead to numerous possible outcomes. Engineers are designing systems to help lower some of these risks and are morally obligated to do so.

In addition, the engineer must weigh and balance any responsibility to represent the client with due diligence. This diligence must be applied to designs and plans for manufacturing processes that limit, reduce, or prevent failures. Ultimately, the engineer participates in means to remedy the problem (i.e., to ameliorate health, environmental, and welfare damages). The evaluation and selection of the best alternative is the stuff of ethical decision making.

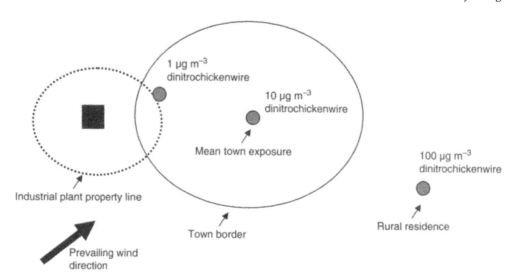

Figure 10.1 Hypothetical approach in protecting public health based on different target populations from emissions of a fictitious toxic substance, "dinitrochickenwire." Protecting the population closest to the source requires the lowest emission ($1 \mu g \ m^{-3}$). Protecting the most people in town who will be exposed allows for an order of magnitude of higher release ($10 \mu g \ m^{-3}$). The very low-exposure scenario of rural residents allows for even higher emissions ($100 \mu g \ m^{-3}$). This approach can also be used in emergency situations, such as how and where to evacuate residents. The concentration would be even lower if the risks are based on a highly sensitive subpopulation (e.g., elderly, infants, or asthmatics), depending upon the effects elicited by the emitted pollutant. For example, if dinitrochickenwire was expected to cause cardiopulmonary effects in babies, an additional factor of safety may push the risk-based controls downward by a factor of 10 to achieve $0.1 \mu g \ m^{-3}$ to protect the maximally exposed, sensitive population. This is akin to the medical process of triage, where the neediest are helped first.

Given this logical and seemingly familiar role of the engineer, why then do disasters and ethical mishaps occur on our watch? What factors cause the engineer to improperly optimize for the best outcome? In part, failures in risk decision making and management are ethical in nature. Sometimes organizational problems and demands put engineers in situations where the best and the most moral decision must be made against the mission as perceived by management. Working within an organization has a way of inculcating the "corporate culture" into professionals. The process is incremental and can "desensitize" employees to acts and policies that an outsider would readily see to be wrong. Much like the proverbial frog placed in water that gradually increases to the boiling pot, an engineer can work in gradual isolation, specialization, and compartmentalization that ultimately changes to immoral or improper behavior, such as ignoring key warning signs that a decision to change a design will have an unfair and disparate impact on certain members of society, that health and safety are being compromised, and that political influence or the "bottom line" of profitability is disproportionately weighted in an engineer's recommendation.[24]

Another reason why optimization is difficult is that an engineer must deal with factors and information that may have not been adequately addressed during formal engineering training or even during career development. Although health decisions must always give sufficient attention to the physical science calculations, these quantitative results are tempered with feasibility considerations.

Engineering risk management is a type of "teleology." The term is derived from the Greek, *teleos*, meaning "outcome." So, what engineers do is based upon something "out there" that we would like

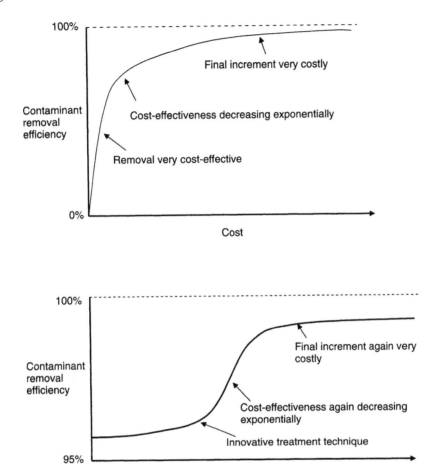

Figure 10.2 Prototypical contaminant removal cost-effectiveness curve. In the top diagram, during the first phase, a relatively large amount of the contaminant is removed at comparatively low costs. As the concentration in the environmental media decreases, the removal costs increase substantially. At an inflexion point, the costs begin to increase exponentially for each unit of contaminant removed, until the curve nearly reaches the steady state where the increment needed to reach complete removal is very costly. The top curve does not recognize innovations that, when implemented, as shown in the bottom diagram, can make a new curve that will again allow for a steep removal of the contaminant until its cost-effectiveness decreases. This concept is known to economists as the law of diminishing returns

to achieve, i.e., a desired outcome, or something we would like to avoid, i.e., an adverse outcome. In particular, engineering is "utilitarian," to the extent that we are called upon "to produce the most good for the most people."[25] However, engineers are also duty bound to our codes of ethics, design criteria, regulations, and standards of practice. In this way, engineering is a type of "deontology."[26] Engineers must, to the best of their abilities, consider all possible design outcomes, planned or otherwise. But this is not easy because the most appropriate "benchmarks" for success in engineering projects are moving targets. There is no assurance of success at the outset, especially for innovative designs. Even if the design is implemented to meet all specifications, the project could be deemed a failure.

At a minimum a designer should consider and ensure the design provides:[27]

1. Overall protection of human health and environment
2. Compliance with applicable or relevant and appropriate requirements
3. Long-term effectiveness and permanence
4. Feedback on progress
5. Short-term effectiveness
6. Ease of implementation
7. Cost considerations
8. Regulatory authority acceptance
9. User/Patient acceptance

The first and fourth criteria are clearly the product of a sound quantitative risk assessment. The other criteria must also include semi-qualitative and qualitative information.

THE HUMBLE ENGINEER

The fool doth think himself wise, but the wise man knows himself to be a fool.

Shakespeare, Touchstone

The Oracle at Delphi told Chaerephon that no one in the world was wiser than Socrates. This genuinely surprised Socrates, so he set out to find those wiser than he. After his quest, Socrates found the Oracle to be correct. He was indeed wisest only because he knew nothing.

Perhaps, a fitting end to this book is a call for humility. Bioethical issues are complex. They can be difficult to discuss in an objective and thoughtful manner. There is much we do not know about the science and engineering aspects of many of the cases and problems mentioned. We must be open to new paradigms and keep an eye peeled toward better ways to solve problems.

I can think of no better advice for the engineer to gain wisdom than to grasp and employ Aristotle's suggested five intellectual virtues,[28] the means by which the soul arrives at the truth:

1. Arrive at scientific knowledge using sound science (deduction and inductive reasoning).[29]
2. Know that technical skill involves the application of proper reasoning.
3. Understand that prudence (practical wisdom) helps us to pursue the "good life."
4. Appreciate that intuition helps us to grasp first principles from which we derive scientific truths.
5. Remind ourselves that wisdom is a combination of scientific knowledge and intuition, which helps us arrive at the highest truths of all.

Putting a bit less philosophically and in engineering vernacular, finding the best ethical approach is much like the testing we require on materials and the performance of devices and systems. According to Michael Davis of the Illinois Institute of Technology, most engineers do not need to know specific moral theories to discern what is right or wrong in a given situation. What is really needed is a systematic means of probing ethical questions, "not a toolbox with one or more theories but one easy-to-use method for guiding discussion, focusing on reasons, and forcing judgments." He has devised a seven-step guide to ethical decision making:[30]

1. *State your concerns.* Davis gives example questions: "There's something about this decision that makes me uncomfortable." or "Do I have a conflict of interest?" Others could be: "This goes against my conscience" or "I do not agree with the definition here of X" (e.g., life, death, acceptable risk, human being, justifiable means, etc.).
2. *Check facts.* Davis contends that "many problems disappear upon closer examination of situation, while others change radically."
3. *Identify relevant factors.* Who are potentially affected, what laws come into play, what does the code of ethics say, what are the practical constraints (e.g., is it realistic to tell everyone or must you rely on an existing communication network?).
4. *Develop list of options.* Davis recommends open-mindedness and imagination. Do not simply consider the best (i.e., positive paradigm) and worst (i.e., negative paradigm) in the circumstances surrounding the decision. This is akin to brainstorming, where even unlikely or farfetched possibilities should at least be considered if they can potentially help.
5. *Test the options.* Davis recommends using a number of tests:

   - *Harm test* – Does this option do less harm than do the other alternatives?
   - *Publicity test* – Would I want my choice of this option be published in the newspaper? (Akin to Kiss's 6:00 o'clock corollary).
   - *Defensibility test* – Could I defend my choice of option before Congressional committee or committee of peers?
   - *Reversibility test* – Would I still think choice of this option good if I were adversely affected by it?
   - *Colleague test* – What will my colleagues say when I describe my problem and suggest this option as my solution?
   - *Professional test* – What might my profession's governing body or ethics committee say about this option?
   - *Organization test* – What does the company's ethics officer or legal counsel say about this?

6. Select the best option based on steps 1–5.
7. Review all the previous steps to see how choosing this option changes to the decision tree or critical path. This can include finding ways to avoid similar ethical problems in the future.

Bioethical issues, more than most ethical challenges, require humility. Nature has a way of teaching engineers to be humble. Engineers are reminded of our limitations by natural disasters, such as the recent devastations of hurricanes, earthquakes and tsunamis, human disasters like the World Trade Center collapses and other acts of terrorism, and the tragedy of actual and potential pandemics, such as the ravages of the disease brought on by human immunodeficiency virus in Africa. We are also humbled when our designs fail and when we see them abused. On the other hand, for centuries, engineers' innovations have provided the front line of preventing and addressing such disasters. The National Academy of Engineering[31] recently declared:

Engineering, through its role in the creation and implementation of technology, has been a key force in the improvement of our economic well-being, health, and quality of life. Three hundred years ago the average life span was 37 years, the primary effort of the majority of humans was focused on provisioning their tables, and the threat of sudden demise due to disease was a lurking reality. Today, human life expectancy is approaching 80 years in many parts of the world as fundamental advances in medicine and technology have greatly suppressed the occurrence of and mortality rates for previously fatal diseases and the efforts of humankind are focused largely on enhanced quality of life.

Indeed, the issues of today are bringing about new challenges. Few can be addressed with a "one size fits all" mindset. The best advice about bioethics may come from our friend Amy in Chapter 1. Recall that she was more concerned in finding a system that provides acceptable results on an ongoing basis than simply solving a short-term problem. Individual cases can be arbitrary and can limit ethical creativity by imposing a forced choice among a few right and wrong answers. We can learn much from Amy's refusal to acquiesce and to follow the path of least resistance. Instead, the ethical engineer must seek an open-ended set of possibilities to arrive at acceptable solutions to bioethical challenges.

Engineers must analyze each possible outcome from a number of perspectives, including those that are not so familiar to the physical scientists, such as the social sciences. When the facts tell us that something is wrong, we must not take the easy way out by acquiescing to the positions of those who fund our research, write our newspaper articles, or who are politically powerful. If they are wrong, sound science dictates that we say so. From this position of honesty, the engineer is surely on the path to understanding and helping to solve the bioethical challenges of our times.

Perhaps the most eloquent expression of our bioethical mandate comes from engineer-President Herbert Hoover:

> To the engineer falls the job of clothing the bare bones of science with life, comfort and hope. . . . The engineer himself looks back at the unending stream of goodness that flows from his successes with satisfactions that few professions may know.

May all our careers be unending streams of goodness.

NOTES AND COMMENTARY

[1] Acts 24: 25 and II Peter 1: 6.

[2] I am so proud to announce that my son, Daniel Joseph Vallero, PE, reached this important milestone in 2006!

[3] I. Kant, *Groundwork of the Metaphysics of Morals*, trans. H.J. Paton (1785; Routledge repr., San Francisco, CA: HarperCollins, 1992).

[4] Personal communications, especially when cofacilitating Duke's annual Responsible Conduct of Research training for Ph.D. students.

[5] L. Kohlberg, *The Philosophy of Moral Development*, vol. 1 (San Francisco, CA: Harper & Row, 1981).

[6] J. Rawls, *A Theory of Justice* (1971; Belknap Press repr., Cambridge, MA, 1999).

[7] D. Irving, "Science, the Formation of Conscience and Moral Decision Making," Proceedings of the Guadalupan Appeal: The Dignity and Status of the Human Embryo, Mexico City, Mexico, 28 October 1999.

⁸ F. Nietzsche, *On the Genealogy of Morals*, trans.Walter Kauffman and R.J. Hollingdale (1989); (New York, NY: Vintage Books, 1887).

⁹ However many scientists, including myself, believe in a supreme cosmic authority as the Author of the commonality of moral principles.

¹⁰ L. Winner, "Engineering Ethics and Political Imagination," in *Broad and Narrow Interpretations of Philosophy of Technology*, ed. P.T. Durbin, 53–64 (The Netherlands: Kluwer Academic Publishers, 1990). (repr., *Ethical Issues in Engineering*, ed. D.G. Johnson, Englewood Cliffs, NJ.: Prentice-Hall, 1991).

¹¹ J.E. Herkert, 2004, "Microethics, Macroethics, and Professional Engineering Societies," in *National Academy of Engineering, Emerging Technologies and Ethical Issues in Engineering: Papers from a Workshop* (14–15 October 2003), 107.

¹² Ibid.

¹³ A.D. Smith (Managing Editor), *The Oxford Dictionary of Biochemistry and Molecular Biology* (Oxford, UK: Oxford University Press, 1997).

¹⁴ J.M. Lackie and J.A.T. Dow, *The Dictionary of Cell Biology* (London, UK: Academic Press Limited, 1995).

¹⁵ R.A. Posner, *Catastrophe: Risk and Response* (New York, NY: Oxford University Press, 2004).

¹⁶ M. Rees, *Our Final Hour: A Scientist's Warning: How Terror, Error, and Environmental Disaster Threaten Humankind's Future in This Century – On Earth and Beyond* (New York, NY, 2003).

¹⁷ Depending on the journal, this can contradict another tenet of scientific research, i.e., the research should be able to be replicated by other researchers, following the methodology described in the article, and derive the same results. However, there is little incentive to replicate research if the likelihood of publication is low; that is, the research is no longer "new" because it was conducted by the original researcher, so the journal may well reject the second, replicate research.

¹⁸ The engineering profession coming to grips with this issue. For example, in emergent "macroethical" areas like nanotechnology, neurotechnology, and even sustainable design approaches. For example, see National Academy of Engineering, 2004, *Emerging Technologies and Ethical Issues in Engineering*, The National Academies Press, Washington, DC.

¹⁹ The Royal Society, *Risk: Analysis, Perception and Management* (London, UK: The Royal Society, 1992).

²⁰ S.L. Derby and R.L. Keeney, "Risk Analysis: Understanding 'How Safe Is Safe Enough?' " *Risk Analysis* 1, no. 3 (1981): 217–24.

²¹ M.G. Morgan, "Probing the Question of Technology-Induced Risk," *IEEE Spectrum* 18, no. 11 (1981): 58–64.

²² Department of the Environment, United Kingdom Government, *Sustainable Development, The UK Strategy*, Cmnd 2426, HMSO, London, UK.

²³ Morgan, "Probing the Question of Technology-Induced Risk," *IEEE Spectrum* 18, no. 11 (1981).

²⁴ For case analyses where engineers have made such unethical decisions, see W.M. Evan and M. Manion, *Minding the Machines: Preventing Technological Disasters* (Upper Saddle River, NJ: Prentice-Hall PTR, 2002).

²⁵ J.S. Mill, 1863, *Utilitarianism*. See M. Martin and R. Schinzinger, *Ethic in Engineering* (New York, NY: McGraw-Hill, 1996), for an excellent discussion of the roles of moral reasoning and ethical theories in engineering decision making.

[26] In previous books, I had shown this term to be of Greek derivation. However, Greek for obligation is "deon". The actual derivation of "deontology" seems to be the prefix "de" added to "on," as in "ontology". Thus, deontology is the opposite(de) of ontology (what things are). Thus, it is the study if what things should be. My thanks to friend and colleague, Panos Georgopoulos, for these insights.

[27] This process follows that called for in National Research Council, *Risk Assessment in the Federal Government: Managing the Process* (Washington, DC: National Academy Press, 1983); and National Research Council, *Issues in Risk Assessment* (Washington, DC: National Academy Press, 1993).

[28] Aristotle, 350 BC, *Nicomachean Ethics*, Book VI.

[29] I recommend knowledge and experience-based intuition as well.

[30] M. Davis, 2005, Teaching Ethics across the Engineering Curriculum, the Online Ethics Center for Engineering and Science at Case Western Reserve University, http://onlineethics.org/essays/education/davis.html (accessed 14 September 2006).

[31] National Academy of Engineering, *The Engineer of 2020: Visions of Engineering in the New Century* (Washington, DC: National Academy Press, 2004).

Epilogue: Practical Bioethics

The Prologue presents a number of shortcomings in applying the prototypical approaches for teaching ethics to engineers. Instead, depending on the subject matter, this book has used a number of methods, including thought experiments and cases to address a range of bioethical issues. Open-ended, process-oriented ethical analysis was applied in hopes that the process of discovery would be similar to the design approach employed by engineers. Indeed, engineers do start with a modicum of information in any design.

SHUTTING DOWN THE PUMP

In the Prologue we considered the hypothetical case entitled "Groupthink and Rugged Individualism" as a foreshadowing of some of the not so obvious ethical issues to be considered in the text. Admittedly, the case is not the most exhilarating. In fact, it may be the least controversial and easiest to solve of any in test. Or, so it seems.

Like many bioethical problems, numbers come into play. Bob the engineer is clearly outnumbered by other practitioners. He is the only engineer. So, the first issue is whether Bob is at the service of the team or of his profession. Is his foremost obligation to the team or to the practice of engineering? Hopefully, it is now clear that the correct answer is the latter.

Another interesting, and often unappreciated, aspect of the problem for Bob is that he has a unique and clearly defined responsibility. This could lead to complacency. The engineering canon does require that an engineer must be competent and perform services only in areas of their expertise. But does that mean that Bob does not have a larger responsibility to point out improprieties, even those where others clearly have greater expertise? Arguably, Bob has a sufficient understanding of informed consent to complain about the aspects of the study with regard to human subjects. In fact, Bob is now culpable for not having raised this issue. The reasonable engineer would likely have little sympathy for Bob's argument that others above him or with "biological" expertise should have taken care of the informed consent issues. As Bob Dylan reminds us in *Subterranean Homesick Blues,* "You dont have to be a weather man to know which way the wind blows".

Engineers have to be concerned about the subject matter. In this case, the study was being done in a terrorist training camp. There are substantial issues associated with this fact. Engineers are not anthropologists! Bob has a paramount duty to the health, safety, and welfare of the public. The fact that he is accompanying a group of scientists in a terrorist camp does not alleviate his moral obligations. Should he have accepted this job? As Michael Davis, philosopher at Illinois Institute of Technology, recently advised me, "It is better to avoid temptation than to risk it." In other words, Bob is possibly putting his career and the public's safety at risk by even agreeing to participate in a study that directly or indirectly increases the threat of terrorism.

On the other hand, his colleagues in the study may believe that there is some greater good and noble end that makes the study worthwhile, such as having a better understanding of the factors that lead to terrorism, which in turn helps to prevent terrorism. Can this be likened to allowing a certain group of people to suffer (e.g., receiving the placebo instead of a known efficacious treatment) for the greater good of advancing the state of medical science (i.e., case-control studies that lead to better drugs)? Or is this simply a publishable and noteworthy endeavor? As Feynman reminds us, it is very easy to fool ourselves and we are very good at rationalizing what we really want to do.

Finally, since most engineers do not publish regularly, they may have not worried much about authorship and confidentiality. But, the arguments about authorship are similar to most team and group projects. Reward and recognition are rarely just and fair. It is clearly unethical not to recognize others' achievements, even if it has been a while since they contributed. Immediacy tends to inflate the perceived contribution. A task done yesterday is easier to recognize than someone's original idea two years ago. However, there is no time limit on intellectual property, at least from an ethical perspective. One ought never to claim another's ideas as one's own. The best policy is full disclosure. No matter how long ago or where the person has since moved, it is still their idea. Also, the rank of the person should not matter. There is an instructive case of a proposal to the National Science Foundation or other granting institutions that was written substantially by a student who graduated and left the institution. After some time, the former student's professor submitted the proposal without attributing it to the student. Such behavior is unethical and probably illegal. The same can be said of any intellectual property, such as patents, so any institution needs a policy where property rights and recognition are agreed upon upfront and implemented justly.

The bottom line is that even a relatively boring case can be loaded with ethical content. And, as my philosopher colleagues keep reminding me, such issues have no final answer. I am not a philosopher by training, but I must agree that most bioethical issues are not conducive to final answers. But, they do benefit from asking and trying to answer them, especially if we give clear reasons for our answers.

OBJECTIVITY AND FINDING TRUTH

Scientists and engineers pride themselves on objectivity. In fact, one of the worst insults one can hurl at a researcher is the lack of objectivity. This may mean sloppiness in carrying out work in the laboratory or in the field. It may mean that the study approach and design were somehow flawed. It may even mean that the researcher has been unfair or even unethical in the way he or she went about my scientific endeavors. That said, the scientific community, which for the most part polices itself in matters of integrity, must be brutally honest about how objective it is being in any matter of importance. From the title of this book, it should be obvious that my assertion is that we as a community of scientific investigators and practitioners must apply systematic thinking in matters of life and personhood.

Engineers have both advantage and disadvantage when comes to matters of bioethics, including respect and personhood. The advantage is that engineers are, first and foremost, problem solvers. We are expected to look coldly at data and facts and to apply knowledge and wisdom to add value to solve a particular societal problem or take advantage of opportunities to improve the world. For example, biomedical engineers apply physics, chemistry, and biology to develop a device that improves motor-muscle response in persons with neurological disorders. The engineer designing such devices, if asked, would probably say that his or her principal value was in the ability to understand the basic sciences and to use models to predict under a controlled set of variables how the device will perform. Many, probably most, engineers would likely not consider it his or her place to decide whether it is right or wrong for such a device to be made, or even how such devices should be distributed equitably to those in need.[1] They perceive it to be the responsibility of the individual client, policy makers, or sanctioning and professional organizations, but not that of the individual engineer.

Some argue that the integrity of the entire profession is simply the aggregation of the moral behavior of its practitioners. This is the so-called bottom-up and microethical view discussed in this text. In other words, if we can prescribe behaviors and prohibit others for each practitioner in a profession, then that is all that is needed to ensure the profession as a whole will have integrity. This is the justification, for example, of the professional codes of ethics. Others argue that the microethical approach is certainly necessary, but is not sufficient to ensure professional and scientific integrity. They argue that, in addition, the profession needs a "macroethical" view. In our motor-muscle device example, our individual engineer may have met the requirements of every canon and principle of the biomedical engineering profession. That is, the engineer did not lie, cheat, or steal. The engineer was careful in applying scientific principles. The engineer followed appropriate steps in publishing results. The engineer had no personal or professional conflicts of interests. However, what if making certain devices were, in and of itself, immoral? What if, for example, there were something inherently questionable about implanting the device, such as any implanted device that improves motor control at the expense of loss or severe change in personality? . . . or one that diminished dignity? . . . or one that compromises freedom?

Such proscriptions of this type of research fall into the category of macroethics. In such cases, the profession is allowing discretion to be taken away from the individual researcher. That is, biomedical engineers, *en masse*, do not want to become "Dr. Frankensteins," either knowingly or unknowingly. Thus, the profession allows itself to invoke restrictions. Some may see this as antiscientific, but most see the need to draw lines between the moral and immoral; acceptable and unacceptable. Sometimes (most times?), the need to draw the line is mutually agreeable, but with great differences in *where* the lines should be drawn.

One disadvantage of looking at bioethical issues from the engineering perspective is that most of the people at the heart of the debate are not engineers. They are medical researchers and care givers. To use an old (and unfortunate) expression, engineers "own few dogs in that fight." This book is one resource for joining the great conversation. I must confess that I have only touched on a few bioethical problems, hopefully those that interest and inform the engineer. Biomedicine, for example, has many issues with more appearing daily. I have avoided many of the patient–physician issues, such as when and if it is ever acceptable to lie to a patient, whether some advance directives are too risky for patients, the morality of direct-to-consumer advertising, the morality of refusing medical treatment for oneself or one's children for religious reasons, military necessity and medical ethics, the ethics of "sham surgeries," and whether alcoholism is a disease. These are important issues to the engineer and non-engineer alike, and I recommend that if the reader is interested, there are numerous biomedical ethics resources that address them. To start, I recommend two readable and short books:

- Carol Levine's *Taking Sides: Clashing Views on Controversial Biomedical Issues*, Eleventh Edition (McGraw-Hill, Des Moines, Iowa, 2006)
- Gregory E. Pence, *The Elements of Bioethics* (McGraw-Hill, New York, New York, 2007).

MORAL COURAGE

Bioethical Question: What is the difference between professional responsibility and moral courage in engineering?

Applying engineering wisdom requires moral courage. Knowing what is the right thing to do is but the first step. As stated at the outset, engineering is active process. One last thought experiment may illustrate the difference[2]:

Thought Experiment

You are walking along a stream and observe a woman flailing in the water, who obviously is in great distress and drowning. You are the only one within the earshot of the woman, so you have to decide whether you will try to save her life. Assume that no forensics or other evidence of your encounter would ever be known.

There are three possible scenarios:

1. You are a lifeguard and it is your legal responsibility to save anyone who is apparently drowning. You are trained and certified to do this. Thus, this decision is arguably an amoral one (it is your job); that is, the decision has no moral components. However, the converse is morally laden. Only if you do not attempt to save the drowning woman in spite of your social responsibility as a lifeguard will this become a moral question, and the lack of action is unquestionably immoral.

2. A second and different scenario is one wherein you recognize the woman to be someone who owes you a large sum of money. If she drowns you will never be paid back. You take the risk to save her. But, the risk is worthwhile and, this is merely a sound business decision. You quickly calculate a benefit-to-cost ratio and find it to be to your financial advantage to save the woman.

3. This leads us to a third scenario. In this case, you are not a particularly good swimmer, and you clearly recognize the drowning woman as a person whom you owe a lot of money. If she drowns you would not have to repay her, so not helping her is clearly to your financial advantage. The morally courageous act is to save the woman's life in the face of risk and in spite of losing a substantial windfall. The action requires *moral courage* – acting ethically even if the result might not be to your personal benefit.[3]

The good news is that when it comes to saving a person's life, most of us, engineers and everyone else, would do so under the conditions of all three scenarios. Only the morally decrepit and the most selfish would fail to do so. And, the professional codes of ethics provide that you do not have to take inordinate risks (if you cannot swim, you would probably end up killing the woman *and* yourself). This is why we are admonished to practice only in our areas of competence. However, in professional responsibility, there are cases where engineers have failed in situations analogous to all three scenarios. For example, many of the case studies of engineering ethics have at least an element of an engineer's failure to do what is professionally and legally required. They let other priorities eclipse those that should have had primacy (e.g., space shuttles that should not have been launched, sky walk designs that should not have been changed, heart valves that should never have reached the market and safety data that were ignored).

The second scenario is seemingly easy. Those decisions that are together morally, financially, and politically expedient are not controversial. The problem is that we are very good at rationalizations, so sometimes we dress up the third situation as if it were the second. We fail to even see a moral issue because we have made up our minds that we have a "win-win" situation. This is often not so much as a direct temptation, but a result of groupthink. We have slowly but progressively become desensitized to issues that prior to acclimation would have set off a warning bell. And, we want to keep being employed, getting grants, or adding clients. In these situations, it takes much moral courage to stand up against an immoral policy (sexual harassment, irresponsible research, early warning signs of failure, etc.). So, moral courage requires not only an objective perspective, but a constant reminder that we are probably less objective about things that fit our bankbooks and bottom lines.

The more complex and important the project is, the more care that is needed to augment the "basics" with additional information and models to address a specific scenario. Such is the case for engineering ethics. Factual premises must be followed by reason. Reason is informed by moral principles. Thus, bioethical behavior depends on grasping the appropriate facts and applying sound reasoning and a moral code to determine the right thing to do or to think about an issue. Unfortunately, most engineering codes (or any other professional codes for that matter) mainly focus on the behavior of an individual, not on the big issues of our time. Thus, the engineering profession has done an admirable job of articulating the ethical framework for the practitioner, the so-called microethics. However, in dealing with many of the societal challenges of the profession, especially the bioethical issues, we are beginning only now to structure our macroethics.

BIOETHICS RESOURCES FOR THE ENGINEER

The seemingly straightforward hypothetical case in the Prologue indicates the range of certainty of ethical decision making. Bioethics is sufficiently complex, complicated, and far-reaching to allow this book to discuss only a small subset of the issues important to engineers. The good news is that engineers by their nature and training possess many of the skills and tools needed to consider these issues in a rational way. Bioethical issues are likely to be associated with too much information rather than not enough. But, the quality of information that is available ranges from "junk science" and whimsy to sound science. The challenge, like most engineering problems, is to apply the right amount and kinds of information to choose the best option.

Sound engineering incorporates sound ethics. Both depend on reliable information. To help, here are a few resources, beginning with suggestions for further reading, followed by some interactive resources.

SUGGESTED READINGS

ETHICS OF EMERGING TECHNOLOGIES

Berne, Rosalyn W. 2005. *Nanotalk: Conversations with Scientists and Engineers about Ethics, Meaning, and Belief in the Development of Nanotechnology.* Mahwah, NJ: Lawrence Erlbaum Associates.

Black, J. 1997. Thinking twice about "tissue engineering". *IEEE Engineering in Medicine & Biology Magazine* 16.4: 102–4.

Black, M.M. and C. Riley. 1973. Moral issues and priorities in biomedical Engineering. *Science Medicine and Man* 1.1: 67–74.

Blank, Robert H. and Andrea L. Bonnicksen. 1993. *Medicine Unbound: The Human Body and the Limits of Medical Intervention. Emerging Issues in Biomedical Policy*, Volume 3. New York: Columbia University Press.

Bledsoe, J. Gary. 2002. Ethical discussions in an undergraduate tissue engineering course. Proceedings of the 2002 IEEE Engineering in Medicine and Biology 24th Annual Conference and the 2002 Fall Meeting of the Biomedical Engineering Society (BMES/EMBS), 23–26 Oct 2002, Houston, TX. *Annual International Conference of the IEEE Engineering in Medicine and Biology – Proceedings* 3: 2672–3.

Brennan, Mark G. and Mark A. Tooley. 2000. Ethics and the biomedical engineer. *Engineering Science and Education Journal*: 5–7.

Brody, Baruch. 1998. *The Ethics of Biomedical Research: An International Perspective.* New York: Oxford University Press.

Brody, Eugene. 1993. *Biomedical Technology and Human Rights.* Aldershot, England: UNESCO.

Bronzino, Joseph D., Ellen J. Flannery, and Maurice Wade. 1990. Legal and ethical issues in the regulation and development of engineering achievements in medical technology – Part I. *IEEE Engineering in Medicine and Biology* 9.1: 79–81.

Bronzino, Joseph D., Ellen J. Flannery, and Maurice Wade. 1990. Legal and ethical issues in the regulation and development of engineering achievements in medical technology – Part II. *IEEE Engineering in Medicine and Biology* 9.2: 53–7.

Bulger, Ruth Ellen, Elizabeth Heitman, and Stanley Joel Reiser (eds). 2002. *The Ethical Dimensions of the Biological and Health Sciences.* Cambridge: Cambridge University Press.

Khushf, G. 2004. The ethics of nanotechnology: visions and values for a new generation of science and engineering, in *Emerging Technologies and Ethical Issues in Engineering.* Washington, DC: National Academies Press, pp. 29–55.

Mnyusiwalla, A., A.S. Dear, and P.A. Singer. 2003. Mind the gap: science and ethics in nanotechnology. *Nanotechnology* 14: R9–R13.

National Academy of Engineering. 2004. *Emerging Technologies and Ethical Issues in Engineering.* Washington, DC: The National Academies Press.

National Academy of Engineering. 2004. *The Engineer of 2020: Visions of Engineering in the New Century.* Washington, DC: The National Academies Press.

National Science and Technology Council. 1999. *Nanotechnology: Shaping the World Atom by Atom.* Washington, DC: National Science and Technology Council.

Pence, G. 2003. *Classic Cases in Medical Ethics: Accounts of Cases that Have Shaped Medical Ethics, with Philosophical, Legal, and Historical Backgrounds.* New York: McGraw-Hill.

Posner, R.A. 2004. *Catastrophe: Risk and Response*, New York: Oxford University Press.

Rees, M. 2003. *Our Final Hour: A Scientist's Warning: How Terror, Error, and Environmental Disaster Threaten Humankind's Future in this Century – On Earth and Beyond*. New York: Basic Books.

Roco, M.C. 2003. Broader societal issues in nanotechnology. *Journal of Nanoparticle Research* 5: 181–9.

Treder, M. 18 February 2006. From heaven to doomsday: seven future scenarios. *Future Brief*.

Wakefield, J. 2004. Doom and gloom by 2100. *Scientific American* 291 (1): 48–9.

Weil, V. 1996. Teaching Ethics in Science, in *Ethics, Values, and The Promise of Science*. Research Triangle Park, NC: Sigma XI Forum Proceedings.

Weil, V. 2001. Ethical issues in nanotechnology, in Roco, M. and W. Bainbridge (eds), *Societal Implications of Nanoscience and Nanotechnology*. Dordrecht, The Netherlands: Kluwer Academic Publishers, pp. 244–51.

Whitbeck, C. 1998. *Ethics in Engineering Practice and Research*. Cambridge University Press.

Wulf, W.A. 2000. Great achievements and grand challenges. *The Bridge* 30 (3–4): 5–10.

Zull, J. 2002. *The Art of Changing the Brain: Enriching the Practice of Teaching by Exploring the Biology of Learning*. Sterling, VA: Stylus Publishing.

ETHICAL ANALYSIS, REASONING, AND DECISION MAKING

Elliott, Deni and Judy E. Stern (eds). *Research Ethics: A Reader*. St Lebanon, NH: University Press of New England.

Fraser, N. 1995. From redistribution to recognition? Dilemmas of justice in a postsocialist age. *New Left Review* 212: 68–93.

Gilligan, C. 1983. *In a Different Voice*. Cambridge, MA: Harvard University Press.

Hampshire, S. 1983. Fallacies in moral reasoning, in A. MacIntyre and S. Hauerwas (eds), *Changing Perspectives in Moral Philosophy*. Notre Dame, IN: University of Notre Dame Press.

Hirschman, A.O. 1970. *Exit, Voice, and Loyalty*. Cambridge, MA: Harvard University Press.

Kidder, R. 1996. *How Good People Make Tough Choices*. New York: Simon and Schuster.

Shapiro, I. 1999. *Democratic Justice*. New Haven, CN: Yale University Press.

Whitbeck, C. 1996. Ethics as design: doing justice to moral problems. *Hastings Center Report* 26 (3): 9–16.

MACROETHICS AND SOCIETAL RISK

Bassingthwaighte, J.B. 2002. The Physiome project: the macroethics of engineering toward health. *The Bridge* 32 (3): 24–9.

Beck, U. 1992. *Risk Society*. London, UK: Sage.

Berman, S., A. Wicks, S. Kotha, and T. Jones. 1999. Does stakeholder orientation matter? *The Academy of Management Journal* (42): 480–506.

Caplan, A. and T. Engelhardt (eds). 1987. Scientific Controversies: Case Studies in the Resolution and Closure of Disputes in Science and Technology. Cambridge: Cambridge University Press.

Collins, H.M. and T. Pinch. 1994. *The Golem: What Everyone Should Know about Science*. Cambridge: Cambridge University Press.

Cram, Nicholas, John Wheeler, and Charles S. Lessard. 1995. Ethical issues of life-sustaining technology. *IEEE Technology and Society Magazine* 14.1: 21–8.

Derr, P., R. Goble, R. Kasperson, and R. Kates. 1982. Worker/public protection: the double standard. *Center for Technology and Development, Reprint 22.* Worcester, MA: Clark University Hazard Assessment Group.

Derr, P., R. Goble, R. Kasperson, and R. Kates. 1984. Responding to worker/public protection: the double standard. *Center for Technology and Development, Reprint 34.* Worcester, MA: Clark University Hazard Assessment Group.

Douglas, M., and A. Wildavsky. 1982. *Risk and Culture.* Berkeley, CA: University of California Press.

Freeman, E. 1984. *Strategic Management: A Stakeholder Approach.* Boston: Pitman.

Funtowicz, S. and J. Ravetz. 1992. Three types of risk assessment and the emergence of post-normal science, in Krimsky, S. and D. Golding (eds), *Social theories of Risk.* Westport, CT: Praeger, pp. 251–74.

Funtowicz, S. and J.R. Ravetz. 1991. A new scientific methodology for global environmental issues, in Costanza, R. (ed.), *Ecological Economics: The Science and Management of Sustainability.* New York: Columbia University Press.

Funtowicz, S. and J. Ravetz. 1993. Science for the post-normal age. *Futures* 25 (7): 739–55.

Funtowicz, S. and J.R. Ravetz. 1994. Emergent complex systems. *Futures* 26 (6).

Funtowicz, S., J.R. Ravetz, I. Shepherd, and D. Wilkinson. 2000. Science and governance in the European Union. *Science and Public Policy* 27 (5).

Hargreaves, I. 2000. *Who's Misunderstanding Whom? Science, Society and the Media.* Economic and Social Research Council. Available at http://www.escr.ac.uk.

Herkert, J.R. 2004. Microethics, macroethics, and professional engineering societies, in *Emerging Technologies and Ethical Issues in Engineering.* Washington, DC: The National Academies Press, pp. 107–14.

Jasanoff, S. 1987. Contested boundaries in policy relevant science. *Social Studies of Science* 17: 195–230.

Kasperson, R.E. and K. Dow. 1993. Hazard perception and geography, in Goring, T. and R. Golledge (eds.), *Behavior and Environment: Psychological and Geographical Approaches, Chapter 8.* Amsterdam, Netherlands: North Holland Press, pp. 193–222.

Kasperson, J.X. and R.E. Kasperson. May 1996. The social amplification and attenuation of risk. *Annals of the American Academy of Political and Social Sciences* 545: 95–105.

Kasperson, R.E. and J.P. Stallen. 1991. *Communicating Risks to the Public: International Perspectives.* Norwell, MA: Reidell.

Kasperson, J.X., R.E. Kasperson, and B.L. Turner, II. 1997. Regions at risk: exploring environmental criticality. *Environment* 39 (10): 4–15, see also pp. 26–29.

Kates, R.W. and J.X. Kasperson. 1983. Comparative risk analysis of technological hazards (a review). *Proceedings of the National Academy of Sciences* 80: 7027–38.

Naser, C.R. 2000. What is life, and what is a machine? The ontology of bioengineering. *Critical Reviews in Biomedical Engineering* 28.3–4: 545–50.

Naurato, N. and T.J. Smith. 2003. Ethical considerations in bioengineering research. *Biomedical Sciences Instrumentation* 39: 573–8.

Renn, O. 1994. *A Regional Concept of Qualitative Growth and Sustainabiltity.* Bericht Nr 2, Akademie für Technikfolgenabschätzung in Baden-Württemberg, Stuttgart.

Renn, O., W. Burns, J. Kasperson, R. Kasperson, and P. Slovic. 1992. The social amplification of risk: theoretical foundations and empirical applications. *Journal of Social Issues* 48 (4): 137–60.

Renn, O., T. Webler, and T. Wiedemann (eds). 1999. *Fairness and Competence in Citizen Participation: Evaluating Models for Environmental Discourse (Technology, Risk and Society)*. New York: Springer.

Roco, Mihail C. and William Sims Bainbridge (eds). June 2002. *Converging Technologies for Improving Human Performance: Nanotechnology, Biotechnology and Cognitive Science*. NSF/DOC-sponsored report, Arlington, VA.

Saha, Subrata. 1990. Ethical questions in biomedical engineering research. *Annual International Conference of the IEEE Engineering in Medicine and Biology Society* 12.5: 1981–2.

Schein, E. 1992. *Organizational Culture and Leadership*, 2nd edition. San Francisco, CA: Jossey-Bass.

Simon, J. and M. Hersch. 2002. An educational imperative: the role of ethical prohibitions in CBW-applicable research. *Minerva* 40 (1).

Stokes, D.E. 1997. *Pasteur's Quadrant – Basic Science and Technological Innovation*. Washington, DC: Brookings Institution Press.

Turner, B.L., W.C. Clark, R.W. Kates, J.F. Richards, J.T. Mathews, and W.B. Meyer. 1993. *The Earth as Transformed by Human Action*. New York: Cambridge University Press.

Vallero, D.A. and P.A. Vesilind. 2006. Preventing disputes with empathy. *Journal of Professional Issues in Engineering and Practice* 132 (3): 272–8.

Vallero, D.A. and P.A. Vesilind. 2006. *Socially Responsible Engineering*. Hoboken, New Jersey: Wiley.

Vesilind, P.A. (ed.). 2005. *Peace Engineering: When Personal Values and Engineering Careers Converge*. Woodsville, New Hampshire: Lakeshore Press.

Weatherall, David. 2003. Problems for biomedical research at the Academia–Industrial Interface. *Science and Engineering Ethics* 9: 43–8.

Wynne, B. 1996. May the sheep safely graze? A reflexive view of the expert-lay knowledge divide, in Lash, S., B. Szerszinski, and B. Wynne (eds), *Risk, Environment and Modernity*. London: Sage, pp. 45–80.

Wynne, B. 2000. Retrieving a human agenda for science. *RSA Journal* 2 (4).

Wynne, B. 2001. Creating public alienation: expert cultures of risk and ethics on GMOs. *Science as Culture* 10 (4): 445–81.

TEACHING ENGINEERING MACROETHICS

Accreditation Board for Engineering and Technology, Inc. (ABET). 2003. *Criteria for Accrediting Engineering Programs: Effective for Evaluations during the 2004–2005 Accreditation Cycle*. Baltimore, MD: ABET.

Association of Schools of Public Health. 2006. Ethics and Public Health: Model Curriculum, http://www.asph.org/document.cfm?page=782.

Bebeau, M., et al. 1995. *Moral Reasoning in Scientific Research: Cases for Teaching and Assessment*. Bloomington, IN: Indiana University.

Boyer, E.L. 1997. *Scholarship Reconsidered: Priorities for the Professoriate*. Menlo Park, CA: The Carnegie Foundation for the Advancement of Teaching.

Casada, Mark E. and James A. DeShazer. 1995. Teaching professionalism, design, & communications to engineering freshmen. *ASEE Annual Conference Proceedings, Volume 1, Investing in the Future*, pp. 1381–6.

Chickering, A. and Z. Gamson. 1991. Applying the seven principles for good practices in undergraduate education. *New Directions for Teaching and Learning*. No. 47, Fall. San Francisco, CA: Jossey-Bass.

Daniels, A.U. 1992. Ethics education at the engineering/medicine interface. *Journal of Investigative Surgery* 5.3: 209–18.

Elliott, D. and J. Stern. 1996. Evaluating teaching and students' learning of academic research ethics. *Science and Engineering Ethics* 2 (3): 345–66.

Glassick, C., M. Huber, and G. Maeroff. 2000. *Scholarship Assessed: Evaluation of the Professoriate*. San Francisco, CA: Jossey-Bass.

Handelsman, J. et al., 2004. Scientific teaching. *Science* 304.

Hofer, B. and P. Pintrich. 1997. The development of epistemological theories: beliefs about knowledge and knowing and their relation to learning, in *Review of Educational Research*, 67: 1. Washington, DC: American Educational Research Association.

Huber, M. and S. Morreale. 2002. *Disciplinary Styles in the Scholarship of Teaching and Learning: Exploring Common Ground*. Washington, DC: American Association of Higher Education and The Carnegie Foundation for the Advancement of Teaching.

Kwarteng, K.B. 2000. Ethical considerations for biomedical scientists and engineers: issues for the rank and file. *Critical Reviews in Biomedical Engineering* 28 (3–4): 517–21.

Mager, R. 1997. *Preparing Instructional Objectives*, 3rd edition. Atlanta, GA: CEP Press.

McKeachie, W. and B. Hofer. 2002. *McKeachie's Teaching Tips: Strategies, Research and Theory for College and University Teachers*, 11th edition. Boston: Houghton-Mifflin.

Menges, R. and M. Weimer. 1996. *Teaching on Solid Ground: Using Scholarship to Improve Practice*. San Francisco, CA: Jossey-Bass.

Mezirow, J. 1991. *Transformative Dimensions of Adult Learning*. San Francisco, CA: Jossey-Bass.

Napper, S.A. and P.N. Hale. 1993. Teaching of ethics in biomedical engineering. *IEEE Engineering in Medicine & Biology Magazine* 12.4: 100–5.

Penslar, R. 1995. *Research Ethics: Cases and Materials*. Bloomington, IN: Indiana University Press.

Rando, W.C. and R.J. Menges. 1991. How practice is shaped by personal theories, in Menges, R.J. and M.D. Svinicki (eds), *New Directions for Teaching and Learning*, Volume 45. San Francisco, CA: Jossey-Bass.

Saha, Subrata. 2002. Teaching bioethics for biomedical engineering students: a case studies approach. Proceedings of the 2002 IEEE Engineering in Medicine and Biology 24th Annual Conference and the 2002 Fall Meeting of the Biomedical Engineering Society (BMES/EMBS), 23-26 Oct 2002, Houston, TX. *Annual International Conference of the IEEE Engineering in Medicine and Biology – Proceedings* 3: 2602.

Schön, D. 1990. *Educating the Reflective Practitioner: Toward a New Design for Teaching and Learning in the Professions*. San Francisco, CA: Jossey-Bass.

Swazey, J. and S. Bird. 1997. Teaching and learning research ethics, in Elliott, D. and J. Stern (eds), *Research Ethics: A Reader*. Lebanon, NH: University Press of New England.

Vallero, D.A. 2002. Teachable moments and the tyranny of the syllabus: 11 September case. *Journal of Professional Issues in Engineering and Practice* 129 (2): 100–105.

Teaching Engineering Microethics

Davis, Michael. 1991. Thinking like an engineer: the place of a code of ethics in the practice of a profession. *Philosophy and Public Affairs* 20.2: 150–67.

Davis, Michael. 1992. Codes of ethics, professions, and conflict of interest: a case study of an emerging profession, clinical engineering. *Professional Ethics* 1.1–2: 179–95.

DeMets, David L. 1999. Statistics and ethics in medical research. *Science and Engineering Ethics* 5: 97–117.

Fielder, J.H. 1991. Ethical issues in biomedical engineering: the Bjork-Shiley heart valve. *IEEE Engineering in Medicine & Biology Magazine* 10.1: 76–8.

Fielder, J.H. 1992. The bioengineer's obligations to patients. *Journal of Investigative Surgery* 5.3: 201–8.

Fielder, J.H. 1999. Ethics and professional responsibility. *IEEE Engineering in Medicine & Biology Magazine* 18.4: 116–7.

Guilbeau, Eric J. and Vincent B. Pizziconi. 1998. Increasing student awareness of ethical, social, legal, and economic implications of technology. *Journal of Engineering Education* 87.1: 35–44.

Heller, J.C. 2000. Beyond a code of ethics for bioengineers: the role of ethics in an integrated compliance program. *Critical Reviews in Biomedical Engineering* 28.3–4: 507–11.

Keefer, M. and K. Ashley. 2001. Case-based approaches to professional ethics: SA systematic comparison of students' and ethicists' moral reasoning, *Journal of Moral Education* 30: 4.

Maguire, L.A. and E.A. Lind. 2003. Public participation in environmental decisions: stakeholders, authorities and procedural justice. *International Journal of Global Environmental Issues* 3 (2): 133–48.

Maguire, L.A. and H. Sondak. 1998. Can using decision analysis and dispute resolution techniques to solve environmental problems help promote equity? in Fletcher, D.J., L. Kavalieris, and B.F.J. Manly (eds), *Statistics in Ecology and Environmental Monitoring 2: Risk Assessment and Decision Making in Biology*. Dunedin, New Zealand: Otago University Press, pp. 97–120.

Maner, Walter. 2006. Heuristic methods for computer ethics. *Metaphilosophy* 33 (in press), http://csweb.cs.bgsu.edu/maner/heuristics/maner.pdf.

Pienkowski, D. 2003. The need for a professional code of ethics in biomedical engineering: a lesson from history. *Critical Reviews in Biomedical Engineering* 28.3–4: 513–6.

Pimple, K. 2001. *Assessing Teaching and Learning in the Responsible Conduct of Research*. Institute of Medicine Committee on Assessing Integrity in Research Environments.

Schrader-Frechette, K. 2000. Ethics and environmental advocacy, in List, P.C. (ed.), *Environmental Ethics and Forestry: A Reader*. Philadelphia, PA: Temple University Press, pp. 209–20.

Schrag, B. 1997–2002. *Research Ethics: Cases and Commentaries*, Volumes 1–6. Bloomington: Association for Practical and Professional Ethics.

Shamoo, A. and D. Resnik. 2003. *Responsible Conduct of Research*. New York: Oxford University Press.

Steneck, N. 2004. *ORI Introduction to the Responsible Conduct of Research*. Washington, DC: US Office of Research Integrity, Dept. of Health and Human Services.

Wueste, Daniel E. 1997. Professions, professional ethics, and bioengineering. *Critical Reviews in Biomedical Engineering* 25.2: 127–49.

Wueste, Daniel E. 1998. Wrongdoing in biomedical research: an ethical diagnosis and prescription. *Critical Reviews in Biomedical Engineering* 26.5–6: 378–80.

USEFUL WEBSITES

Societal and Ethical Issues of Nanotechnology: http://www.sei.nnin.org/readings.html.
Links to

- Nano SEI Discussions
- Reports and Surveys
- Nano SEI Specific Areas
- Social and Ethical Dimensions of Science and Technology
- Alternative Visions
- What is 'nano'?
- Video & Multimedia Resources

Currently includes the following discussions:

- "Nanotechnology and Societal Transformation" (2001) Crow, Michael M. & Sarewitz, Daniel. In *Final Report from the Workshop Held at the National Science Foundation*, Edited by Mihail Roco & William Bainbridge, 28–29 Sept. 2000. National Science Foundation (NSF).
- "Of Chemistry, Love and Nanobots" (2001) Smalley, R.E. *Scientific American*, 285, pp. 76–77
- "The Promises and Perils of Nanoscience and Nanotechnology: Exploring Emerging Social and Ethical Issues" (2003) Sweeney, Aldrin E., Seal, Sudipta & Vaidyanathan, Pallavoor. *Bulletin of Science, Technology & Society*, 23 (4), pp. 236–245.
- "Zeroing In on Ethical Issues in Nanotechnology" (2003) Weil, V. *Proceedings of the IEEE*, 91 (1), pp. 1976–1979.
- " 'Mind the Gap': Science and Ethics in Nanotechnology" (2003) Mnyusiwalla, Anisa, Daar, Abdallah S. & Singer, Peter A. *Nanotechnology*, 14, pp. R9–R13.
- "Grilichesian Breakthroughs: Inventions of Methods of Inventing and Firm Entry in Nanotechnology" Darby, Michael R. & Zucker, Lynne G. NBER Working Paper No. 9825.
- "The Societal Implications of Nanotechnology" (2003) Langdon Winner's testimony to the Committee on Science of the US House of Representatives on The Societal Implications of Nanotechnology, Wednesday, 9 April 2003.
- "Broader Societal Issues of Nanotechnology" (2003) Roco, Mihail. *Journal of Nanoparticle Research*, 5, pp. 181–189.
- " 'Plenty of room at the bottom': Towards an Anthropology of Nanoscience" (2003) Johansson, Mikael. *Anthropology Today*, 19 (6), pp. 3–6.
- "Social and Ethical Issues in Nanotechnology: Lessons from Biotechnology and Other High Technologies" (2003) Wolfson, Joel Rothstein. *Biotechnology Law Report*, 22 (4), pp. 376–396.
- "Bio-to-Nano? Learning the Lessons, Interrogating the Comparison" (2004) Grove-White, Robin, Kearnes, Matthew, Miller, Paul, Macnaghten, Phil, Wilsdon, James & Wynne, Brian. A Working Paper by the Institute for Environment, Philosophy and Public Policy, Lancaster University and Demos.

- " 'Societal and Ethical Implications of Nanotechnology': Meanings, Interest Groups, and Social Dynamics" (2004) Schummer, Joachim. *Techne: Research in Philosophy and Technology*, 8 (2), pp. 56–87 [Online].
- "Analyzing the European Approach to Nanotechnology" (2004) Michelson, Evan. Occasional Paper on Nanotechnology. Woodrow Wilson International Center for Scholars – Foresight and Governance Project.
- "Bibliography of Studies on Nanoscience and Nanotechnology" (2004) Schummer, Joachim. Special edition on "Nanotech Challenges". *HYLE – International Journal for Philosophy of Chemistry*, 10 (2).
- "Discovering the Nanoscale" (2004) Book edited by Davis Baird, Alfred Nordmann & Joachim Schummer. Amsterdam: IOS Press.
- "Nano Hyperbole and Lessons from Earlier Technologies" (2004) Toumey, Chris. *Nanotechnology Law & Business Journal*, 1 (4), Article 9.
- "The Nanotech Land Grab" (2004) *Corporate Legal Times*, pp. 32–39
- "Nanotech Challenges (Part I)" (2004) [html] *Techne*: Research in Philosophy and Technology. Special Issue on "Nanotech Challenges". Jointly published with *Hyle*, 8 (2).
- "Nanotech Challenges (Part II)" (2005 Spring) *Techn*: Research in Philosophy and Technology. Special Issue on "Nanotech Challenges". Jointly published with *Hyle*, 8 (3).
- "What Counts as a 'Social and Ethical Issue' in Nanotechnology?" (2005)[pdf] Lewenstein, Bruce V. *HYLE – International Journal for Philosophy of Chemistry*, 11 (1), pp. 5–18.
- "To Be Nano or Not to Be Nano?" (2005) Joachim, Christian. *Nature Materials*, 4 (2), pp. 107–109.
- "The Salience of Small: Nanotechnology Coverage in the American Press, 1986–2004" (2005) Lewenstein, Bruce V., Gorss, Jason & Radin, Joanna. Paper submitted to the International Communication Association for conference, 26–30 May 2005. [Contact authors for a copy.]
- "A Short, Short Citizens' Guide to 'The Next Big Idea' " (no date) Macoubrie, Jane. Woodrow Wilson International Center for Scholars.
- "Nanotechnology: Risks and the Media" (Winter 2005). Friedman, Sharon M. & Egolf, Brenda P. Jane. *IEEE Technology and Society Magazine*, 24 (4), pp. 5–11.

United Nations Educational, Scientific and Cultural Organization, Ethics of Science & Technology: http://portal.unesco.org/shs/en/ev.php-URL_ID=1373&URL_DO=DO_TOPIC&URL_SECTION=201.html.

Includes a site dedicated to nanotechnology ethics: http://portal.unesco.org/shs/en/ev.php-URL_ID=6314&URL_DO=DO_TOPIC&URL_SECTION=201.html.

Foresight Nanotech Institute: http://www.foresight.org/

The website states that Foresight is a member-supported organization. Our membership, including over 14 000 individuals and a growing number of corporations, is diverse demographically and geographically. They are interested in ensuring that the future of nanotechnology unfolds for the benefit of all. These concerned individuals include scientists, engineers, business people, investors, publishers, artists, ethicists, policy makers, interested laypersons, and students from grammar school to graduate level.

Ethical Issues in Nanotechnology: http://www.ethicsweb.ca/nanotechnology/

The EthicsWeb is a collection of ethics-related websites, run by philosopher–ethicist Chris MacDonald. This link is dedicated to ethics associated with nanotechnology. It includes an annotated bibliography at http://www.ethicsweb.ca/nanotechnology/bibliography.html.

New York Academy of Sciences, Peter Singer discusses "Nanotechnology Ethics" at http://www.nyas.org/ebriefreps/main.asp?intSubsectionID=1121.

The Nano Ethics Group: http://www.nanoethics.org/

The site states that the Nanoethics Group is a non-partisan and an independent organization that studies the ethical and societal implications of nanotechnology.

Procedural Ethics Chronological Site Index: http://csweb.cs.bgsu.edu/maner/heuristics/toc.htm

This site states that it is a collection of step-by-step ethical reasoning procedures, taken from a variety of sources that may prove useful to computer professionals engaged in ethical decision-making. Some of these procedures are examined in "Heuristic Methods for Computer Ethics," which will appear in final form *Metaphilosophy*, Volume 33 (cited in this bibliography).

One of these may be a good match for a particular situation or, more likely, it will be necessary to combine elements from several. It is always possible that a good decision would emerge from a *gestalt* impression, without doing any step-by-step analysis, but it would be difficult to defend such a decision.

Although these procedures have limitations, they provide a fertile source of ideas and can reveal elements missing from the decision procedure you currently use. In addition, it is possible to map elements from these procedures into a uniform set of 12 stages.

NOTES AND COMMENTARY

[1] This is changing. I attended a meeting of engineering and science educators sponsored by the National Science Foundation in January 2007. Many shared new ways of addressing large, societal issues in addition to the practitioner-specific needs.

[2] This thought experiment is adapted from P.A. Vesilind, L. Heine and J. Hendry. *TRAMES: A Journal of the Humanities and Social Sciences*. Issue No. 1, 2006, 22–31. Vesilind is my colleague and co-author of our recent book, *Socially Responsible Engineering* (Hoboken, NJ: John Wiley & Sons, 2006).

[3] R. Kidder, *Moral Courage* (New York, NY: William Morrow, 2005).

Appendix 1

National Society of Professional Engineers Code of Ethics for Engineers

PREAMBLE

Engineering is an important and learned profession. As members of this profession, engineers are expected to exhibit the highest standards of honesty and integrity. Engineering has a direct and vital impact on the quality of life for all people. Accordingly, the services provided by engineers require honesty, impartiality, fairness, and equity, and must be dedicated to the protection of the public health, safety, and welfare. Engineers must perform under a standard of professional behavior that requires adherence to the highest principles of ethical conduct.

I. FUNDAMENTAL CANONS

Engineers, in the fulfillment of their professional duties, shall:

1. Hold paramount the safety, health, and welfare of the public.
2. Perform services only in areas of their competence.
3. Issue public statements only in an objective and truthful manner.
4. Act for each employer or client as faithful agents or trustees.
5. Avoid deceptive acts.
6. Conduct themselves honorably, responsibly, ethically, and lawfully so as to enhance the honor, reputation, and usefulness of the profession.

II. RULES OF PRACTICE

1. Engineers shall hold paramount the safety, health, and welfare of the public.

   a. If engineers' judgment is overruled under circumstances that endanger life or property, they shall notify their employer or client and such other authority as may be appropriate.

b. Engineers shall approve only those engineering documents that are in conformity with applicable standards.

c. Engineers shall not reveal facts, data, or information without the prior consent of the client or employer except as authorized or required by law or this Code.

d. Engineers shall not permit the use of their name or associate in business ventures with any person or firm that they believe are engaged in fraudulent or dishonest enterprise.

e. Engineers shall not aid or abet the unlawful practice of engineering by a person or a firm.

f. Engineers having knowledge of any alleged violation of this Code shall report thereon to appropriate professional bodies and, when relevant, also to public authorities, and cooperate with the proper authorities in furnishing such information or assistance as may be required.

2. Engineers shall perform services only in the areas of their competence.

a. Engineers shall undertake assignments only when qualified by education or experience in the specific technical fields involved.

b. Engineers shall not affix their signatures to any plans or documents dealing with subject matter in which they lack competence, nor to any plan or document not prepared under their direction and control.

c. Engineers may accept assignments and assume responsibility for coordination of an entire project and sign and seal the engineering documents for the entire project, provided that each technical segment is signed and sealed only by the qualified engineers who prepared the segment.

3. Engineers shall issue public statements only in an objective and truthful manner.

a. Engineers shall be objective and truthful in professional reports, statements, or testimony. They shall include all relevant and pertinent information in such reports, statements, or testimony, which should bear the date indicating when it was current.

b. Engineers may express publicly technical opinions that are founded upon knowledge of the facts and competence in the subject matter.

c. Engineers shall issue no statements, criticisms, or arguments on technical matters that are inspired or paid for by interested parties, unless they have prefaced their comments by explicitly identifying the interested parties on whose behalf they are speaking, and by revealing the existence of any interest the engineers may have in the matters.

4. Engineers shall act for each employer or client as faithful agents or trustees.

a. Engineers shall disclose all known or potential conflicts of interest that could influence or appear to influence their judgment or the quality of their services.

b. Engineers shall not accept compensation, financial or otherwise, from more than one party for services on the same project, or for services pertaining to the same project, unless the circumstances are fully disclosed and agreed to by all interested parties.

c. Engineers shall not solicit or accept financial or other valuable consideration, directly or indirectly, from outside agents in connection with the work for which they are responsible.

d. Engineers in public service as members, advisors, or employees of a governmental or quasi-governmental body or department shall not participate in decisions with respect to services solicited or provided by them or their organizations in private or public engineering practice.

  e. Engineers shall not solicit or accept a contract from a governmental body on which a principal or an officer of their organization serves as a member.

5. Engineers shall avoid deceptive acts.

  a. Engineers shall not falsify their qualifications or permit misrepresentation of their or their associates' qualifications. They shall not misrepresent or exaggerate their responsibility in or for the subject matter of prior assignments. Brochures or other presentations incident to the solicitation of employment shall not misrepresent pertinent facts concerning employers, employees, associates, joint venturers, or past accomplishments.

  b. Engineers shall not offer, give, solicit or receive, either directly or indirectly, any contribution to influence the award of a contract by public authority, or which may be reasonably construed by the public as having the effect of intent to influencing the awarding of a contract. They shall not offer any gift or other valuable consideration in order to secure work. They shall not pay a commission, percentage, or brokerage fee in order to secure work, except to a bona fide employee or bona fide established commercial or marketing agencies retained by them.

III. PROFESSIONAL OBLIGATIONS

1. Engineers shall be guided in all their relations by the highest standards of honesty and integrity.

  a. Engineers shall acknowledge their errors and shall not distort or alter the facts.

  b. Engineers shall advise their clients or employers when they believe a project will not be successful.

  c. Engineers shall not accept outside employment to the detriment of their regular work or interest. Before accepting any outside engineering employment they will notify their employers.

  d. Engineers shall not attempt to attract an engineer from another employer by false or misleading pretenses.

  e. Engineers shall not promote their own interest at the expense of the dignity and integrity of the profession.

2. Engineers shall at all times strive to serve the public interest.

  a. Engineers shall seek opportunities to participate in civic affairs; career guidance for youths; and work for the advancement of the safety, health, and well-being of their community.

  b. Engineers shall not complete, sign, or seal plans and/or specifications that are not in conformity with applicable engineering standards. If the client or employer insists on such unprofessional conduct, they shall notify the proper authorities and withdraw from further service on the project.

  c. Engineers shall endeavor to extend public knowledge and appreciation of engineering and its achievements.

3. Engineers shall avoid all conduct or practice that deceives the public.

     a. Engineers shall avoid the use of statements containing a material misrepresentation of fact or omitting a material fact.

     b. Consistent with the foregoing, engineers may advertise for recruitment of personnel.

     c. Consistent with the foregoing, engineers may prepare articles for the lay or technical press, but such articles shall not imply credit to the author for work performed by others.

4. Engineers shall not disclose, without consent, confidential information concerning the business affairs or technical processes of any present or former client or employer, or public body on which they serve.

     a. Engineers shall not, without the consent of all interested parties, promote or arrange for new employment or practice in connection with a specific project for which the engineer has gained particular and specialized knowledge.

     b. Engineers shall not, without the consent of all interested parties, participate in or represent an adversary interest in connection with a specific project or proceeding in which the engineer has gained particular specialized knowledge on behalf of a former client or an employer.

5. Engineers shall not be influenced in their professional duties by conflicting interests.

     a. Engineers shall not accept financial or other considerations, including free engineering designs, from material or equipment suppliers for specifying their product.

     b. Engineers shall not accept commissions or allowances, directly or indirectly, from contractors or other parties dealing with clients or employers of the engineer in connection with work for which the engineer is responsible.

6. Engineers shall not attempt to obtain employment or advancement or professional engagements by untruthfully criticizing other engineers, or by other improper or questionable methods.

     a. Engineers shall not request, propose, or accept a commission on a contingent basis under circumstances in which their judgment may be compromised.

     b. Engineers in salaried positions shall accept part-time engineering work only to the extent consistent with policies of the employer and in accordance with ethical considerations.

     c. Engineers shall not, without consent, use equipment, supplies, laboratory, or office facilities of an employer to carry on outside private practice.

7. Engineers shall not attempt to injure, maliciously or falsely, directly or indirectly, the professional reputation, prospects, practice, or employment of other engineers. Engineers who believe others are guilty of unethical or illegal practice shall present such information to the proper authority for action.

     a. Engineers in private practice shall not review the work of another engineer for the same client, except with the knowledge of such engineer, or unless the connection of such engineer with the work has been terminated.

     b. Engineers in governmental, industrial, or educational employ are entitled to review and evaluate the work of other engineers when so required by their employment duties.

    c. Engineers in sales or industrial employ are entitled to make engineering comparisons of represented products with products of other suppliers.

8. Engineers shall accept personal responsibility for their professional activities, provided, however, that engineers may seek indemnification for services arising out of their practice for other than gross negligence, where the engineer's interests cannot otherwise be protected.

    a. Engineers shall conform with state registration laws in the practice of engineering.

    b. Engineers shall not use association with a nonengineer, a corporation, or partnership as a "cloak" for unethical acts.

9. Engineers shall give credit for engineering work to those to whom credit is due, and will recognize the proprietary interests of others.

    a. Engineers shall, whenever possible, name the person or persons who may be individually responsible for designs, inventions, writings, or other accomplishments.

    b. Engineers using designs supplied by a client recognize that the designs remain the property of the client and may not be duplicated by the engineer for others without express permission.

    c. Engineers, before undertaking work for others in connection with which the engineer may make improvements, plans, designs, inventions, or other records that may justify copyrights or patents, should enter into a positive agreement regarding ownership.

    d. Engineers' designs, data, records, and notes referring exclusively to an employer's work are the employer's property. The employer should indemnify the engineer for use of the information for any purpose other than the original purpose.

    e. Engineers shall continue their professional development throughout their careers and should keep current in their specialty fields by engaging in professional practice, participating in continuing education courses, reading in the technical literature, and attending professional meetings and seminars.

—As Revised January 2003

"By order of the United States District Court for the District of Columbia, former Section 11(c) of the NSPE Code of Ethics prohibiting competitive bidding, and all policy statements, opinions, rulings or other guidelines interpreting its scope, have been rescinded as unlawfully interfering with the legal right of engineers, protected under the antitrust laws, to provide price information to prospective clients; accordingly, nothing contained in the NSPE Code of Ethics, policy statements, opinions, rulings or other guidelines prohibits the submission of price quotations or competitive bids for engineering services at any time or in any amount."

STATEMENT BY NSPE EXECUTIVE COMMITTEE

In order to correct misunderstandings, which have been indicated in some instances since the issuance of the Supreme Court decision and the entry of the Final Judgment, it is noted that in its decision of 25 April 1978, the Supreme Court of the United States declared: "The Sherman Act does not require competitive bidding."

It is further noted that as made clear in the Supreme Court decision:

1. Engineers and firms may individually refuse to bid for engineering services.
2. Clients are not required to seek bids for engineering services.
3. Federal, state, and local laws governing procedures to procure engineering services are not affected, and remain in full force and effect.
4. State societies and local chapters are free to actively and aggressively seek legislation for professional selection and negotiation procedures by public agencies.
5. State registration board rules of professional conduct, including rules prohibiting competitive bidding for engineering services, are not affected and remain in full force and effect. State registration boards with authority to adopt rules of professional conduct may adopt rules governing procedures to obtain engineering services.
6. As noted by the Supreme Court, "nothing in the judgment prevents NSPE and its members from attempting to influence governmental action . . . ."

NOTE: In regard to the question of the application of the Code to corporations vis-à-vis real persons, business form or type should not negate nor influence conformance of individuals to the Code. The Code deals with professional services, which services must be performed by real persons. Real persons in turn establish and implement policies within business structures. The Code is clearly written to apply to the Engineer and items incumbent on members of NSPE to endeavor to live up to its provisions. This applies to all pertinent sections of the Code.

Appendix 2

Biomedical Engineering Society Code of Ethics

Approved February 2004

Biomedical engineering is a learned profession that combines expertise and responsibilities in engineering, science, technology, and medicine. Since public health and welfare are paramount considerations in each of these areas, biomedical engineers must uphold those principles of ethical conduct embodied in this Code in professional practice, research, patient care, and training. This Code reflects voluntary standards of professional and personal practice recommended for biomedical engineers.

BIOMEDICAL ENGINEERING PROFESSIONAL OBLIGATIONS

Biomedical engineers in the fulfillment of their professional engineering duties shall:

1. Use their knowledge, skills, and abilities to enhance the safety, health, and welfare of the public.
2. Strive by action, example, and influence to increase the competence, prestige, and honor of the biomedical engineering profession.

BIOMEDICAL ENGINEERING HEALTH CARE OBLIGATIONS

Biomedical engineers involved in health care activities shall:

1. Regard responsibility toward and rights of patients, including those of confidentiality and privacy, as their primary concern.
2. Consider larger consequences of their work in regard to cost, availability, and delivery of health care.

BIOMEDICAL ENGINEERING RESEARCH OBLIGATIONS

Biomedical engineers involved in research shall:

1. Comply fully with legal, ethical, institutional, governmental, and other applicable research guidelines, respecting the rights of and exercising the responsibilities toward colleagues, human and animal subjects, and the scientific and general public.
2. Publish and/or present properly credited results of research accurately and clearly.

BIOMEDICAL ENGINEERING TRAINING OBLIGATIONS

Biomedical engineers entrusted with the responsibilities of training others shall:

1. Honor the responsibility not only to train biomedical engineering students in proper professional conduct in performing research and publishing results, but also to model such conduct before them.
2. Keep training methods and content free from inappropriate influence from special interests.

Glossary of Terms
Likely to Be Encountered
in Bioethical Decision Making[1]

A posteriori knowledge (Latin: What comes after) – Knowledge gained from experience. Modern science is found for acquiring such knowledge. Contrast with *a priori* knowledge.

A priori knowledge (Latin: What comes before) – Knowledge independent of experience. Contrast with *a posteriori* knowledge.

ABET (Accreditation Board for Engineering and Technology, Inc.) – Entity that accredits United States college and university programs in applied science, computing, engineering, and technology.

Abortion – Premature exit of the fetus from the uterus. A medically-induced abortion can be completed surgically or chemically (e.g., hormonal drugs).

Acceptable engineering practice –

1. Amount of engineering-related work needed, in years, to meet one of the minimal criteria to sit for the Professional Engineer (PE) examination.
2. Reasonably expected professional performance by an engineer demonstrating competence, especially adhering to codes for design, construction, and operation.

Accreditation – Nongovernmental, peer review of an institution to ensure educational quality. Institutional accreditation evaluates overall institutional quality. One form of institutional accreditation is regional accreditation of colleges and universities. Specialized accreditation examines specific programs of study, rather than an institution as a whole. This type of accreditation is granted to specific programs at specific levels. Architecture, nursing, law, medicine, and engineering programs are often evaluated through specialized accreditation.

Act utilitarianism – Form of utilitarianism wherein the test of utility maximization is applied directly to single acts. In act utilitarianism, an action is evaluated on a case-by-case basis for practical usefulness. Compare to rule utilitarianism.

Ad hominem – Fallacious argument attacking the holder of a view rather than the position itself.

Adult stem cell – Undifferentiated cell found in a differentiated tissue that can renew itself and (with certain limitations) differentiate to yield all the specialized cell types of the tissue from which it originated.

Advanced directive – A document informing the health providers and physicians the type and extent of care the patient requests in the event that the patient becomes unable to make medical decisions (e.g., in a coma).

Amoral – Lacking any moral characteristics. An amoral act is neither morally good nor morally bad; it simply exists. Contrast with moral or immoral.

Analogy – Comparison of similarities between two things to a conclusion about an additional attribute common to both things. This is a type of inductive reasoning (see Inductive Reasoning).

Animal welfare ethics – Moral consideration of animals in research, food supply, and other human endeavors.

Antagonism – Effect from a combination of two agents is less than the sum of the individual effects from each agent $(1 + 1 < 2)$.

Anthropocentrism – Philosophy- or decision framework-based human beings. View that all and only humans have moral value. Nonhuman species and abiotic resources have value only in respect to that associated with human values. Contrast with biocentrism and ecocentrism.

Anthropogenic – Made, caused, or influenced by human activities. Contrast with biogenic.

Applied mathematics – Mathematical techniques typically used in the application of mathematical knowledge to domains beyond mathematics itself.

Archetype – Symbolic imagery. Psychologist Carl Jung applied the term to the generalized patterns of images that form the world of human representations in recurrent motifs, passing through the history of all culture.

Association – Relationship, not necessarily causal, between two variables. The antecedent variable comes before and is associated with an outcome; however, it may or may not be the cause of the outcome. For example, mean birth weight of minority babies is less than that of babies of the general population. Ethnicity is an antecedent of low birth weight, but not the cause. Other factors, e.g., nutrition, smoking status, and alcohol consumption, may be the causal agents.

Autonomy – Individual's capability of deliberation about personal goals and of acting under the direction of such deliberation. To respect autonomy is to give weight to autonomous persons' considered opinions and choices while refraining from obstructing their actions unless they are clearly detrimental to others.

Autosome – Any chromosome other than the sex chromosomes or the mitochondrial chromosome.

Baby –

1. Very young (e.g., birth to one year of age) child.
2. A child that has not yet been born (see Fetus). For example, expectant mothers often refer to their baby's kicking and other movements.

Bacteria – Unicellular microorganisms that exist either as free-living organisms or as parasites, ranging from beneficial to harmful to humans.

Bayesian – Statistical approach that addresses probability inference, named after Thomas Bayes (*An Essay towards Solving a Problem in the Doctrine of Chances*, 1763), to decision making and inferential statistics that deals with probability inference (i.e., using knowledge of prior events to predict future events). In a Bayesian network, priors are updated by additional data that yield posterior probabilities that are often more robust than classical probabilities.

Beatitudes – Declarations of blessedness. Often associated with Christ's Sermon on the Mount (Matthew 5: 1–7:29).

Behaviorism – Viewpoint founded by psychologist John Watson (*Psychology as the Behaviorist Views It*, 1913) holding that behaviors can be conditioned and measured. Behaviorism is based solely on observable behaviors, considering cognition and mood to be overly subjective.

Belmont Report – "Ethical Principles and Guidelines for the Protection of Human Subjects of Research" by the National Commission for the Protection of Human Subjects of Biomedical and Behavioral Research. Released on 18 April 1979. Articulated the basic ethical principles and guidelines that should assist in resolving the ethical problems that surround the conduct of research with human subjects. Expounded three basic ethical principles: respect for persons, beneficence, and justice.

Beneficence – Promotion of good for others and contribution to their welfare. Obligation to protect subjects from harm. The principle of beneficence requires that probable risks to research subjects be minimized and the potential for benefit to the subjects and/or to society as a whole be maximized. One of three principles stemming from the Belmont Report (see Belmont Report).

Benefit–cost analysis (or cost–benefit analysis) – Method designed to determine the feasibility or utility of a proposed or existing project. Yields a benefit–cost ratio (see Benefit–Cost Ratio (BCR)).

Benefit–cost ratio (BCR) – Weighted benefits divided by weighted costs; used to compare and differentiate among project alternatives. Gross BCR <1 is undesirable. The greater the BCR, the more acceptable the alternative.

Benevolence –

1. Disposed to do good, to act charitably, to possess virtue.
2. Goodness and moral excellence. Compare to beneficence.

Best practice –

1. Optimal service to the client.
2. Treatment is appropriate, accepted, and widely used according to expert consensus; embodies an integrated, comprehensive, and continuously improving approach to care (medicine). It is morally obligatory that health care practitioners provide patients with the best practice (also known as standard therapy or standard of care).

Bias –

1. Unfair influence (sociology).
2. Systematic error in one direction; such as the positive bias of a scale that reads 1 mg too high (instrument error) or the negative bias in interpretations of lesions reported by a physician performing the procedure (operator bias) that consistently miss some lesions. Bias makes the reported values less accurate.

Bio – Prefix indicating "life" (Greek).

Biocentrism – View that all life has moral value. Contrast with anthropomorphism.

Bioethics –

1. Inquiry into ethical implications of biological research and applications.
2. Ethical inquiry into matters of life, especially biomedical and environmental ethics (preferred definition of this text).

Biogenic – Made, caused, or influenced by natural processes. Contrast with anthropogenic.

Bioinformatics – Management and analysis of data using advanced computing techniques applied to biological research and inquiry.

Biological engineering – Combination of biomedical and biosystem engineering (see Biomedical Engineering and Biosystem Engineering) to develop useful biology-based technologies that can be applied across a wide spectrum of societal needs, including diagnosis, treatment, and prevention of disease, design and fabrication of materials, devices, and processes, and enhancement and sustainability of environmental quality.

Biological response – Manner and type of effect in an organism (e.g., disease, change in metabolism, and homeostasis).

Biologically Based Dose Response (BBDR) model – Predictive model that describes biological processes at the cellular and molecular level linking the target organ dose to the adverse effect.

Biomarker – A chemical, physical, or biological characteristic that indicates biological condition. The biomarker may be a chemical to which an organism is exposed (e.g., lead in blood), a metabolite of the chemical (e.g., cotinine in blood as an indication of exposure to nicotine), or a biological response (e.g., an increase in body temperature as a result of exposure to a pathogen).

Biomedical engineering – Application of engineering principles to medicine, including drug delivery systems, therapeutic systems, and medical devices.

Biomedical ethics –

1. Application of moral reasoning to address questions that emerge in health care and clinical situations.
2. Study of morality in biomedical situations.

Biomedical testing – Investigations to determine whether a change in a body function might have occurred because of exposure to a hazardous substance.

Biostatistics – Application of statistical tools to interpret biological and medical data.

Biosystem – Living organism or a system of living organisms that are able to interact with other organisms directly or indirectly.

Biosystem engineering (bioengineering) –

1. Application of biological sciences to achieve practical ends.
2. Integration of physical, chemical, or mathematical sciences and engineering principles for the study of biology, medicine, behavior, or health to advance fundamental concepts, to create knowledge for the molecular to the organ systems levels, and to develop innovative biologics, materials, processes, implants, devices, and informatics approaches for the prevention, diagnosis, and treatment of disease, for patient rehabilitation, and for improving health (from the National Institutes of Health).

Biota – Any living creature, plant (flora), animal (fauna), or microbial.

Biotechnology – Use of living creatures to produce things of value to humans (e.g., drugs and food).

Bioterrorism – Use of living agents to cause intentional harm to people (e.g., anthrax spores or pathogenic viruses) and society (e.g., agricultural pests).

Blastocyst – Preimplantation embryo of about 150 cells that make up a sphere of an outer layer of cells (i.e., trophoblast) that will form the placenta, a fluid-filled cavity (blastocoel), and the inner cell mass, i.e., a cluster of cells on the interior from which the stem cells are derived. Human embryonic stem cells are frequently taken from four- to five-day-old blastocysts.

Body burden – Total amount of a substance in the body.

Bottom-Up – View where fundamental components are first considered, working upward to larger perspectives. Contrast with top-down.

Butterfly effect – Sensitive dependence on initial conditions. Metaphor for the extreme sensitivity of chaotic systems (see Chaos), in which small changes or perturbations lead to drastically different outcomes. The phrase is derived from a butterfly flapping its wings in California, and thereby initiating a change in weather patterns that results in the formation of a thunderstorm in Nebraska (from Edward Lorenz in his 1963 article, "Deterministic Nonperiodic Flow," *Journal of Atmospheric Sciences* 20: 130–41; although in his presentation to the New York Academy, it was not a butterfly but a seagull's flapping of the wing that posited as the initial condition. Later in 1972, Lorenz used the butterfly in the example).

Cancer – Disease of heritable, somatic mutations affecting cell growth and differentiation, characterized by an abnormal, uncontrolled growth of cells. Malignant growth or tumor caused by abnormal and uncontrolled cell division.

Carcinogen – Physical or chemical agent that induces cancer.

Cardinal virtues – Four paramount virtues in classical philosophy: justice, prudence, fortitude, or temperance.

Case-control study – An epidemiologic study contrasting those with the disease of interest (cases) to those without the disease (controls). The groups are then compared with respect to exposure history, to ascertain whether they differ in the proportion exposed to the chemical(s) under investigation.

Categorical imperative – Central theme of Immanuel Kant's deontological ethics (see Deontology or deontological ethics) that sets one principle from which all specific moral imperatives are derived: "Act only according to that maxim by which you can at the same time will that it should become a universal law" (*Groundwork of the Metaphysic of Morals* [*Grundlegung zur Metaphysik der Sitten*], 1785).

Causation, causality – Relationship between causes and effects. Contrast with association.

Casuistry –

1. Analyzing actual cases by applying moral principles to inform ethical decision making (Cicero is one of the first to apply this method).
2. Derogatory term to denote special and subtly deceptive argumentation.

Cell – Basic unit of life; autonomous, self-replicating unit that either constitutes a unicellular organism or is a subunit of a multicellular organism; the lowest denomination of life.

Central nervous system – Portion of the nervous system that consists of the brain and the spinal cord.

Chaos theory – Exposition of the apparent lack of order in a system that nonetheless obeys specific rules. Condition discovered by the physicist Henri Poincaré around the year 1900 that refers to an inherent lack of predictability in some physical systems (i.e., Poincaré's concept of dynamical instability).

Character – Description of a person's attributes, abilities, and traits, especially those related to moral strength. Contrast with personality.

Charity – Virtue manifested by impartial love. In Christianity, coexists with faith and hope.

Chi-square (χ^2) test – Statistical test to determine the probability that an observed deviation from the expected event or outcome occurs solely by chance.

Child –

1. Human being less than eighteen years of age (United Nation's Convention on the Rights of the Child). Subcategories include:
   - Infant (baby) (ages 0–1.5)
   - Neonate (in the first month after birth)
   - Toddler (ages 1.5–4)
   - Middle childhood (age 4–11)
     o Prepubescence (near age 10–11)
     o Preadolescence (near age 11–13)
   - Adolescence and puberty (age 14–20)
2. Human beings after conception until eighteen years of age (same categories as above, but additional category of "unborn child" with subcategories of zygote, embryo, and fetus (see Zygote, Embryo, Fetus)).

Childhood – Time of rapid development of human personality and morality.

-cide – Suffix indicating act of killing (Latin: *caedere* – to cut)

Chromosome – Structure within a cell's nucleus consisting of strands of deoxyribonucleic acid (DNA) coated with specialized cell proteins, and duplicated at each mitotic cell division. Chromosomes transmit the genes of the organism from one generation to the next.

Client – A person or entity (e.g., company) that seeks service from a professional. The engineer, for example, is required to be a faith agent to the client, meaning that the engineer is responsible for representing the best interests of the client. Contrast with customer.

Clone – Line of cells genetically identical to the originating stem cell.

Code of ethics – Established set of moral expectations of a group, especially of professional societies.

Code of Hammurabi – One of the first known collections of 282 moral laws probably written between 2100 BC and 1800 BC, but credited to Babylonian King Hammurabi (1792–1750 BC).

Coefficient of determination (r^2) – Proportion of the variance of one variable predictable from another variable. The ratio of the explained variation to the total variation, which represents the percentage of the data nearest to the line of best fit. For example, if $r = 0.90$, then $r^2 = 0.81$, meaning that 81% of the total variation in one variable (y) can be explained by the linear relationship between the two variables (x and y) as described by the regression equation. Thus, the remaining 19% of the total variation is unexplained.

Coherence – Criterion for causality based on the amount and degree of agreement among studies linking cause to effect; especially among various types of studies (e.g., animal testing, human epidemiological investigations, and *in vitro* studies).

Cohort study – Epidemiologic study comparing those with an exposure of interest to those without the exposure. These two cohorts are then followed over time to determine the differences in the rates of disease between the exposure subjects. Also called a prospective study.

Coma – Prolonged period of unconsciousness following traumatic brain injury. In this sleep-like state, there is no speech, the eyelids are usually closed, and there is no response to commands. A number of scales (e.g., the Glasgow Coma Scale, Disability Rating Scale, and the Rancho Los Amigos Scale

(RLAS) are used to characterize the severity of coma. The RLAS, for example, uses eight levels in descending order of severity (Source: Miller School of Medicine, University of Miami)):

- Level I – No response to any stimuli; indicates coma.
- Level II – Generalized response, i.e., patient reacts inconsistently and nonpurposefully to stimuli in a nonspecific manner, such as eye blinking, changes in breathing rate, gross body movement, and vocalization; indicates coma.
- Level III – Localized response, i.e., patient reacts specifically but inconsistently to stimuli, such as turning head toward a sound or focusing on an object presented and following simple commands in an inconsistent, delayed manner; not considered coma, but stimulation techniques appropriate through Levels III.
- Level IV – Confused and agitated, i.e., patient is in a heightened state of activity with severely decreased ability to process information. The patient is detached from the present and responds primarily to his/her own internal confusion. Behavior is often bizarre.
- Level V – Confused, inappropriate, nonagitated, i.e., patient appears alert and is able to respond to simple commands fairly consistently, but responds to more complex commands in a non purposeful, random manner and is agitated by external stimuli.
- Level VI – Confused and appropriate, i.e., the patient shows goal-directed behavior, but is dependent on external input for direction. He/she follows simple directions and shows carryover for tasks that have been relearned, such as self-care activities. Responses may be incorrect due to memory problems, but they are appropriate to the situation.
- Level VII – Automatic and appropriate, i.e., the patient appears appropriate and oriented, but goes through daily routines automatically, and has shallow recall of what he/she has been doing. The patient shows increased, but superficial awareness of self and other people, demonstrates decreased judgment and problem-solving abilities, lacks realistic planning for the future, and requires at least minimal supervision for learning and safety purposes. Judgment and other higher-level cognitive abilities remain compromised.
- Level VIII – Purposeful and appropriate, i.e., the patient is alert and oriented, and is able to recall and integrate past and recent events, is aware of and responsive to the environment, and needs no supervision once learning has occurred. He/she may continue to show decreased reasoning, tolerance for stress, judgment in emergencies or unusual circumstances, and decreased social, emotional, and intellectual capacities.

Common sense – Innate, sound judgment, not specially resulting from formal training. However, engineers nurture common sense by blending formal, technical training with experience and mentorship.

Compartmentalization – Viewing a system by its individual components. This can be problematic when an engineer does not consider the system as a whole (e.g., when the structural engineer and soil engineer do not collaborate on selecting the best and safest combination of materials and structures suited to a soil type, or when a biomedical engineer does not work closely with various specialized health care professionals in a clinical setting to adapt a realistic device to the comprehensive needs of the patient). Compartmentalization can be good when it allows the engineer to focus adequate attention on the components (see Bottom-Up), so long as the design is properly built into a system.

Competence –

1. Skill in practice. For professionals, competence is requisite to ethical practice.
2. Sufficient velocity for a fluid to carry a load (especially a fluid's ability to carry solids).

Complexity – Relative measure of uncertainty in achieving functional requirements or objectives. Designers are frequently expected to reduce the complexity of engineered systems.

Confidence, confidence level –

1. Client's trust in a professional.
2. Amount of certainty that a statistical prediction is accurate. Physical sciences may differ from social sciences in what is considered acceptable confidence, e.g., the former may require 99% while social scientific research may consider 95% to be acceptable. Depending on the application, engineering research ranges in acceptable confidence level (e.g., structural fatigue research may require higher confidence levels than environmental research).

Conflict of interest – Situation wherein professional or personal obligations or personal or financial interests detract from a professional's ability to carry out one's responsibilities and duties fairly.

Confounder – Factor that distorts or masks the true effect of risk factors in an epidemiologic study. A condition or variable that is a risk factor for disease and is associated with an exposure of interest. This association between the exposure of interest and the confounder (a true risk factor for disease) may make it falsely appear that the exposure of interest is associated with disease. For example, a study of low birth weight children in low-income families must first address the confounding effects of tobacco smoking before ascribing the actual risk associated with income.

Conscience –

1. Set of ethical principles held by a person that govern one's thoughts, motivations, and actions.
2. Awareness of these principles.
3. In psychoanalytical theory, the part of the superego (see Superego) that informs the ego (see Ego) on what should be done.

Consequentialism – Ethical theory with the perspective that the value of an action derives solely from the value of its consequences. Consequentialists hold that the consequences of a particular action form the basis for any valid moral judgment about that action, so that a morally right action is an action that produces good consequences. One of three major theories of normative ethics (see Normative Ethics), along with virtue ethics and deontological ethics (see Deontology or deontological ethics).

Contingent probability – Probability that an event will occur as a result of one or more previous events. Also known as conditional probability.

Contractarianism – Ethical theory based on adherence to the social contract. Includes libertarianism and Rawlsism (see Libertarianism and Rawlsism).

Control group – Group used as the baseline for comparison in epidemiologic studies or laboratory studies. This group is selected because it either lacks the disease of interest (case-control group) or lacks the exposure of concern (cohort study). Also known as a reference group.

Cooking – Irresponsible data manipulation of retaining only those results that fit the study objectives, hypothesis, or theory and discarding others. This is done in an attempt to improve statistical significance and confidence, e.g., a better chi-square or correlation coefficient (r), or coefficient of determination (i.e., r^2).

Copyright – Legal documentation of written intellectual property.

Correlation coefficient (r) – Statistical measurement of the strength and the direction of a linear relationship (association) between two variables.

Creativity – Ability to create knowledge, to see patterns and make connections so as to approach problem solving and opportunities in an inventive and novel way.

Critical path – Systems engineering of activities, decisions, and actions that must be completed on schedule and at a sufficient level of quality for the entire project to be successful.

Cross-sectional study – Epidemiological study of observations representing a particular point in time. Contrast with longitudinal study.

Customer – Buyer of goods and services.

Data – Plural of datum. Gathered facts from which conclusions can be drawn.

Death –

1. Cessation of life (see Life); legal definition is "the cessation of life; permanent cessations of all vital functions and signs" (Black's Law Dictionary).

2. Whole body death occurs with complete cessation of breathing and heartbeat, leading to anoxia (no oxygen), then ischemia (no blood flow), with death occurring with ischemia in the brain.

3. Brain death, which has generally displaced whole body death as technology has temporarily prevented brain ischemia even with cessation of cardiovascular function (e.g., during a transplant operation), occurs when the brain ceases to function as indicated by lack of response to external stimuli, absence of body movements, cessation of spontaneous breathing, and two isoelectric (flat) electroencephalogram (EEG) readings separated by 24 hours (Ad Hoc Committee of the Harvard Medical School to Examine the Definition of Brain Death, published in 1968 in the *Journal of the American Medical Association*, 205 (337)).

Deductive reasoning – A conclusion is necessitated by previously known facts. If the premises are true, the conclusion must be true. Starting from general knowledge and moving to specifics (e.g., from cause to effects). Contrast with inductive reasoning.

Deep ecology – Environmental movement initiated in 1972 by Norwegian philosopher, Arnie Naess, that advocates radical measures to protect the natural environment irrespective of their effect on the welfare of humans (opposite of anthropocentrism).

Demand – Quantity of a good or service that society chooses to buy at a given price.

Deontology or deontological ethics (meaning opposite of ontology) – Ethical theory basing right and wrong on duty rather than seeking virtue or moral ends.

Descriptive ethics – Study of what a group actually holds to be right and wrong, and maintains as ideals. Contrast with normative ethics.

Developing nation – Nation at the lower end of socioeconomic and technological advancement; formerly known as third world nation.

Developmental toxicity – Adverse effects on developing organism that may result from exposure prior to conception (either parent), during prenatal development, or postnatally until the time of sexual maturation. The major manifestations of developmental toxicity include death of the developing organism, structural abnormality, altered growth, and functional deficiency.

Device (medical) – Instrument, apparatus, implement, machine, contrivance, implant, *in vitro* reagent, or other similar or related article, including any component, part, or accessory, that is intended for use in the diagnosis of disease or other conditions, or in the cure, mitigation, treatment, or prevention of disease, in man or other animals, or intended to affect the structure of any function of the body and that does not achieve its primary intended purpose through chemical action and that is not depended upon being metabolized for the achievement of its primary intended purposes (Federal Food,

Drug and Cosmetics Act, 1938). The last sentence of this definition helps to distinguish a device from a drug. Both are regulated, but differently, by the US Food and Drug Administration.

Dignity – A person's worthiness of respect, privacy, autonomy, and self-worth.

Dilemma – Choice between two equally unattractive options.

Disaster (Latin: *dis* and *astrum* meaning "bad star") – A relative term meaning a catastrophic event that wreaks great destruction. However, the term is not exclusive to large-scale events, such as hurricanes or earthquakes, but can also include small-scale events with highly negative consequences, such as an engineering or medical failure where one or a few people are impacted but that has other implications (malpractice, bad publicity, blame, etc.).

Disease – Abnormal and adverse condition in an organism (although sometimes applied metaphorically, such as the "disease" of poverty).

Disparate effect – Health outcome, usually negative, that is disproportionately high in certain members of a population, such as an increased incidence of certain cancers in minority groups.

Disparate exposure – Exposure (see Exposure) to a physical, chemical, or biological agent that is disproportionately high in certain members in a population, such as the higher than average exposure of minority children to lead.

Disparate susceptibility – Elevated risk of certain members of a population (e.g., genetically predisposed) to the effects of a physical, chemical, or biological agent; can lead to disparate effects (see Disparate Effect).

Distributive justice – Just allocation of goods, services, and utility within society. Focuses more on outcome than on procedures (i.e., procedural justice).

Doctrine of the mean – Confucian, 23-chapter book on usefulness of a golden way (Tao) to gain virtue.

Dose – Amount of a substance available for interactions with metabolic processes or biologically significant receptors after crossing the outer boundary of an organism. Potential dose is the amount ingested, inhaled, or applied to the skin. Applied dose is the amount presented to an absorption barrier and available for absorption (although not necessarily having yet crossed the outer boundary of the organism). Absorbed dose is the amount crossing a specific absorption barrier (e.g., the exchange boundaries of the skin, lung, and digestive tract) through uptake processes. Internal dose is a more general term denoting the amount absorbed without respect to specific absorption barriers or exchange boundaries. The amount of the chemical available for interaction by any particular organ or cell is termed the delivered or biologically effective dose for that organ or cell.

Dose–response – Relationship between a quantified exposure (dose) and the proportion of subjects demonstrating specific biologically significant changes in incidence and/or in degree of change (response).

Drug – Substance intended for use in the diagnosis, cure, mitigation, treatment, or prevention of disease, which is regulated by the US Food and Drug Administration. Contrast with device and nutritional supplement (see Device and Nutritional Supplement).

Dry lab –

1. *In silico* research (contrast with wet lab).
2. Walkthrough prior to actual laboratory work (step preceding wet lab).
3. Unethical practice of forging (making up) data.

Dual use –

1. Science, engineering, and technology designed to provide both military and civilian benefits.

2. Research and technology that simultaneously benefit and place society at risk (e.g., biotechnological advances that improve vaccines but also increase the risks of bioterrorism).

Dynamical instability – See Chaos Theory.

Ecocentrism – Philosophy or decision framework based on the whole ecosystem rather than a single species. Contrast with anthropocentrism.

Ecofeminism – View that oppression of women and environmental degradation are linked. French feminist Françoise D'Eaubonne coined the term in 1974 in an attempt to characterize how male-dominated societies have inflected violence on both women and nature.

Educated men and women – Term used to distinguish college-educated persons, included in certain graduation ceremonies.

Effectiveness – Measure of the extent and degree to which a design achieves a goal (i.e., effectiveness of design is measured with respect to design specifications and objectives).

Efficacy – Ability or capacity to generate a desired effect.

Efficiency – Ratio of total energy or mass output to total energy or mass input, expressed as a percentage. Treatment or removal efficiency is the product of the contaminant mass prior to treatment (I) times the contaminant mass after treatment (E) divided by I. To express efficiency as a percentage, these values are multiplied 100 times: $\frac{I \times E}{I} \times 100$.

Ego – The self. In psychoanalytic theory, the ego represents the id (see Id) to the outer world; mediating between the id and superego (see Superego).

Egoism – Theoretical framework holding that self-interest dictates ethical decision making (also known as psychological egoism or ethical egoism).

Elasticity –

1. Ability of a body to return to its original shape after being stressed (engineering).
2. Degree to which an item's price change results from a unit change in supply ("supply elasticity") or a unit change in demand ("demand elasticity") (economics).

Elitism – Sense of entitlement based on class and socioeconomic status.

Embryo –

1. Human being from the time of fertilization until the end of the eighth week of gestation, thereafter considered to be a fetus.
2. Developing organism from the time of fertilization until significant differentiation has occurred, when the organism becomes known as a fetus. An organism in the early stages of development.

Embryonic stem cell – Undifferentiated cells from the embryo with the potential to become all cell types found in the body. Embryonic stem cells are found in the inner cell mass of blastocysts (see Blastocyst).

Embryonic stem cell line – Embryonic stem cells that have been cultured *in vitro* that allow proliferation without differentiation for months to years.

Emotive theory of values – Theory holding that judgments about values are simply emotional expressions based on the inclinations and attitudes of the individual doing the judging.

Empathy – Ability to place someone in another person's (or other species') mental or spiritual position.

Empiricism – Position of modern science that all knowledge about the physical world is based on observation and sensory experiences. See *A posteriori* knowledge.

Endocrine system – Chemical messaging system in organisms used for regulation by secretion of hormones by glands that are sent through the circulatory system to cells where the hormones bind to receptors.

Ends justifying means – Rationale for excusing unethical approaches to achieve worthy ends (at least in the view of those undertaking the actions).

Engineering –

1. Application of scientific and mathematical principles to practical ends, especially design, manufacture, and operation of structures, machines, processes, and systems.
2. The profession that implements (1).

Environmental engineering – Subdiscipline of engineering (usually civil engineering) concerned with applications of scientific principles and mathematics to improve the condition of the environment.

Environmental equity – Fair distribution of environmental risks across population groups.

Environmental justice – Fair treatment of all people regardless of race, color, national origin, or income with respect to the development, implementation, and enforcement of environmental laws, regulations, and policies.

Environmental racism – Disproportionately negative treatment of members of ethnic and racial minority groups in decisions regarding polluting agents in their neighborhoods and communities. Term was coined in 1982 by the then director of the United Church of Christ's Commission for Racial Justice, the Reverend Ben Chavis.

Environmental science – Systematic study of the environment and its components and processes (e.g., nutrient cycling, pollutant transport, and adverse effects).

Environmentalism –

1. Advocacy in the protection of the environment.
2. Philosophy underpinning this advocacy. Such advocacy may or may not be scientifically based (i.e., differs from environmental science and environmental engineering).

Epidemiology – Study of the causes, distribution, and control of disease in populations.

Epistemology – Philosophical inquiry into the nature of knowledge.

Error –

1. Mistake.
2. In statistics, the difference between a reported value and the actual value (see Bias).

Ethics –

1. Set of moral principles.
2. Study of morality and moral decision making.

Eugenics – Study of hereditary (genetic) improvement of the human race by controlled selective breeding. It has been used as a rationale for genocide and other immoral acts (e.g., in Nazi Germany).

Euthanasia – Deliberate use of medical means to cause death.

Event – Set of outcomes that are preceded and linked to an earlier set of outcomes (probability theory).

Evil – An immoral act (noun) or morally bad (adjective). Opposite of good.

Ex situ – Moved off-site (e.g., contaminated soil transported to an incinerator for treatment).

Ex vivo – Outside the body, frequently the equivalent of *in vitro* (see *In vitro*).

Experiment – Investigation to support or reject a hypothesis or to increase knowledge about a phenomenon.

Exposure – Contact with a substance through various pathways, especially oral, dermal, and inhalation.

Extraordinary means – Treatments and medical care that are unduly burdensome or sorrowful, such as amputation, or beyond the economic means of the person, or which only prolong the suffering of a dying person. If such measures are considered to be morally extraordinary, they are not obligatory. Recent advances in medicine, however, complicate the decision whether to undergo or forego medical treatment, since medicine can now save many people who would simply have been allowed to die in the past. Thus, the moral permissibility and obligation are not always straightforward, wherein some may consider ordinary means to be extraordinary and vice versa.

Fact – That which can be shown to be true, to exist, or to have occurred.

Failure – Lack of success as indicated by design specifications and measures of success.

Faith –

1. Belief beyond that based on logical proof or material evidence.
2. Christian virtue of secure belief in God and trusting acceptance of God's will.

Faithful agent – A trustee or representative authorized by a person who is entrusted with that person's best interests. Thus, a faithful agent will avoid conflicts of interest (see Conflict of Interest), will maintain appropriate confidentiality, and in other ways guard the trust given by the client.

Fallacy of *non sequitur* – Invalid argument wherein the conclusion does not follow from the premises.

False negative – Finding of the absence of a condition (e.g., disease) in a test when, in fact, the disease is present (e.g., a lung cancer screen shows that the patient has no cancer, but at a level of detection below the screen, the cancer has cancer cells in the lung). Type II error (see Type II Error).

False positive – Positive finding of a test when, in fact, the true result was negative (e.g., A drug screen shows that a person has used opiates, even though the person has not). Type I error (see Type I Error).

Fetus (foetus) – Developing mammal (including human) subsequent to embryonic stage and preceding birth.

Forging – Inventing or concocting research results and reporting methods and findings of fictitious or exaggerated experiments. Synonymous with dry lab (definition 3).

Futile care – Treatment that is unlikely to restore quality of life, improve a disease, decrease pain, or ameliorate suffering, and could render more harm than benefit to the patient's wellbeing.

Fuzzy logic – System dealing with the partial truths, assigning with values ranging from completely true to completely false.

Game theory – Decision making under conditions of uncertainty and interdependence; taking into account the characteristics of players, strategies, actions, payoffs, outcomes at equilibrium.

Gamete – Reproductive cell (i.e., an egg or a sperm).

GANTT chart – Graphical display of a project or program showing each task as a horizontal bar with its length being proportional to time needed for completion.

Gene – Ordered sequence of nucleotides located in a particular position on a particular chromosome, representing the fundamental unit of heredity.

Genetics – Scientific investigation of heredity.

Genome – Total genetic composition of an individual.

Genomics – Study of genes, including their functions.

Germ theory – Paradigm that diseases are caused by singular, proximate, pathogenic microbes. Displaced miasma theory (see Miasma) in late nineteenth century.

Gestalt theory (German: Form) – View that perception and other psychological phenomena must be understood for their overall patterns and forms, as opposed to the individual components.

Gestation – Time from fertilization of the ovum to birth. Also known as uterogestation.

Golden mean – Optimal position between extremes (deficiency and excess) expressed by Aristotle and other Ancient Greeks.

Golden Rule – Empathetic view; ethics of reciprocity: Do unto others as you would have them do unto you (Matthew 7: 12).

Good – Moral excellence; an action or behavior representing such excellence. Opposite of evil (see Evil).

Good Samaritan – Helpful and sympathetic person modeled after the person in Jesus Christ's parable (Luke 10: 25–37).

Grand rounds – Training sessions for physicians and other health care providers covering health topics likely to be encountered in practice.

Green engineering – Design, commercialization, and use of processes and products, which are feasible and economical while minimizing the generation of pollution at the source and the risk to human health and the environment (US Environmental Protection Agency).

Green hospital – Health care facility that incorporates sustainable practices into its operations (e.g., recycling, green purchasing, and minimization of wastes).

Gross domestic product (GDP) – Total monetary value of all final goods and services produced within a nation's borders during a given period of time.

Harm – Damage to another person or creature.

Harm principle – John Stuart Mill's recommendation that utilitarianism's (see Utilitarianism) premise of greatest good is restricted (e.g., by law) when others are harmed.

Hazard – Potential source of harm.

Healthy worker effect – Situation where there are fewer deaths for workers in an industry compared with the US population, usually because severely ill and chronically disabled are excluded from employment. The general population may not be the best comparison group for workers.

Hedonism – Belief that pleasure is the greatest good and highest aspiration of humankind. In early utilitarian thinking, hedonism was the basis for utility or good.

Hierarchy of needs – Psychological construct proposed by Abraham Maslow (*A Theory of Human Motivation*, 1943) holding that as humans meet their basic needs, they are successively able to meet higher (growth) needs, until finally reaching self actualization (which includes morality and creativity).

Hill's criteria – Minimal conditions necessary to establish causal relationship between two items; presented by British medical statistician Sir Austin Bradford Hill (1897–1991) as a means of finding causal links between a specific factor (e.g., exposure to air pollution) and specific adverse effects (e.g., asthma). These criteria, originally recommended for occupation setting but now applied to numerous health and environmental problems, are meant to be guidelines rather than inviolable rules of epidemiology (see Epidemiology).

Holism –

1. Theory that living matter is made up of organic or unified wholes that are greater than the simple sum of their parts. In this view, wholes have some priority over the elements, members, individuals, or parts composing them. Holistic views propose that the meaning and truth of claims cannot be assessed separately, but must be assessed as part of theories, bodies of theory, or worldview.
2. Holistic investigation or system of treatment.

Homeostasis – Ability of an organism to self-regulate functions; inherent trend toward stability (e.g., risk homeostasis).

Honesty – Virtue of acting with truthfulness, openness, and fairness.

Hope –

1. To anticipate a positive outcome.
2. In Christianity (1 Corinthians 13), the desire and search for future goodness that is difficult but possible to attain with God's help. Coexists with faith and love (charity).

Hormone – Chemical released by glands of the endocrine system (see Endocrine System).

Humors – Ancient Greek's explanation of personality, especially how when the four fluids become unbalanced. Galen (c. 190) extended these four fluids to characterize temperament: blood (sanguine); black bile (melancholic); yellow bile (choleric); phlegm (phlegmatic). These characterizations remain today, although the connection to the fluids is no longer supported (e.g., a happy person is considered to be sanguine and a sad person is melancholic, and an unpleasant person is considered to be bilious).

Hypothetico-deductive method – Method of logical deduction, attributed to Karl Popper (*The Logic of Scientific Discovery*, 1934), limiting scientific discovery to that which is testable; requiring an approach that formulates hypotheses, *a priori*, with the intent of rejecting these hypotheses. The method assumes that a hypothesis can never truly be proved, but at best can be corroborated.

Hysteresis – Failure to return to previous condition, such as due to an energy loss that always occurs under cyclical loading and unloading of a spring, proportional to the area between the loading and unloading load-deflection curves within the elastic range of a spring (engineering), or the failure of a variable to return to its initial equilibrium after a temporary shock (economics).

Iatrogenic – Caused by an act of the treating physician (e.g., prescribed drug's side effects lead to a stroke).

Id – In psychoanalytic theory, the most primitive of the three divisions of the psyche, having to be mediated by the ego and superego (see Ego and Superego).

Ideal utilitarianism – Form of utilitarianism holding the view that the value of an action derives solely and completely from the value of the action's consequences.

Immoral – Contrary to established moral principles.

In silico – Based on information, usually using computational methods, rather than using actual materials being studied. To some extent, *in silico* research is an alternative to *in vivo* and *in vitro* research (see *In vivo* and *In vitro*), which is desirable in the case of limiting animal research and in reducing risks in humans who undergo *in vivo* procedures.

In situ – Taking place where it is found (e.g., cleaning up hazardous waste sites where they exist, rather than moving the materials off-site).

In utero – In the womb (e.g., fetal alcohol syndrome results from the unborn child's exposure to alcohol and its metabolites during gestation (see Gestation)).

In vitro – Outside of the organism (literally: "in glass," i.e., in a test tube).

In vitro fertilization – Union of an egg and sperm outside the body and in an artificial environment.

In vivo – Inside the organism (e.g., experiments within a rat to observe biochemical responses to a chemical dose).

Incidence or incidence rate – Number of new cases in a defined population within a period of time.

Induced termination of pregnancy – Synonym of abortion (see Abortion).

Inductive reasoning – Starting from a specific experience and drawing inferences from the specific set of facts or instances to a larger number of instances (generalization). Conclusions are drawn from the perspective that all individuals of a kind have a certain character on the basis that some individuals of the kind have that character. Contrast with deductive reasoning.

Inference – Reasoning that one statement (the conclusion) is derived from one or more other statements (the premises). See Syllogism.

Informatics – Application of computational and other technologies to access and to enhance information; one means of turning data into information (see Data and Information).

Information – Processed and organized data (see Data). Value-added data as a step toward knowledge.

Information processing theory – Cognitive processes consisting of stages of encoding, storage, and retrieval of data.

Institutional review board – Group of scientists and nonscientists who review proposed studies to ensure that the rights of study participants are protected.

Instrumental value – Worth based on usefulness. In biomedical ethics, the perspective of whether a human life has value that depends on usefulness (Will the baby be loved? – Will the elderly person continue to enjoy life?) is an instrumental viewpoint. In environmental ethics, use of the term environmental resource implies that ecological value is based on the utility of the ecosystem (e.g., wetlands as breeding area for game fish, as retention areas to prevent floods, and as sinks for carbon to prevent global warming). Contrast with intrinsic value.

Integrity – Moral soundness based on honesty and upright character (see Character).

Intellectual property – Proprietary knowledge protected by law, e.g., patents and copyright.

Intervention – Direct involvement of corrective action to change existing condition for the better (i.e., both an engineering and a medical concept).

Intrinsic value – Worth based on existence, not usefulness. All humans have intrinsic value in contemporary morality. In biomedical ethics, however, there is no unanimity of thought about the intrinsic value of an embryo or a fetus, or a person nearing end of life. Those subscribing to sanctity of life viewpoints see intrinsic value of any human being (beginning with the human zygote and ending in natural death). In environmental ethics, there is no unanimity of thought about nonhuman species. For example, the loss of a species is morally wrong based on the value of the existence of the species, not its actual or potential value (e.g., as a cure for cancer or as a food source for a food species). Contrast with instrumental value.

Intuition – Direct perception of meaning without conscious reasoning. Compare to deductive and inductive reasoning.

Junk science – Term applied to questionable data or information used to support advocacy positions; or factually correct data and information misapplied to support advocacy positions.

Justice – The fair and equal treatment of people. Providing what is deserved. Distribution of goods and services based on fairness and what is deserved.

Knowledge – Familiarity, awareness, or understanding gained through experience or study. A necessary step toward wisdom.

Land ethic – Ethical view proposed by Aldo Leopold (*A Sand County Almanac*, 1949) that "reflects the existence of an ecological conscience, and this in turn reflects a conviction of individual responsibility for the health of land." This is a precursor to ecocentrism (see Ecocentrism).

Latency – Period of time between disease occurrence and detection, sometimes used interchangeably with induction.

Law of diminishing returns – Economic principle espoused by Thomas Malthus (*Essay on the Principle of Population*, 1798) stating that when a fixed input is combined in production with a variable input, using a given technology, increases in the quantity of the variable input will eventually depress the productivity of the variable input. Malthus proposed this as a law from his pessimistic idea that population growth would force incomes down to the subsistence level.

Law of supply and demand – Economic principle stating that, in equilibrium, prices are determined so that demand equals supply; thus changes in prices reflect shifts in the demand or supply curves.

Left brain/right brain thinking – Differentiation between where incoming stimuli are processed in the brain. Verbal, sorting, and detail-oriented processing occurs in the left, and the spatial, intuitive, and nonverbal processing in the right.

Leviathan – Treatise (full title: *Leviathan or the Matter, Form and Power of a Common-wealth Ecclesiastical and Civil*) written by Thomas Hobbes in 1651, expounding the theory that absolute government was the exclusive vehicle needed to balance personal interests and desires with their rights of life and property. Thus, members of society enter into a social contract (see Social Contract) with the sovereign.

Libertarianism – Form of contractarianism (see Contractarianism) holding that individuals ought to have complete freedom of action, provided that these actions do not infringe on the same freedom of others. Thus, government is circumscribed.

Liberty – Freedom of choice; autonomy; immunity, especially from government's arbitrary exercise of authority.

Life – Period from onset (i.e., conception) to end (i.e., death) of a unique organism. Antonym can be either death (see Death) or nonliving.

Linearity – Following the mathematical equation for a line ($/y = mx + b$; where m is the slope and b is the y intercept). Also used to describe the degree to which data points approximate the line of best fit (linear regression).

Logic – Branch of philosophy addressing inference (e.g., using a syllogism (see Syllogism) to determine the validity of an ethical argument).

Logical positivism – Twentieth-century philosophical view that all meaningful statements are either analytic or can be verified conclusively using observation and experimentation, and that nonverifiable (e.g., metaphysical) theories have no meaning (also known as logical empiricism). Even for the physical sciences (let alone, theology), logical positive comes undone, since much of scientific theory is not analytic, or even empirically verifiable (e.g., Newton's general law of gravitation).

Longitudinal study – Epidemiological study using data gathered at more than one point in time, e.g., after an exposure or a medical intervention. Contrast with cross-sectional study.

Love – In theological thought, the divinely infused habit to cherish the Creator for his own sake above all things, and fellow humans for the sake of the Creator. The greatest of the three Christian virtues (also known as charity (see Charity)).

Low birth weight – Weight at birth of human baby below what is considered the ideal range, often less than 5 pounds and 8 ounces (2500 grams). Compare to very low birth weight.

Lowest-Observed-Adverse-Effect Level (LOAEL) – Lowest exposure level at which there are biologically significant increases in frequency or severity of adverse effects between the exposed population and its appropriate control group.

Macroethics – Expectations of an entire profession, e.g., the engineering profession's positions regarding emerging technologies, social justice, or sustainability.

Maleficence – Causing evil or harm to another.

Malfeasance – Wrongdoing, misconduct, or misbehavior.

Malevolence – Evil intent.

Malpractice – Unprofessional activity, especially that which harms a patient (medicine) or client (law or design).

Malthusian population theory – View first articulated by Reverend Thomas Robert Malthus (1766–1834) that the human population would increase at a geometric rate and the food supply at an arithmetic rate. The view was overly pessimistic and did not account for technological advances (e.g., the Green Revolution).

Mendelism – Heredity theory underlying classical genetics, proposed by Roman Catholic monk and scientist Gregor Mendel in 1866.

Mental model – Template in one's mind, based on past experience and current knowledge, to decipher and use sensory information.

Mentor – Guide or advisory guide to a more junior colleague.

Metabolism – Act of a living organism converting and degrading a substance from one form to another (known as a metabolite).

Metabonomics – Study of total metabolite pool (especially using computational methods).

Metaphor – Figurative comparison of known things to other, more difficult to understand concepts.

Miasma – Dominant disease theory from the seventeenth century through much of the nineteenth century that posited the cause of disease to be from foul emanations from air, soil, and water. Displaced by germ theory (see Germ Theory) as standard disease paradigm in the nineteenth century.

Microethics – Expectations of the individual professional practitioner or researcher. Compare to macroethics.

Minimax theorem – Key convention of game theory holding that lowest maximum expected loss in a two-person zero-sum game equals the highest minimum expected gain. It is a useful technique to address uncertainties in decision making.

Model – Mathematical function with parameters that can be adjusted so the function closely describes a set of empirical data. A mechanistic model usually reflects observed or hypothesized biological or physical mechanisms, and has model parameters with real-world interpretation. In contrast, statistical or empirical models selected for particular numerical properties are fitted to data; model parameters may or may not have real-world interpretation. When data quality is otherwise equivalent, extrapolation from mechanistic models (e.g., biologically based dose–response models) often carries higher confidence than extrapolation using empirical models.

Moral –

1. Pertaining to the judgment of goodness or evil of human action and character.
2. Often, an adjective for goodness or ethically acceptable actions (opposite of immoral).

Moral development – Progression of morality and ethics in an individual, growing from one based on rewards and punishment to principled conscience.

Moral hazard – Actuarial risk resulting from uncertainty about the honesty of the insured, especially the risk that one party to a contract will change behavior to the detriment of the other party once the contract has been concluded (important to the insurance industry and the legal profession).

Moral objectivism – View that moral principles are absolute, reflect universal moral truths and are derived from reason, divinely inspired, or both (contrast with moral relativism).

Moral relativism – View that moral principles are derived from personal, cultural, historical, or social sources, so they are neither universal nor absolute (contrast with moral objectivism).

Moral sense – Motivation deriving logically from ethical or moral principles that rule a person's judgments and actions.

Morality – Distinction between what is right and wrong.

Morbidity – State of disease.

Mores – Morally significant norms of a society.

Mortality rate – Proportion of a population that dies during a specified time period. Also called death rate.

Multipotent stem cells – Stem cells that can give rise to several other cell types, but those types are limited in number. Hematopoietic cells (i.e., blood stem cells that can develop into several types of blood cells) are examples. Scientists have long held the opinion that differentiated cells cannot be altered or caused to behave in any way other than the way in which they have been naturally committed, but recent findings have called this assumption into question (see Nanog).

Nanog – Protein central to the unique properties of embryonic stem cells. The protein, named after the mythological Celtic land of the ever-young *Tir nan Og*, is required for the special ability of stem cells to multiply without limit while remaining able to make many different types of cell. Researchers are investigating whether transcription can produce nanog in adult stem cells, making them pluripotent. That is, they will behave just like embryonic stem cells.

Naive realism – Describes the belief that physical objects continue to exist when they are no longer perceived. Contrast with solipsism.

Nanotechnology – Science and engineering addressing the design and production of extremely small (<100 nanometers diameter) devices and systems fabricated from individual atoms and molecules.

Nash equilibrium – Solution concept of game theory wherein equilibrium is reached so that each player's action is optimal given the actions of the other players; that is, all players are executing dominant strategies.

Natural law – View that natural order exists in the human world, which is good; so people must not violate this order.

Negative paradigm – Most unethical action or case possible. In line-drawing, the negative paradigm is the polar opposite of the positive paradigm (see Positive Paradigm).

Negative utilitarianism – Permutation of utilitarianism (see Utilitarianism) where the metric of morality is the prevention of greatest evil or the promotion of the least harm for the greatest number.

Nerve – Enclosed, cable-like bundle of nerve fibers or axons.

Neurotechnology – Techniques and apparatus used to analyze and treat the nervous system, especially the brain (e.g., implantation of devices in the brain to allow paralyzed patients to move a robotic arm).

New Testament – Second half of the Christian Bible that includes the teachings of Jesus Christ and that of his followers.

Nicomachean ethics – Aristotle's exposition of virtue ethics; ten books based on his lectures at the Lyceum (either edited by or dedicated to his son, Nicomachus).

Nongovernmental organization (NGO) – Entity that advocates or represents positions, including scientific, legal, and medical perspectives, without a governmental mandate. Examples include Doctors without Borders, Resources for the Future, and Engineers without Borders.

No-Observed-Adverse-Effect Level (NOAEL) – Highest exposure level where there are no biologically significant increases in the frequency or severity of adverse effect between the exposed population and its appropriate control.

Nonmaleficence – Affirmation of doing no harm or evil.

Normative ethics – Study of ethics concerned with classifying actions as right and wrong. Contrast with descriptive ethics.

Norms – Expectations of behavior of individuals within society.

Nosocomial – Contracted in the hospital (e.g., infection from an airborne microbial exposure while being treated for a broken arm).

Nutritional supplement – A dietary supplement intended to be ingested in pill, capsule, tablet, or liquid form, not represented for use as a conventional food or as the sole item of a meal or diet, and labeled accordingly. It is regulated by the US Food and Drug Administration as a food and not as a drug.

Ockham's Razor – Principle espoused by medieval nominalist William of Ockham that entities are not to be multiplied beyond necessity. The principle encourages asking whether any proposed kind of entity is necessary.

Odds ratio – Ratio of the odds of disease among the exposed compared with the odds of disease among the unexposed. For rare diseases, such as cancer, the odds ratio can provide an estimate of relative risk.

Omics – Shorthand term for computational, biological subfields for describing very large-scale data collection and analysis, all with the suffix "omics" (e.g., genomics, proteomics, and metabonomics)

Onus – Burden of responsibility.

Opportunity risk – Likelihood that a better opportunity will present itself after an irreversible decision has been made (e.g., prohibiting research in an emerging technology may prevent exposure to a toxic substance to a few, but in the process the cure for a disease may be lost).

Optimal range – Range of success, below and above which are unacceptable (e.g., trivalent chromium must be taken within the optimal range because intake at too low a dosage leads to a nutritional deficiency and too high a dosage leads to toxicity).

Optimization – Selecting the best design for the conditions. The "best" is determined by the designer based on one or more variables (e.g., a heart valve may have three key variables: flow rate, reliability, and durability – the engineer would design the valve by optimizing these three variables to achieve the best performance).

Organ – Completely differentiated unit of an organism that provides a certain, specialized function.

Organism – A living entity; consisting of one or more cells (unicellular and multicellular organisms, respectively).

Outlier – Value that is markedly smaller or larger than other values in a data set. Can be problematic for researchers since it decreases the coefficient of determination (i.e., r^2).

Pandora's box – Metaphor for a prolific source of problems that, when opened, lead to a cascade of miseries (Greek: All gifted; from mythology of a box given to Pandora by Zeus who ordered that she not open it; Pandora succumbed to her curiosity and opened it; all the miseries and evils flew out to afflict humankind).

Paradox – Argument appearing to justify a self-contradictory conclusion by using valid deductions from acceptable premises.

Parametrics – Descriptors of an entire population, without the need of inference. Compare to statistics.

Pareto efficient – Resource allocation wherein there is no rearrangement that can make anyone better off without making someone else worse off.

Partial birth abortion – Late-term abortion, clinically known as intact dilation and extraction (IDX or Intact D&X).

Pathogen – Microbe capable of producing disease.

Patient – Individual receiving medical attention, such as treatment.

Peace engineering – Application of technical skills to promote peace.

Pedagogy – Instruction techniques used to promote learning.

Pentateuch, The (Greek: five tools) – First five books of Judeo-Christian Scripture; also known as the Law or the Torah. Moses is credited as the author.

Perception – Information and knowledge gained through the senses.

Persistent vegetative state (PVS) – Condition wherein a person loses higher cerebral powers, but maintains sleep–wake cycles with at least partial hypothalamic and brain stem autonomic functions. Compare to coma.

Person – Human being.

Personality – Set of traits possessed by a person that uniquely influences cognitions, motivations, and behaviors. Contrast with character.

Personhood – Status of having moral rights. Some restrict the status of personhood only to individuals who are self aware and capable of rational thought and moral agency. However, this limitation of status is opposed in that some individuals lose these capacities temporarily and because persons at various stages are yet to attain some or all of these capacities or have lost them.

PERT chart – "Program Evaluation Review Technique" diagram depicting project tasks and their interrelationships.

Pesticide – Substance used to control pesticide by using its toxic properties.

Pharmacodynamics – Manner in which a substance exerts its effects on living organisms (also known as toxicodynamics if it involves a toxic substance).

Pharmacokinetics – Behavior of substances within an organism, especially by absorption, distribution, bio-transformation, storage, and excretion (also known as toxicokinetics if it involves a toxic substance).

Philosophy – Pursuit of wisdom (Greek: *philos* for love and *sophia* meaning wisdom).

Physician – One who practices medicine, especially a licensed medical practitioner.

Physiologically Based Pharmacokinetic (PBPK) Model – Model used to characterize pharmacokinetic behavior of a chemical. Available data on blood flow rates and on metabolic and other processes that the chemical undergoes within each compartment are used to construct a mass-balance framework for the PBPK model.

Pica – Ingestion of non-food substances (e.g., paint chips, soil) that can account for disparate exposures (see Disparate Exposure) in certain subpopulations (e.g., minority children).

Plagiarism – Unethical practice of presenting someone else's written works as one's own.

Pluripotent stem cells – Cells that can produce all the cell types of the developing body, such as the cells in the inner cell mass of the blastocyst (see Blastocyst).

Point of departure – The dose–response (see Dose–Response) point that marks the beginning of a low-dose extrapolation. This point is most often the upper bound on an observed incidence or on an estimated incidence from a dose–response model.

Polychlorinated biphenyl (PCB) – Highly toxic molecule of two benzene rings bonded to each other with chlorine substituents (209 structural variations, known as congeners); presently banned but manufactured by Monsanto for much of the twentieth century.

Polycyclic aromatic hydrocarbon – Class of products of incomplete combustion consisting of fused aromatic rings. A number of them are suspected carcinogens (e.g., benzo(a)pyrene).

Positive paradigm – Most ethical action or case possible. In line drawing, the positive paradigm is the polar opposite of the negative paradigm (see Negative Paradigm).

Precautionary principle – Risk management (see Risk) approach taken when scientific knowledge is incomplete and the possible consequences are devastating (e.g., global climate change). The principle holds that scientific uncertainty must not be accepted as an excuse to postpone cost-effective measures to prevent a significant problem.

Prescriptive statement – Statement that says how things should be, as opposed to a descriptive statement, which says how things are. Contrast with proscriptive statement. Codes of ethics contain both prescriptive and proscriptive statements.

Prevalence – Proportion of cases that exist within a population at a specific point in time, relative to the number of individuals within that population at the same point in time.

Price elasticity of demand – Percentage change in quantity demanded of a good as the result of a given percent change in price (the percentage change in quantity demanded divided by the percentage change in price).

Price elasticity of supply – Percentage change in quantity supplied of a good as the result of a given percent change in price (the percentage change in quantity supplied divided by the percentage change in price).

Prisoner's dilemma – Game theory scenario wherein two parties each seek self-interests non-cooperatively, yielding a result that leaves them both worse off. Tragedy of the Commons (see Tragedy of the Commons) is an example.

Probability – Measurement of the likelihood that an event will occur; ranging from 0 (no likelihood whatsoever) to 1 (absolute certainty).

Profession – Group (e.g., physicians or engineers) with a common mission, requiring substantial education and training, self determination of professional requirements to enter, organized into an identifiable professional body, and requiring the adherence to standards of conduct.

Proscriptive statement – Prohibition against certain acts. Codes of ethics contain both proscriptive and prescriptive statements (see Prescriptive Statement).

Proteomics – Study of proteins in the body, especially the protein complement of the genome (see).

Psychiatry – Medical specialty that addresses the origin, diagnosis, prevention, and treatment of mental disorders.

Psychology – Study of human behavior.

Public, The – Whole collection of people comprising a society.

Punishment – Penalty inflicted on a person for committing an act unacceptable to society, something meant to deter an unacceptable act.

Pursuit of happiness – One of the unalienable rights articulated in the US Declaration of Independence; along with life and liberty, this pursuit sets the stage of limited government and self-determination of citizens.

Racism – Belief that one group of human beings is inferior to another as a result of their race.

Rationality – Philosophical theories concerned with truth, reason, and knowledge.

Rawlsism – Form of contractarianism (see Contractarianism) based on the view that "a practice is just if it is in accordance with the principles that all who participate in it might reasonably be expected to propose or to acknowledge before one another when they are similarly circumstanced and required to make a firm commitment in advance" (John Rawls, 1957, *Justice as Fairness*, 54 (22), pp. 659–60). A key means of achieving this is through the veil of ignorance (see Veil of Ignorance).

Reasonable person standard – Position (legal, engineering, etc.) expected to be held by a hypothetical person in society who exercises average care, skill, and judgment in conduct.

Reasoning – Derivation of a conclusion from premises.

Reinforcement – Linking a stimulus and response with a reward or punishment, leading to an increased or decreased likelihood that the response will be repeated subsequently when the stimulus is reintroduced (key concept of behaviorism (see Behaviorism)).

Relative risk – Ratio of the risk of disease or death among the exposed segment of the population to the risk among the unexposed.

Reliability – Probability that a device or system will perform its specified function, without failure under stated environmental conditions, over a required lifetime.

Religion – Structured belief system, centered around the supernatural and sacred, and including expectations (tenets) for morality consistent with beliefs.

Renaissance – Time from about AD 1450 to 1600, characterized by advances in the humanities and learning, and the birth of modern science in Western civilization.

Respect for nature – View expounded by Paul Taylor (*Respect for Nature*, 1986) defending biocentrism (see Biocentrism), based on a systematic and comprehensive account of moral relationships between humans and other living things. This requires an acceptance that all living things have inherent moral value, so that respect for nature is the "ultimate moral attitude."

Respect for persons – Principle expounded in the Belmont Report (see Belmont Report) that incorporates at least two ethical convictions: first, that individuals should be treated as autonomous agents, and, second, that persons with diminished autonomy are entitled to protection. The principle of respect for persons thus divides into two separate moral requirements: the requirement to acknowledge autonomy and the requirement to protect those with diminished autonomy.

Responsible conduct of research (RCR) –

1. Ethical approach to research, especially performing research honestly and objectively, with integrity, and with full consideration of the possible effects (e.g., protecting human subjects, animal welfare).
2. Requirement by public research agencies that grantees and contractors (e.g., National Institutes of Health) properly train their researchers (e.g., Ph.D. students) in research ethics.

Reverence for life – View introduced by Albert Schweitzer (1875–1965) based on "world and life affirmation," based on *ehrfurcht vor dem leben* (German: Attitude of awe and wonder tempered with fear, summed up by Schweitzer as "I am life which wills to live, in the midst of life which wills to live.").
Right of conscience – Moral prerogative to exercise professional judgment.
Right of conscientious refusal – Moral prerogative of a professional to refuse to engage in activity that he or she considers to be unethical.
Right to die – Political and philosophical position that it is the individual's right to decide to live, especially when confronted with a terminal illness or extremely serious, progressive disease; often opposed on the basis of sanctity of life (see Sanctity of Life).
Right to life – Viewpoint holding that human life has a paramount value; central feature of debates concerning abortion, capital punishment, embryonic stem cell research, euthanasia, and assisted suicide. As such, it is a central focus of pro-life advocates.
Rights ethics – View that an act is morally right when it is the preferred way to respect the rights of others. John Locke (*Essays on the Law of Nature*, 1664) advocated rights ethics.
Risk – Likelihood of an adverse outcome.
Risk–benefit analysis – Comparison of risks of various options to their benefits (e.g., health and wildlife risks of applying a pesticide compared to the benefits of crop protection).
Risk shifting – Taking an action that changes and reduces the risk of one population, but increases the risk in a different population (e.g., banning DDT reduces the risk of cancer in developed nations but increases the risk of malaria in tropical and subtropical developing countries).
Risk trade-off – Eliminating or reducing one risk, but introducing or increasing a countervailing risk (e.g., reducing the pain of a headache by taking aspirin, but increasing the risk of Reye's syndrome; or removing mold in buildings may increase worker exposure to asbestos).
Rule utilitarianism – Form of utilitarianism wherein the test of utility maximization is applied to acts only indirectly through some other suitable object of moral assessment; that is, to rules of conduct. Rule utilitarianism bases the morality of an action on whether or not that action follows a universal rule that would have good consequences if everyone followed it (compare to the duty-based concept, categorical imperative).

Sanctity of life – View that human life is precious from conception to natural death, since it is created in the image of the Creator.
Scale – Spatial extent from very small (molecular) to very large (planetary).
Science – Systematic investigation, through experiment, observation, and deduction, in an attempt to produce reliable explanations of the physical world and its processes.

Scientific method – Progression of inquiry: (1) to identify a problem you would like to solve, (2) to formulate a hypothesis, (3) to test the hypothesis, (4) to collect and to analyze the data, and (5) to draw valid conclusions.

Senescence – Organism's aging process.

Sensitivity – Ability of a test to detect a condition when it is truly present. Compare to specificity.

Seven virtues – The four cardinal virtues (see Cardinal Virtues) plus the three theological virtues, faith, hope, and love (charity), given in 1 Corinthians 13.

Slope factor, cancer – Dose–response curve for a substance indicating cancer potency (units = mass per body mass per time; e.g., $mg\,kg^{-1}\,day^{-1}$).

Social contract – Agreement among the members of a society and its government delineating and limiting the rights and duties of each. Thomas Hobbes argued that the power of a sovereign is justified by an implied contract wherein the people agree to obey the government in return for a guarantee of peace and security, which is not possible in the "state of nature" (see State of Nature) conceived to exist without such contract.

Social ecology – View that environmental problems are firmly rooted in human social interactions. Social ecologists believe that an ecologically sustainable society can still be socially exploitative.

Society – Collection of human beings that is distinguished from other groups by shared institutions and a common culture.

Solipsism – Idea that the only one's existence is a certainty, and that true knowledge of anything else is impossible. Contrast with naive realism.

Somatic cell nuclear transfer (SCNT) – Method of cloning by transferring the nucleus from a donor somatic cell into an enucleated egg to produce a cloned embryo.

Spatial scale – Geographic extent of a resource or problem. Global scale examples include pandemics, changes in climate, or nuclear threats. Continental scale examples are shifting biomes and border control between nations. Regional scale examples include the contamination of rivers or polluting the air. Local scale examples include crime, job loss, hazardous waste sites, and landfills.

Specialization – Degree to which an individual professional concentrates his or her practice into a narrow range of expertise and activities.

Specificity – Ability of a test to exclude the presence of a condition when it is truly not present. Compare to sensitivity.

Standard deviation – Measure of the spread in a data set; wider spread means larger standard deviation.

Standard mortality ratio (SMR) – Ratio of the number of deaths observed in the study group to the number of deaths expected based on rates in a comparison population, multiplied by 100.

State of nature – According to Thomas Hobbes (*Leviathan*, 1651) human beings would behave "badly" toward one another in the absence of a social contract (see Social Contract) with government. Hobbes believed that such a state would lead to a "war of every man against every man" and make life "solitary, poor, nasty, brutish, and short."

Statistics – Mathematics concerned with collecting and interpreting quantitative data and applying probability theory to estimate conditions of a universe or population from a sample.

Stimulus–response theory – Basic concept of behaviorism (see Behaviorism) based on the stimulus–response theory that a stimulus will cause a response either by pairing a response with a reflective trigger (e.g., Pavlov's famous experiment of inducing salivation in dogs by ringing a bell that was paired with food) or by rewarding a response in the presence of a stimulus (e.g., a rat pressing a bar that releases a food pellet).

Strain – Geometrical expression of deformation caused by the action of stress (see Stress) on a physical body.

Stratified random sample – Separation of a sample into several groups and randomly assigning subjects to those groups.

Stress –

1. Applied force or system of forces that are apt to strain (see Strain) or deform a physical body.
2. The internal resistance of a physical body to such applied force or system of forces.

Subjectivism – Philosophical view that knowledge and value are dependent on and limited by one's subjective experience.

Superego – In psychoanalytic theory, component of human psyche that has incorporated the values and standards of society in conflict with the primitive selfishness of the id (see Id). This conflict is mediated by the ego (see Ego).

Supererogatory – Morally good beyond what is obligatory (duty-based).

Supply – Quantity of a good or service that a seller would like to sell at a particular price.

Supply curve – Relationship between a good's quantity supplied and the good's price.

Susceptibility – Extent to which an individual is prone to the effects of an agent.

Sustainability – Processes and activities that are currently useful and that do not diminish these same functions for future generations.

Sustainable design – Application of principles of sustainability (see Sustainability) to structures, products, and system; an aspect of green engineering (see Green Engineering).

Sword of Damocles – A constant and impending hazard or peril (from the Greek moral anecdote wherein King Damocles required a guest to sit precariously under a large sword held only by a horse hair).

Syllogism – Argument according to Aristotle's logical theory that includes a major premise, a minor premise, and a conclusion. Ethical syllogisms have factual premise, a connecting premise, an evaluative premise, and a moral conclusion.

Sympathy – Compassion; sharing the feelings of others. Compare to empathy.

Synderesis (or synteresis) – Immediate, intuitive grasp of the fundamental principles of morality.

Synergy – Effect from a combination of two agents is greater than the sum of the additive effects from both agents $(1+1>2)$.

System –

1. Combination of organized elements comprising a unified whole.
2. In thermodynamics, a defined physical entity containing boundaries in space, which can be open (i.e., energy and matter can be exchanged with the environment) or closed (no energy or matter exchange).

Tao – "Method" or "way" described in the writings of Tao-te ching, Chuang-tzu, and others. Continuous activity between the *yin* and the *yang* (see *Yin* and *Yang*).

Tautology – Proposition that is already true by definition, in no need of logical deduction (e.g., the statement "An octagon has eight sides" is a tautology).

Taxonomy – Classification system.

Technology –

1. Application of scientific knowledge.
2. The apparatus that results from such applications.

Teleology – See Consequentialism.

Temporal scale – Range of complexity associated with time. Extremely short temporal scale may be measured in nanoseconds, e.g., nuclear reactions, whereas long temporal scales may be measured in millions of years, e.g., fossilization of plants to coal.

Temporality – Criterion for causality requiring that the cause (e.g., exposure to an infectious agent) precede the effect (e.g., disease).

Terrorism – Unlawful use of, or threatened use of, force or violence against individuals or property to coerce or intimidate governments or societies, often to achieve political, religious, or ideological objectives (US Department of Defense).

Thermodynamics, Laws of – Physical principles addressing the physical relationships between energy and matter, especially those concerned with the conversion of different forms of energy.

Thought experiment – Conception of the consequences of an intervention in the world without actually intervening. In some instances, an actual experiment is preferred but not possible or feasible in practice, or perhaps not even in theory.

Tissue – Collection of interconnected cells that carry out a similar function in an organism.

Top-down – Starting at the upper levels of organization and working downward to the details. Contrast with bottom-up.

Torah – Hebrew law as recorded in the Pentateuch, i.e., the first five books of the Old Testament.

Total quality management (TQM) – Structured management approach, introduced by engineer W. Edwards Deming, based on continued refinements to improve the quality of products and services informed by ongoing feedback from customers and clients.

Toxicity – Extent and degree of biological harmfulness of a chemical, physical, or biological agent, ranging from acute (short-term harm) to chronic (long-term harm) toxicity.

Tragedy of the Commons – Term coined by Garrett Hardin (*Science*, volume 162 (1968)) characterizing the degradation of commonly held resources as a result to selfish self-interest in maximizing utility of each individual using the resource.

Transgenic – Modified genetically.

Trimming – Irresponsible data manipulation of smoothing of irregularities to make observational data appear more accurate and precise than is the case. This is done in an attempt to improve the coefficient of determination (i.e., r^2).

Truth – Conformity to fact. Compare to honesty.

Type I error – Error of rejecting a true null hypothesis. Contrast with type II error.

Type II error – Error of accepting (not rejecting) a false null hypothesis. Contrast with type I error.

Uncertainty – Difference between what is known and what is actually the truth. Scientific uncertainty includes error and unknowns, such as those resulting for selecting variables, undocumented variability, and limitations in measurements and models. In science, there is almost always uncertainty. The goal is to decrease uncertainty and to document known uncertainties.

Utilitarianism – Theory proposed by Jeremy Bentham (*Principles of Morals and Legislation*, 1789) and James Mill (*Utilitarianism*, 1863) that action should be directed toward achieving the greatest happiness for the greatest number of people (if applied to nonhuman species, this is referred to merely as "greatest number").

Utility –

1. Condition of being useful.
2. Useful outcome.
3. Level of enjoyment an individual attains from choosing a certain combination of goods.

Valuation – Quantifying or otherwise placing value on goods and services. Monetized valuation uses monetary currency (e.g., gross domestic product), whereas many environmental and quality of life resources are not readily conducive to monetized valuation, e.g., old growth forests have nonmonetized value (e.g., habitat, ecological diversity) but little monetized value since they are not used for timber.

Value –

1. Principle, standard, or quality that are good for a person to hold.
2. Worth.

Value engineering (VE) – Systematic application of recognized techniques by a multidisciplined team to identify the function of a product or service, establish a worth for that function, generate alternatives through the use of creative thinking, and provide the needed functions to accomplish the original purpose of the project, reliably, and at the lowest life-cycle cost without sacrificing safety, necessary quality, and environmental attributes of the project (US Department of Transportation).

Value of life –

1. Economic or moral worth of a human being.
2. Marginal cost of preventing death (social sciences).

Variability – True heterogeneity or diversity (e.g., among a population that is exposed to airborne pollution from the same source and with the same contaminant concentration, the risks to each person as a result of breathing the polluted air will vary).

Veil of ignorance – Notion introduced by John Rawls (*A Theory of Justice*, 1971) to determine morality of an action supposing that societal roles can be completely reshaped and redistributed, and that the ethical decision maker does not know his or her reassigned role. This is an attempt to reach distributive justice (see Distributive Justice).

Very low birth weight – Weight of human baby at birth less than 3 pounds and 5 ounces (1500 grams). Compare to low birth weight.

Vice – Immorality; negative moral trait. Opposite of virtue.

Virtue – Moral excellence or goodness. Opposite of vice.

Virtue ethics – Ethical theory that emphasizes the virtues, or moral character, in ethical decision making. Focuses on what makes a good person, rather than what makes a good action. One of three major theories of normative ethics (see Normative Ethics), along with consequentialism and deontology (see Consequentialism and Deontology or Deontological Ethics).

Vulnerability – Condition of an individual or population determined by physical, social, economic, and environmental variables, wherein susceptibility to a hazard increases (e.g., asthmatics are more vulnerable to the effects of some air pollutants than is the average person).

Way of knowing – Method of learning. Discovery and reason hold primacy as scientific ways of knowing. Intuition (see Intuition), to scientists, is an advanced, integrated means of knowing. All are considered *a posteriori* knowledge (see *A posteriori* knowledge). However, theological ways of knowing also include *a priori* knowledge (see *A priori* knowledge) given by the Creator. Thus, intuition can be a combination of *a posteriori* and *a priori* knowledge.

Weight of evidence – Strength of data and information supporting a conclusion. When little reliable data are available the weight of evidence is lacking, whereas when numerous studies provide reliable information to support a particular position (e.g., exposure to a chemical associated with a health effect), the weight of evidence is strong.

Whistleblower – Informant who exposes wrongdoing within an organization or to an outside authority (e.g., government agency) that has or appears to have the ability to stop the misconduct.

Wilderness – A large, natural (or nearly natural) region, not controlled by humans.

Willingness to pay – Economic concept meaning the most money that people will give for a good or service; depicted by the total area under a demand curve.

Wisdom – Insight, erudition, and enlightenment resulting from the accumulation of knowledge and the ability to discern what is meaningful from what is not.

Yin and *Yang* – Concept in Chinese philosophy describing two opposing and equal forces in the universe needed for harmony. *Yang* includes such principles as strength, liveliness, and brightness, whereas *yin* includes weakness, passivity, and darkness.

Zygote – Diploid cell that results from the fertilization of an egg cell by a sperm cell.

NOTE AND COMMENTARY

[1] These terms are the author's operational definitions. The sources are numerous and many terms have been modified from their original definitions. However, here are the resources used to augment this glossary:

1. The Online Ethics Center Glossary: http://onlineethics.org/glossary.html
2. University of Nebraska Medical Center Glossary: http://www.unmc.edu/ethics/words.html
3. L.M. Hinman's "Ethics Update – Glossary": http://ethics.sandiego.edu/Glossary.html.
4. R.N. Johnson's "A Glossary of Standard Meanings of Common Terms in Ethical Theory": http://web.missouri.edu/~philrnj/eterms.html.
5. US Department of Health and Human Services, Office of Research Integrity: http://ori.dhhs.gov/education/products/rcradmin/glossary.shtml.
6. US Department of Health and Human Services, Agency for Toxic Substances and Disease Registry: http://www.atsdr.cdc.gov/glossary.html.
7. US Environmental Protection Agency, "Terms of Environment: Glossary, Abbreviations and Acronyms": http://www.epa.gov/OCEPAterms/.
8. US Environmental Protection Agency, "Glossary of IRIS Terms": http://www.epa.gov/iris/gloss8.htm.
9. Fox Chase Cancer Center, "Glossary of Ethics Terms": http://www.fccc.edu/ethics/Glossary_of_Ethics_Terms.html.

In addition, the sources cited in the Notes and Commentary following each chapter of this book contain useful definitions of terms used in these specific instances.

Name Index

Subject Index

Page numbers in **bold** refer to definition of the term

Printed and bound by CPI Group (UK) Ltd, Croydon, CR0 4YY

03/10/2024

01040335-0004